**NORTH CAROLINA
STATE BOARD OF COMMUNITY COLLEGES
LIBRARIES
ASHEVILLE-BUNCOMBE TECHNICAL COMMUNITY COLLEGE**

DISCARDED

JUN 19 2025

D1709210

RCRA REGULATORY COMPLIANCE GUIDE

RCRA REGULATORY COMPLIANCE GUIDE

by

Mark S. Dennison

Attorney at Law

NOYES PUBLICATIONS
Park Ridge, New Jersey, U.S.A.

Copyright © 1993 by Mark S. Dennison
 No part of this book may be reproduced or utilized in
 any form or by any means, electronic or mechanical,
 including photocopying, recording or by any information storage and retrieval system, without permission
 in writing from the Publisher.
Library of Congress Catalog Card Number: 92-32509
ISBN: 0-8155-1321-6
Printed in the United States

Published in the United States of America by
Noyes Publications
Mill Road, Park Ridge, New Jersey 07656

10 9 8 7 6 5 4 3 2 1

Library of Congress Cataloging-in-Publication Data

Dennison, Mark S.
 RCRA regulatory compliance guide / by Mark S. Dennison.
   p.   cm.
 Includes bibliographical references and index.
 ISBN 0-8155-1321-6
 1. Hazardous wastes--Management. 2. United States. Resource
Conservation and Recovery Act of 1976. I. Title. II. Title:
Resource Conservation and Recovery Act regulatory compliance guide.
TD1030.D45   1992
363.72'8756'0973--dc20   92-32509
                                                       CIP

To Tracey, my one and only love and happiness

Preface

This book has been written especially for companies as a legal compliance guide to safe, legal methods of handling hazardous wastes. The book is organized in a practical format to answer questions faced by companies that produce hazardous wastes in their day-to-day business operations. Although the book is a useful guide, it should be remembered that it is not intended as a substitute for the advice normally provided by legal counsel or environmental consultants. The book is prepared in a concise manner and describes the most frequently confronted issues in a general manner. The book is intended to explain the Resource Conservation and Recovery Act, as well as other environmental laws and regulations, in a way that the company can understand and apply to its business operations. The company should first read the book from beginning to end to acquire a working knowledge of the law that applies and then use the information to properly manage generated wastes.

The book explains how to identify hazardous wastes at the company, what generator classification the company will fall under, how to safely handle the wastes, how to store hazardous wastes, and how to properly dispose of the wastes. The practical aspects of hazardous waste management are examined along with the body of laws and regulations the company needs to comply with in order to avoid costly liability and penalties. Special issues concerning recordkeeping and reporting, worker safety, and environmental liability insurance are also closely examined. This book should be useful to any company that generates hazardous wastes. I have attempted to explain a complex regulatory scheme as concisely as possible using a practical step-by-step approach to the subject. Whenever in doubt about a specific problem, the company can always contact the state or federal hazardous waste regulatory authority or seek the advice of a trained professional.

Mark S. Dennison
August, 1992

Acknowledgments

I wish to express my appreciation to everyone who played a role in publication of this book.

Special thanks go to Robert Noyes, President of Noyes Publications, for believing in the project and giving me the opportunity to publish with his company. Thanks also to Janet Paul and the excellent staff at Noyes for their quality editing and production of the book.

My thanks to several people who helped develop my interest and commitment to environmental law and writing: Billy Want, author and attorney at law in Charleston, SC; Michael Perlin, author and professor of law at New York Law School; Owen Smith, author and professor of law at Long Island University; Dan Mandelker, author and professor of law at Washington University School of Law; Dan Tarlock, author and professor of law at IIT Chicago-Kent School of Law.

I am especially grateful to my long-time friend, Randy Masters, who always believed I should write; and to my friend, Henry Seligson, who helped me to see that I could.

Most importantly, I must thank the woman I love, Tracey Huff, for standing by me through the years. Thank you for always believing in me, and for giving me all your encouragement, love and support. Most of all, thank you for being the daily source of inspiration in my life.

About the Author

Mark S. Dennison is an attorney and author or co-author of numerous books and articles on environmental, land use and zoning law issues. He writes a weekly environmental column for *Environmental Protection News* and *HazMat News,* and is the current editor of *Remediation* journal. He was previously the managing editor for the Clark Boardman Environmental Law Series and the acquisitions editor for the Wiley Law Environmental Law Library. He is in private practice in Ridgewood, New Jersey, specializing in environmental, land use and zoning law. He is admitted to practice in New Jersey and New York. Mr. Dennison earned his law degree at New York Law School, and is currently completing an LL.M. degree in environmental law at Pace University School of Law.

Condensed Contents

 1. Regulatory Framework 1
 2. The Resource Conservation and Recovery Act 11
 3. Hazardous Waste Identification and Classification 23
 4. Hazardous Waste Generators 37
 5. Hazardous Waste Storage and Disposal 43
 6. Hazardous Waste Reporting and Recordkeeping 55
 7. Hazard Communication 71
 8. Transporting Hazardous Wastes 87
 9. Hazardous Waste Liability Insurance 101
10. Hazardous Waste Penalties, Costs and Compliance Audits 117
Appendix 1: U.S. EPA Regional Offices 127
Appendix 2: Federal Hazardous Waste Regulatory Offices 131
Appendix 3: State Hazardous and Solid Waste Regulatory Offices (HW/SW) . 139
Appendix 4: RCRA Hazardous Wastes 151
Appendix 5: EPA Notification of Hazardous Waste Activity 171
Appendix 6: Tier One Form 191
Appendix 7: Tier Two Form 197
Appendix 8: Appendix E to Hazard Communication Standard (Advisory)—
  Guidelines for Employer Compliance 205
Appendix 9: Sample Material Safety Data Sheet 219
Appendix 10: Toxic Chemical Release Inventory Reporting Form R 223
Appendix 11: EPA Catalogue of Hazardous and Solid Waste Publications ... 289
List of Regulations and Statutes Referenced in Text 345
Index 349

Notice

To the best of the Publisher's knowledge the information contained in this publication is accurate; however, the Publisher assumes no responsibility nor liability for errors or any consequences arising from the use of the information contained herein. Final determination of the suitability of any information, procedure, or product for use contemplated by any user, and the manner of that use, is the sole responsibility of the user.

The book is intended for informational purposes only. The reader is warned that caution must always be exercised when dealing with chemicals, products, or procedures which might be considered hazardous. Expert advice should be obtained at all times when implementation is being considered.

Mention of trade names or commercial products does not constitute endorsement or recommendation for use by the Publisher.

Contents

1. **REGULATORY FRAMEWORK** 1
   1.1 **Introduction** 1
   1.2 **Environmental Agencies—Federal** 1
   1.3 **Environmental Agencies—State** 2
   1.4 **Hazardous Waste Statutes and Regulations** 3
       1.4.1 Resource Conservation and Recovery Act (RCRA) 4
       1.4.2 Comprehensive Environmental Response, Compensation
             and Recovery Act (CERCLA) 6
       1.4.3 Federal Water Pollution Control Act (FWPCA) 8
       1.4.4 Toxic Substances Control Act (TSCA) 9
       1.4.5 Clean Air Act (CAA) 10
       1.4.6 Occupational Safety and Health Act (OSHA) 10

2. **THE RESOURCE CONSERVATION AND RECOVERY ACT** 11
   2.1 **Introduction** 11
   2.2 **RCRA's Scope and Purpose** 11
   2.3 **Wastes Regulated Under RCRA** 12
   2.4 **Subtitle C Fundamentals** 14
   2.5 **Hazardous Waste Generators** 16
   2.6 **Hazardous Waste Transportation** 18
   2.7 **Recordkeeping and Reporting** 19
   2.8 **Land Disposal Restrictions** 19

3. **HAZARDOUS WASTE IDENTIFICATION AND CLASSIFICATION** 23
   3.1 **Introduction** 23
   3.2 **Definition of Hazardous Waste** 23
   3.3 **Federal Laws Governing Hazardous Substances** 24
   3.4 **Terminology Relating to Hazardous Waste** 24
   3.5 **RCRA Hazardous Wastes** 25
   3.6 **"Listed" and "Characteristic" Hazardous Wastes** 28
   3.7 **Hazardous Waste Classification** 29
   3.8 **EPA Hazard Classes** 29
   3.9 **Common Hazardous Wastes** 30

xii CONTENTS

      3.9.1 Petroleum Products 31
      3.9.2 Chlorinated Hydrocarbons 31
      3.9.3 Polychlorinated Biphenyls (PCBS) 32
      3.9.4 Asbestos 33
      3.9.5 Sulfuric Acid 34
      3.9.6 Heavy Metals 35

4. HAZARDOUS WASTE GENERATORS 37
  4.1 Introduction 37
  4.2 Hazardous Waste Generator Categories 38
      4.2.1 Conditionally Exempt Small Quantity Generator 38
      4.2.2 Small Quantity Generator (SQG) 39
      4.2.3 Large Quantity Generator (LQG) 40
  4.3 Determining the Waste Generator Category 40

5. HAZARDOUS WASTE STORAGE AND DISPOSAL 43
  5.1 Introduction 43
  5.2 Hazardous Waste Storage 43
  5.3 Labeling 44
  5.4 Emergency Response Plans 44
  5.5 Recordkeeping and Reporting Requirements 45
  5.6 Land Disposal Restrictions 46
  5.7 RCRA Hazardous and Solid Waste Amendments 47
  5.8 LDR Treatment Standards 49
  5.9 LDR Variances 50
  5.10 Waste Mixtures 50
  5.11 Contaminated Soils 50
  5.12 Waste Minimization 51
  5.13 Sample Waste Minimization Program 52

6. HAZARDOUS WASTE REPORTING AND RECORDKEEPING 55
  6.1 Introduction 55
  6.2 RCRA Hazardous Waste Spill Reporting 55
  6.3 Underground Storage Tank Releases 57
  6.4 UST Corrective Actions 58
  6.5 CERCLA Hazardous Waste Spill Reporting 60
  6.6 Clean Water Act Hazardous Waste Spill Reporting 62
  6.7 Developing a Contingency Plan 63
  6.8 Emergency Coordinator 63
  6.9 Hazardous Waste Recordkeeping 64
      6.9.1 RCRA Recordkeeping Requirements 64
      6.9.2 UST Recordkeeping Requirements 65
      6.9.3 CERCLA Recordkeeping Requirements 67
  6.10 Hazardous Waste Recordkeeping Checklist 68

7. HAZARD COMMUNICATION 71
  7.1 Introduction 71
  7.2 OSHA Hazard Communication Standard Requirements .. 72
  7.3 Material Safety Data Sheets 72

CONTENTS

- 7.4 Written Hazard Communication Plan ... 73
- 7.5 Employee Training Program ... 74
  - 7.5.1 Exempt Substances ... 75
- 7.6 Emergency Planning and Community Right-to-Know Requirements ... 75
- 7.7 Reporting Spills or Releases ... 76
  - 7.7.1 Annual Reporting ... 76
  - 7.7.2 Tier One Form ... 77
  - 7.7.3 Tier Two Form ... 78
  - 7.7.4 Form R ... 79
  - 7.7.5 Reporting Requirements Added to 1992 Form R ... 80
  - 7.7.6 Exempt Substances ... 81
- 7.8 OSHA Asbestos Standard ... 81
- 7.9 Monitoring of Asbestos Exposure ... 82
  - 7.9.1 Employee Notification ... 82
  - 7.9.2 Record of Exposure Measurements ... 83
  - 7.9.3 Requirements for Companies Exceeding the .1 Threshold ... 83
  - 7.9.4 Medical Surveillance ... 84
  - 7.9.5 Medical Surveillance Records ... 85

8. TRANSPORTING HAZARDOUS WASTES ... 87
- 8.1 Introduction ... 87
- 8.2 Hazardous Materials Transportation Act ... 88
- 8.3 Resource Conservation and Recovery Act ... 89
- 8.4 Hazardous Waste Shipments Preparation ... 90
- 8.5 Hazard Class ... 91
- 8.6 DOT Shipment Number ... 91
- 8.7 Packaging Requirements ... 91
- 8.8 Used Batteries ... 92
- 8.9 Labels and Placards ... 93
- 8.10 Markings ... 94
- 8.11 Choosing a Reputable Transporter ... 95
- 8.12 Hazardous Waste Manifest Requirements ... 96
- 8.13 Manifest Exception Reports ... 98
- 8.14 Manifest Exemptions ... 98
- 8.15 Transporting Hazardous Wastes in Company Vehicles ... 99

9. HAZARDOUS WASTE LIABILITY INSURANCE ... 101
- 9.1 Introduction ... 101
- 9.2 Types of Coverage ... 102
- 9.3 The Current Problem ... 104
- 9.4 Standard CGL Policy ... 104
- 9.5 Duty to Defend ... 105
- 9.6 Types of Damages Covered by the CGL Policy ... 106
- 9.7 Definition of Occurrence ... 108
- 9.8 Trigger of Coverage ... 108
- 9.9 "Sudden and Accidental" Pollution Exclusion ... 110
- 9.10 The "Absolute Pollution Exclusion" ... 112
- 9.11 The "Owned Property" Exclusion ... 114
- 9.12 Insuring Against Environmental Risks ... 115

10. HAZARDOUS WASTE PENALTIES, COSTS AND COMPLIANCE AUDITS 117
   10.1   Introduction 117
   10.2   Hazardous Waste Penalties 117
   10.3   Hazardous Waste Management Costs 119
   10.4   Hazardous Waste Audits 120
   10.5   Preparing for an Audit 122
   10.6   Performing an Audit 122

APPENDIX 1: U.S. EPA REGIONAL OFFICES 127

APPENDIX 2: FEDERAL HAZARDOUS WASTE REGULATORY OFFICES 131

APPENDIX 3: STATE HAZARDOUS AND SOLID WASTE REGULATORY OFFICES (HW/SW) 139

APPENDIX 4: RCRA HAZARDOUS WASTES 151

APPENDIX 5: EPA NOTIFICATION OF HAZARDOUS WASTE ACTIVITY 171

APPENDIX 6: TIER ONE FORM 191

APPENDIX 7: TIER TWO FORM 197

APPENDIX 8: APPENDIX E TO HAZARD COMMUNICATION STANDARD (ADVISORY)—GUIDELINES FOR EMPLOYER COMPLIANCE 205

APPENDIX 9: SAMPLE MATERIAL SAFETY DATA SHEET 219

APPENDIX 10: TOXIC CHEMICAL RELEASE INVENTORY REPORTING FORM R 223

APPENDIX 11: EPA CATALOGUE OF HAZARDOUS AND SOLID WASTE PUBLICATIONS 289

LIST OF REGULATIONS AND STATUTES REFERENCED IN TEXT 345

INDEX 349

1

Regulatory Framework

1.1 INTRODUCTION

Before any company can fully understand the various laws and regulations governing hazardous wastes, it must gain a working knowledge of the governmental agencies that are responsible for administering and enforcing those laws. Generally speaking, all environmental regulation begins at the federal level. Congress enacts an environmental statute and authority to administer it is usually delegated to a particular federal agency, such as the U.S. Environmental Protection Agency (EPA) or the Department of Transportation (DOT). Often, the federal statute will permit or require that state governments institute programs modeled after the federal scheme. In such cases, the state program will control activities carried out in the state. The state laws must be at least as stringent as the federal law in order to be valid. For this reason, it is crucial that any company first understand the federal law and then determine whether its state law has additional requirements.

1.2 ENVIRONMENTAL AGENCIES - FEDERAL

The primary federal agencies in charge of oversight and enforcement in the hazardous waste area are the EPA, DOT and OSHA. In many instances the federal agencies may have overlapping responsibilities in regard to the hazardous waste laws. For example, EPA has developed a hazardous waste list pursuant to RCRA, and DOT has a list of hazardous wastes and hazardous materials. Any company will need to know which wastes are listed on both the EPA and DOT lists in order to comply with the regulations issued by each agency. The EPA list is crucial for identifying waste quantities at the company, whereas the DOT list is crucial for understanding the shipping requirements for a particular waste. Each federal statute will designate one or more federal agencies to implement the statute.

Appendix 1 contains a listing of EPA Regional offices. Appendix 2 contains a listing of other important federal governmental offices and agencies.

PRIMARY FEDERAL AGENCIES REGULATING HAZARDOUS WASTE
* Environmental Protection Agency (EPA)
* Department of Transportation (DOT)
* Occupational Safety and Health Administration (OSHA)

FEDERAL AGENCY RESPONSIBILITIES
* EPA - RCRA hazardous waste list
* DOT - hazardous waste shipping requirements
* OSHA - monitors worker exposure to hazardous substances

1.3 ENVIRONMENTAL AGENCIES - STATE

State agencies are in charge of oversight and enforcement of state hazardous waste laws. Although the federal government maintains the primary role in hazardous waste regulation, states have implemented their own programs to ensure tighter control over hazardous waste activities within state boundaries. The federal government encourages the states to implement their own programs. Most states have a hazardous waste management authority that oversees compliance with state laws, which are usually tougher than the federal statutes on certain issues. For example, the states may include substances on their list of hazardous wastes that are not included on EPA's list. Appendix 3 contains a listing of state agencies and offices.

1.4 HAZARDOUS WASTE STATUTES AND REGULATIONS

Although the Resource Conservation and Recovery Act is the main focus of this book, companies should be aware that many other federal and state environmental laws regulate hazardous waste activities to some degree. An overview of the primary federal statutes regulating hazardous wastes is provided in this section. A full discussion of each of these laws is beyond the scope of this book. Detailed information about these laws should be sought from other sources.

PRIMARY FEDERAL LAWS THAT THE COMPANY NEEDS TO KNOW
* Resource Conservation and Recovery Act (RCRA)
* Comprehensive Environmental Response, Compensation, and Liability Act (CERCLA)
* Federal Water Pollution Control Act (FWPCA)
* Toxic Substances Control Act (TSCA)
* Clean Air Act (CAA)
* Occupational Safety and Health Act (OSHA)

1.4.1 RESOURCE CONSERVATION AND RECOVERY ACT (RCRA)

The most comprehensive environmental law statute to address the problem of hazardous wastes is the Resource Conservation and Recovery Act of 1976.[1] The statute is administered primarily by EPA.

RCRA is the most significant statute for any company to understand. The primary goal of RCRA is to protect water, land and air from contamination from solid wastes.[2] A secondary goal of the statute is to control hazardous wastes and set standards for the proper identification, listing and handling of hazardous wastes from generation to final disposal. EPA has established several criteria for determining whether a waste is hazardous under RCRA.[3]

Two subtitles are of primary importance to most companies, Subtitle C and Subtitle I. Subtitle C provides the framework for hazardous waste management. The key provisions relate to listing of hazardous wastes, EPA identification number requirements for hazardous waste activities, storage requirements and hazardous waste manifest requirement for waste

---

[1] 42 USC Section 6901 *et seq.*
[2] It is crucial for the company to understand that the term solid waste, although fairly clearly defined in the statute, is quite misleading. The language of the definition referring to "solid, liquid, semisolid, or contained gaseous material," indicates that solid wastes do *not* need to be in solid physical form. In fact, most solid wastes are in liquid or semi-liquid form.
[3] 40 CFR Part 261, Identification and Listing of Hazardous Waste, contains the elements of the hazardous waste definition. Chapter 3 discusses the procedures for identifying whether wastes are considered hazardous under RCRA.

shipments. Subtitle I contains the underground storage tank regulations.

RCRA authorizes EPA to set specific waste management practices for generators, transporters, and treatment, storage, and disposal facilities (TSDFs). Although RCRA only requires TSDFs to obtain permits, RCRA's recordkeeping, manifesting, and reporting requirements for all other generators and transporters are intended to create a framework that will account for all hazardous wastes.

When RCRA was ammended in 1984, Congress added several new programs to complete the hazardous waste management framework. First, Congress directed EPA to promulgate restrictions on land disposal of hazardous waste (commonly known as the Land Ban).[4] Second, Congress directed EPA to issue regulations imposing stringent groundwater monitoring requirements on all existing land disposal facilities. Third, Congress added a new corrective action program to RCRA that gives EPA the authority to order generators and TSDFs to perform cleanups of hazardous waste releases and to undertake investigations and corrective actions at permitted facilities.

Subtitle I of RCRA, the Underground Storage Tank (UST) part of the statute, was enacted in 1984. The statute requires EPA to set national standards for the installation and operation of new USTs and the upgrading and operation of existing USTs. EPA promulgated UST regulations in December 1988, which set technical standards and financial requirements for UST owners and operators. In addition, the statute requires

---
[4]*See* Chapter 5.

the reporting of all leaks and spills from USTs and the remediation of all contaminated property resulting from such leaks and spills. Regulation of USTs is complex and a proper discussion would take an additional volume. Therefore, Subtitle I cannot be covered in this book.

RCRA
* RCRA Subtitle C and EPA regulations contain requirements for identifying, handling and disposing of hazardous wastes
* 40 CFR Part 261 contains the EPA list of hazardous wastes
* RCRA Subtitle I and EPA regulations contain underground storage tank requirements

---

1.4.2 COMPREHENSIVE ENVIRONMENTAL RESPONSE, COMPENSATION AND RECOVERY ACT (CERCLA)

Congress enacted the Comprehensive Environmental Response, Compensation and Liability Act ("CERCLA")[5] in 1980 to facilitate hazardous waste cleanup whenever hazardous substances are released into the environment, and to hold responsible parties liable for such cleanup. CERCLA establishes a National Priority List of the worst sites that are remediated by the government with monies from the so-called "Superfund."[6] Money is

---

[5] 42 USC Section 9601 *et seq.*
[6] There are currently approximately 1,200 sites on the CERCLA National Priority List. The average cost of cleaning up a site is about $30 million. *See* "Superfund and Transaction Costs," No. R-4132-ICJ, RAND Institute (April 24, 1992).

recouped from potentially responsible parties (PRPs) in cost recovery actions. The company might become a PRP if it contributed hazardous waste to a cleanup site. Therefore, it is essential that the company knows that its wastes are shipped to a licensed disposal facility and are not going into an unlicensed dumpsite.

A business could also be found to be owning or operating its own site if hazardous wastes were improperly disposed of on the premises. The surest way to avoid the costly liability of cleanup is to assure that hazardous wastes are sent to a licensed disposal facility or properly and safely contained at the company.

CERCLA identifies four classes of responsible parties: (1) present owners and operators of a site, (2) past owners and operators (if they owned or operated the site at the time the hazardous substance came to be located at the site), (3) generators of the hazardous substance found at the site, and (4) persons who arranged for the hazardous substance disposal at the site. A responsible party's liability is strict, joint, several, and retroactive.[7]

Where the company is most likely to become concerned with CERCLA is if it spills or accidentally

---

[7]This means that the government does not need to prove that a PRP caused the hazardous condition. All the government need prove is that the PRP contributed some quantity of the hazardous waste to the site requiring cleanup. A company could contribute a single drum of waste to the site and still be liable for the entire amount of cleanup. Retroactive simply means that even if the company improperly disposed of some hazardous waste before the CERCLA statute was enacted in 1980, it would be liable despite the fact that the disposal was in compliance with the law that existed prior to 1980.

releases hazardous wastes into the air, ground or water.[8] In such a situation, an emergency response is required to abate any contamination and cleanup the spill. Largely for this reason, Congress passed the Emergency Planning and Community Right-To-Know Act (EPCRA) in 1986 as an independent component of CERCLA.[9] EPCRA requires all states to create Emergency Planning Committees at state and local levels and to develop emergency plans in the event of a release of a hazardous material. Companies are subject to notification and reporting requirements and must make information concerning spills and use or manufacture of hazardous and toxic chemicals available to local, state, and federal officials.

CERCLA
* Governs spills or releases of hazardous substances into the environment
* Hazardous waste sites are cleaned up and PRPs pay for it
* A PRP is anyone who owns or operates a site; generated waste found at the site; or arranged for the disposal of waste found at the site
* The company must report all hazardous waste spills or releases to the Emergency Response Center

---

1.4.3 FEDERAL WATER POLLUTION CONTROL ACT (FWPCA)

The primary objective of the FWPCA, also known as the "Clean Water Act," is to "restore and maintain the

---

[8] See Chapter 6.
[9] See Chapter 7.

chemical, physical, and biological integrity of the Nation's waters."[10] FWPCA Section 311 governs discharges of oil and hazardous substances into U.S. waters.

The current form of the Act was enacted in 1972 and amended in 1977 and 1987. Through the National Pollution Discharge Elimination System (NPDES) permit process, the Act regulates the direct discharge of pollutants into surface waters of the United States. The Environmental Protection Agency (EPA) is delegated the authority to establish effluent limitations for industrial discharges and to require monitoring and reporting on the quantity and quality of the effluent discharged.[11]

1.4.4 TOXIC SUBSTANCES CONTROL ACT (TSCA)

TSCA[12] was enacted to regulate chemical substances that pose an unreasonable risk of injury to health and the environment. TSCA is administered primarily by the EPA. It imposes specific requirements on the use, storage and disposal of different chemicals. TSCA section 6(e) specifically addresses disposal of polychlorinated biphenyls (PCBs). TSCA section 7 deals with imminent hazards presented by chemical substances.[13] A company's primary concern with TSCA

---

[10] FWPCA Section 101, 33 USC Section 1251(a).
[11] A detailed discussion of the Clean Water Act is beyond the scope of this handbook, however, the company must be aware that a spill or release of a hazardous substance into U.S. waters will expose the company to possible penalties and cleanup costs under the Clean Water Act.
[12] 15 USC Section 2601 *et seq*.
[13] TSCA Section 7, 15 USC Section 2606.

should be the proper maintenance of PCB containing equipment, and proper disposal of any PCB waste.

1.4.5 CLEAN AIR ACT (CAA)

The CAA provides a legal framework for air pollution control. The Act sets out primary and secondary ambient air quality standards to protect human health, safety and the environment.[14] In addition, the Act contains air quality standards for "hazardous air pollutants (HAPs)."[15] With the passage of the 1990 Amendments to the Clean Air Act, the EPA must now develop regulations for up to 190 different substances identified in the statute. Previously, EPA had only identified eight HAPs. The company should review this list to determine whether it is complying with the emissions standards for any hazardous pollutant it may be discharging into the air.

1.4.6 OCCUPATIONAL SAFETY AND HEALTH ACT (OSHA)

The Occupational Safety and Health Act[16] contains provisions which regulate toxic and hazardous substance safety in the workplace. The regulations include minimum standards for the prevention of harmful exposure of employees to chemical and hazardous emissions. OSHA also requires that employers maintain current health records on employees who have been exposed to hazardous substances and communicate workplace hazards to employees. OSHA regulations are discussed more fully in Chapter 7.

---

[14] CAA Sections 108-109, 42 USC Sections 7408-7409.
[15] CAA Section 112, 42 USC Section 7412.
[16] OSHA, 29 USC Sections 651-678.

2

The Resource Conservation and Recovery Act

2.1 INTRODUCTION

This chapter serves as a general overview and reference to the Resource Conservation and Recovery Act (RCRA), which is the primary focus of this book. It should be remembered that other environmental laws regulate hazardous waste in addition to RCRA. Those laws were briefly discussed in Chapter 1, and are beyond the scope of this book. This chapter will familiarize the reader with the hazardous waste regulation under RCRA, pointing out how it affects companies that generate and dispose of hazardous waste, and focusing on effective ways to comply with the statute and its regulations. Following a basic discussion of RCRA and its regulations, a helpful checklist is provided which summarizes the various RCRA regulatory requirements. This checklist should prove useful to the company as a quick reference guide when reviewing RCRA compliance. The checklist also provides a working outline of the issues that will be discussed in more detail throughout the book.

2.2 RCRA'S SCOPE AND PURPOSE

The Resource Conservation and Recovery Act originally became law in 1976. RCRA established a comprehensive regulatory scheme for hazardous waste management. Its primary focus is on waste minimization and safe treatment, storage and disposal of hazardous wastes.[1] The statute was amended in 1984 by the Hazardous and Solid Waste Amendments of 1984 ("HSWA"), which expanded RCRA's scope with comprehensive hazardous waste management requirements.[2] RCRA basically consists of three subtitles. Subtitle C covers Hazardous Waste Management, Subtitle D covers State or Regional Solid Waste Plans, and Subtitle I covers Regulation of Underground Storage Tanks. A separate volume would be necessary to adequately discuss each of RCRA's three subtitles. The individual requirements of Subtitle C are addressed in this book.

2.3 WASTES REGULATED UNDER RCRA

RCRA provides the "cradle to grave" scheme for regulating hazardous wastes. The regulatory framework is intended to track waste from generation to disposal. More specifically the statute regulates "solid waste." The exact meaning of solid waste causes some confusion because as the term is defined in the statute and regulations, a waste need not be in solid form to be considered a solid waste.[3] Liquids, gases, semi-solid

---

[1] RCRA's purpose is expressed in the statute as: "[W]herever feasible, the generation of hazardous waste is to be reduced or eliminated as expeditiously as possible. Waste that is nevertheless generated should be treated, stored, or disposed of so as to minimize the present and future threat to human health and the environment." RCRA Section 1003(b).
[2] Pub. L. No. 98-616, 98 Stat. 3221 (1984).
[3] Solid waste is defined in the statute as "garbage, refuse, sludge . . . or other discarded material . . . from industrial, commercial, mining, and agricultural operations." 42 USC 6903(27).

and solid materials can all fall within the definition of "solid waste."[4]

The solid waste definition is even more perplexing because it is not always clear which materials are considered wastes. A waste is generally understood to mean a material that no longer has a use. EPA regulations provide tests for determining what materials are considered "wastes," however, these tests are far from clear. In general terms, the EPA test looks at the type of material and the type of management activity involved to determine whether a material is a product or a waste. A product can be put to some use and is therefore exempt from RCRA regulation, whereas a material that is no longer useable is considered a waste, which may be regulated under RCRA. Recycled materials may be thought of as a product exempt from RCRA regulation, however, recycled materials may be considered solid wastes in many instances. EPA specifies several types of recycled materials as being subject to regulation as solid wastes under RCRA.[5] The courts continue to debate which recycling activities are covered by RCRA.[6]

---

[4] See 50 Fed. Reg. 614 (Jan. 4, 1985) for a detailed description of EPA's definition of the term "solid waste."

[5] These include recycled materials that are: (1) used in a manner constituting disposal (40 CFR Section 261.2(c)(1)); (2) Burned for energy recovery (40 CFR Section 261.2(c)(2)); (3) Reclaimed (40 CFR Section 261.2(c)(3)); or (4) Accumulated speculatively (40 CFR Section 261.2(c)(4)).

[6] See, e.g., American Mining Congress v. EPA, 824 F.2d 1177 (D.C. Cir. 1987); American Petroleum Institute v. EPA, 906 F.2d 729 (D.C. Cir. 1990).

2.4 SUBTITLE C FUNDAMENTALS

Aside from determining whether a material fits the definition of a "solid waste," if it is not also a hazardous waste, it will not be subject to regulation under Subtitle C of RCRA. Obviously, a material may be a waste but not hazardous, or hazardous but not a waste. In neither of these instances would the material be regulated under RCRA Subtitle C. Therefore, the proper inquiry involves first determining whether a material is a waste, and then deciding whether or not it is hazardous. Determining whether a solid waste is "hazardous" is normally the key inquiry for purposes of Subtitle C.

Two situations arise where a waste is considered hazardous under RCRA. First, the waste may be found on an EPA list of hazardous wastes. These are RCRA "listed" hazardous wastes.[7] Second, a waste may exhibit certain hazardous characteristics, which will subject the waste to RCRA regulation as a hazardous waste. These wastes are commonly referred to as "characteristic" hazardous wastes. A solid waste is a characteristic hazardous waste if it exhibits any of

---

[7]The lists are found in 40 CFR Part 261 and are as follows: (1) Hazardous waste from non-specific sources (40 CFR Section 261.31); (2) Hazardous waste from specific sources (40 CFR Section 261.32); (3) Discarded commercial chemical products, off-specification species, container residues, and spill residues thereof (40 CFR Section 261.33).

four characteristics: Ignitability,[8] Corrosivity,[9] Reactivity,[10] or Toxicity.[11]

In addition to listed and characteristic hazardous wastes, two more categories of hazardous wastes are regulated under RCRA Subtitle C: solid waste that has been mixed with a hazardous waste, and solid waste generated from a treatment, storage or disposal facility. The first type is regulated under EPA's "mixture rule,"[12] and the second type is governed by EPA's "derived from rule."[13] Wastes that have been mixed with a characteristic hazardous waste are only subject to regulation if the waste mixture exhibits one of the four hazardous characteristics.

Certain materials are exempt from classification as solid wastes and some materials are exempt from classification as RCRA hazardous wastes. Excluded materials include: materials that are disposed of in public sewer systems, industrial discharges that are subject to Clean Water Act permitting requirements, residues from fossil fuel combustion, and certain "mining" wastes.[14]

---

[8] See 40 CFR Section 261.21 for a description of this characteristic.
[9] See 40 CFR Section 261.22 for a description of this characteristic.
[10] See 40 CFR Section 261.23 for a description of this characteristic.
[11] See 40 CFR Section 261.24 for a description of this characteristic. This category was formerly known as EP Toxicity, named after EPA's former test for toxicity. The earlier test was replaced by the new Toxicity Characteristic Leaching Procedure (TCLP test) on June 1, 1990. See 55 Fed. Reg. 11798 (Mar. 29, 1990) for a full descripton of the new testing procedure.
[12] See 40 CFR Section 261.3(a)(2)(iii).
[13] See 40 CFR Section 261.3(c).
[14] See 40 CFR Section 261.4.

2.5 HAZARDOUS WASTE GENERATORS

Generators of hazardous wastes are regulated to different degrees depending on the amount of wastes they generate. Part 262 of Title 40 of the Code of Federal Regulations contains detailed requirements for hazardous waste generators.

All generators are required to identify and keep track of their hazardous wastes.[15] Each generator must apply for an EPA hazardous waste identification number.[16] This number is used for various reporting and recordkeeping purposes. Treatment, storage or disposal of hazardous waste is prohibited without the EPA I.D. number.

For a company that carries out waste generating activities, it is crucial that it keep careful track of all wastes generated on the premises, and implement an effective hazardous waste management program to assure RCRA compliance. Whenever possible, waste minimization should be sought. Waste minimization, such as recycling activities, can be a cost-effective means of managing company wastes, and by minimizing the quantity of waste generated, the company may be subject to less stringent regulation. Above all, the company should exercise care to ensure that it never becomes classified as a treatment, storage and disposal facility (TSDF) because TSDFs are the most heavily regulated class of generators under RCRA. TSDFs have specific permitting requirements, must meet certain technical standards, and need high amounts of environmental liability insurance. It is easy enough for most companies to

---

[15] 40 CFR Section 262.11.
[16] 40 CFR Section 262.12.

avoid being classified as a TSDF as long as they are careful about how and for how long they store any wastes, and make sure that any on-site treatment falls outside the scope of TSDF status.[17] Because it is unlikely that most companies will become classified as TSDFs, the specific requirements for TSDFs are not considered in this book. These requirements are extensive and are beyond the scope of this volume. Suffice it to say that TSDF status is something any company would rather not have to know about.

A generator may store hazardous waste on-site for 90 days without a permit if it complies with the following requirements:

* Each container of waste is clearly labeled with the words "Hazardous Waste" and the date of the start of the accumulation period.[18]
* The waste is stored in containers or tanks that meet the standards set out in 40 CFR 265 Subparts I and J.
* The generator performs waste analysis and trial tests required by 40 CFR 265.200.[19]
* The generator has trained personnel in certain areas of hazardous waste management[20] and has a contingency plan and emergency procedures in place.[21]

---

[17] For example, a company may choose to reclaim certain spent chemicals on-site to minimize the amount of waste generated. Although reclamation may be considered "treatment" of hazardous waste (which triggers TSDF status), if properly done, the process would be considered a recycling activity that is exempt from RCRA regulation. 40 CFR Section 261.6(c)(1).
[18] 40 CFR Section 262.34(a)(2), (3).
[19] 40 CFR Section 262.34(a)(1).
[20] 40 CFR Section 262.34(a)(4).
[21] See 40 CFR Section 265 Subpart C (Preparedness and Prevention regulations) and 40 CFR Section 265 Subpart

RCRA regulations provide for total exemption from both the permit requirements and the 90 day rule for limited "satellite" accumulation of hazardous waste provided that the generator does not accumulate more than 55 gallons of a hazardous waste or one quart of acutely hazardous waste. Some additional requirements for satellite accumulation are set forth in the regulations, however, these quantity limitations are the most important prerequisites for the satellite accumulation status.[22]

Small quantity generators (SQGs) of hazardous waste are also exempt from RCRA permitting requirements provided that certain conditions are met. A SQG generates more than 100 kilograms but less than 1,000 kilograms of hazardous waste in a calendar month.[23] A SQG may store hazardous waste on-site for up to 180 days without a permit.[24] Additional requirements for SQGs are set forth in the regulations,[25] which are discussed in more detail in Chapter 4.

2.6 HAZARDOUS WASTE TRANSPORTATION

All generators are subject to specific requirements when transporting hazardous wastes off-site for treatment, storage or disposal. RCRA requires that each shipment of hazardous waste be accompanied by a hazardous waste manifest.[26] In addition, to the

---

D (the Contingency Plan and Emergency Procedures regulations).
[22] 40 CFR Section 262.34((c)(1).
[23] 40 CFR Section 262.34(d).
[24] 40 CFR Section 262.34(d).
[25] 40 CFR Section 262.34(d)(1)-(5).
[26] 40 CFR Section 262.20 - 262.22.

manifest requirement, the generator must properly package, label and mark each container according to Department of Transportation (DOT) regulations. These requirements are explained in detail in Chapter 8.[27]

2.7 RECORDKEEPING AND REPORTING

RCRA also contains specific recordkeeping and reporting requirements for all generators of hazardous wastes. Generally speaking, each generator must keep copies of all hazardous waste manifests, biennial reports and other types of records for at least three years.[28] Reporting requirements include: filing of biennial reports,[29] manifest exception reports;[30] and other information.[31] The reporting and recordkeeping requirements are explained in detail in Chapter 6.

2.8 LAND DISPOSAL RESTRICTIONS

The Hazardous and Solid Waste Amendments of 1984 (HWSA) instituted new requirements for land disposal of hazardous wastes. These land disposal restrictions (LDRs) are commonly referred to as the "Land Ban." Regulations promulgated by EPA to implement the LDRs went into effect in phases, the last phase taking effect on May 8, 1990. The purpose of the land disposal restrictions is to ban hazardous waste disposal on or

---

[27]The primary DOT regulations concerning hazardous waste transport are: Packaging (40 CFR Section 262.30); Labeling (40 CFR Section 262.31); Marking (40 CFR Section 262.32); and Placarding (40 CFR Section 262.33).
[28]40 CFR Section 262.40(a)-(c).
[29]40 CFR Section 262.41
[30]40 CFR Section 262.42
[31]40 CFR Section 262.43. Specific reporting requirements for SQGs are found at 40 CFR Section 262.44.

into land except after wastes have been treated according to specific EPA treatment standards. In general terms, the EPA treatment standards usually require that, before land disposal, wastes must be treated to meet a certain concentration level or be treated by a specified technology such as incineration. In certain instances, a company may be able to obtain a variance from the land disposal restrictions. The company submits an application to EPA for a treatability variance,[32] delisting, extension, or other type of relief from the LDRs.[33] The land disposal restrictions are discussed in detail in Chapter 5.[34]

RCRA QUICK REFERENCE CHECKLIST

Step 1. Identify Hazardous Wastes
___ Check EPA Hazardous Waste List (40 CFR 261.31-.34)
___ Check Wastes That Have Been Mixed With Listed Wastes
___ Check Wastes for Hazardous Characteristics (40 CFR 261.21-.24)
  * Ignitability
  * Corrosivity
  * Reactivity
  * Toxicity
___ Check Waste Mixtures for Hazardous Characteristics

---

[32] 40 CFR Section 268.44.
[33] Other options include no migration petitions and petitions for equivalent treatment methods.
[34] EPA has published a useful overview of the land disposal restrictions. Overview of RCRA Land Disposal Restrictions (LDRs), U.S. EPA, Office of Solid Waste and Emergency Response, Directive: 9347-3-01FS (July 1989).

RESOURCE CONSERVATION AND RECOVERY ACT

___ Check to See if Any Waste Exemption Applies (40 CFR 261.4)

Step 2. Hazardous Waste Storage

___ Apply for EPA Hazardous Waste I.D. Number
___ Use Safe Containers Compatible with Waste Being Stored
___ Properly Label All Containers
___ Keep Hazardous Waste Log

Step 3. Determine Generator Category

___ Satellite Accumulation
  * No more than 55 gallons of hazardous waste
  * No more than 1 quart of acutely hazardous waste
  * Storage for no more than 90 days
___ Small Quantity Generator
  * Between 100 and 1,000 kilograms of hazardous waste per month
  * Storage of no more than 180 days
  * 270 day storage if waste not transported more than 200 miles
  * Trained employee on-site or on call to handle emergencies
___ Large Quantity Generator
  * More than 1,000 kilograms of waste per month
  * Storage of no longer than 90 days
  * Subject to permitting and other requirements (40 CFR Part 264)

Step 4. Hazardous Waste Disposal

___ Must Use Manifest When Transporting Off-Site
___ Must Comply With DOT Hazardous Waste Requirements

22 RCRA REGULATORY COMPLIANCE GUIDE

      * Proper Packaging (40 CFR 262.30)
      * Proper Labeling (40 CFR 262.31)
      * Proper Marking (40 CFR 262.32)
      * Proper Placarding (40 CFR 262.33)
___ Check Compliance With Land Disposal Restrictions
      * Wastes below EPA mandated concentration levels
      * Wastes treated according to EPA specifications (40 CFR 268)
      * Variances obtained when necessary

Step 5. Hazardous Waste Recordkeeping and Reporting
___ Hazardous Waste Manifests
___ Test Results
___ Biennial Report
___ Manifest Exception Report
___ Training Records
___ Contingency Plan

3

Hazardous Waste Identification and Classification

3.1 INTRODUCTION

It is the responsibility of each company to determine whether hazardous wastes are present on the premises. First, the company will need to make an assessment of which wastes are present, and then determine whether the waste is regulated as a "listed" or "characteristic" waste. Second, the company will need to determine what quantities of the waste are present. Finally, the company will need to determine how to comply with all applicable federal and state laws and regulations governing the hazardous waste.

3.2 DEFINITION OF HAZARDOUS WASTE

In order to properly manage hazardous wastes found at the company, it is essential that the company first understand what wastes are considered "hazardous" under the various laws and regulations. The terms "hazardous waste," "hazardous substance," "hazardous material," and "chemical substance" are the common legal terms referring to substances that are regulated because of their hazardous nature. This book will generally refer to the term "hazardous waste" since it is the term used in the Resource Conservation and Recovery Act (RCRA), which is the primary environmental law governing hazardous waste management. The other terms will be used only when the more specific term is required to explain other environmental laws that use a different term.

3.3 FEDERAL LAWS GOVERNING HAZARDOUS SUBSTANCES

Resource Conservation and Recovery Act (RCRA) - administered by EPA, uses "hazardous waste" and generally means a material with hazardous characteristics that is no longer useable.

Hazardous Materials Transportation Act (HMTA) - administered by DOT, uses the term "hazardous material" to refer to materials of a hazardous nature that have a functional use; once the material has been used, it then becomes a hazardous waste.

Comprehensive Environmental Response, Compensation, and Liability Act (CERCLA) - administered by EPA, uses the term "hazardous substance" to refer primarily to both hazardous materials and hazardous wastes that have been released into the environment.

Toxic Substances Control Act - administered by EPA uses the term "chemical substance" to refer to substances that pose an unreasonable risk of injury to health and the environment.

3.4 TERMINOLOGY RELATING TO HAZARDOUS WASTE

Hazardous Material - a substance that has a functional use, such as gases, fertilizers, fuels, acids, solvents and cleaning compounds, which poses an unreasonable risk of harm to health, safety or property.

Hazardous Waste - the by-product of a used material

that is no longer useable. Hazardous wastes are dangerous to handle, or to dispose of, and may cause death, illness, or a hazard to health or the environment when improperly handled.

Toxic Waste - due to its chemical make-up, a toxic waste has properties that are capable of killing, injuring or damaging an organism. It is a type of hazardous waste.

Hazardous Substance - this is the broadest term and includes hazardous materials, hazardous wastes and toxic wastes.

3.5 RCRA HAZARDOUS WASTES

There are several environmental statutes that define hazardous wastes and materials. The Resource Conservation and Recovery Act is the primary law governing handling of hazardous wastes and is the focus of this book. The U.S. Environmental Protection Agency (EPA) has authority under RCRA to list as hazardous those wastes exhibiting certain hazardous properties. These wastes are known as "listed wastes" because they are listed in EPA regulations.[1] The company should identify its wastes and then determine whether any of its wastes are on the EPA list. If the waste is unlisted, it may still be regulated, however. Unlisted wastes that exhibit hazardous characteristics are regulated as "characteristic wastes."[2]

---

[1] The RCRA hazardous waste list is reproduced from 40 CFR Part 261 as Appendix 4 at the end of this book.
[2] A listed waste need not be checked for hazardous wastes characteristics since EPA has already concluded that such wastes are hazardous. Moreover, under 40 CFR Section 262.11(b), (c), a waste generator must first determine whether a waste is listed, and only if

The company must be careful to check all its wastes. If unlisted, the company should have samples tested if there is any possibility that the waste could be a characteristic waste. When in doubt, the company may call the EPA RCRA/Superfund hotline for advice at (800) 424-9346, or contact the appropriate EPA Regional Office. Addresses and phone numbers for the various Regional Offices are listed in Appendix 1.

RCRA defines a hazardous waste as any solid waste or combination of solid wastes: "which because of its quantity, concentration, or physical, chemical, or infectious characteristics may (A) cause or significantly contribute to an increase in mortality or an increase in serious irreversible, or incapacitating reversible illness; or (B) pose a substantial present or potential hazard to human health or the environment when improperly treated, stored, transported, or disposed of, or otherwise managed."[3]

RCRA hazardous wastes are regulated by the EPA. EPA has established several criteria to determine whether a particular waste is hazardous within the meaning of the statutory definition.[4] A prerequisite to a waste's regulation under RCRA is that it be a solid waste. It is crucial for the company to understand that the term "solid waste" is quite misleading. Solid wastes can, in fact, be in solid, liquid, or semi-solid

---

unlisted, does it determine whether the waste exhibits a hazardous waste characteristic.
[3]RCRA Section 1004(5); 42 USC Section 6903.
[4]Consult 40 CFR Part 261, "Identification and Listing of Hazardous Waste," for the elements of the regulatory definition.

form, and may even contain gaseous material.[5] Most solid wastes are actually in liquid or semi-liquid form.

The first place for the company to look for EPA listed wastes is in Part 261 of Volume 40 to the Code of Federal Regulations (CFR). Appendix 4 presents the RCRA list of hazardous wastes. EPA concentrates most on these listed wastes.

If the waste is unlisted, the company needs to make a further determination on whether a waste may be a characteristic waste. A qualified professional should be consulted, and if necessary, waste samples tested for hazardous characteristics. EPA basically classifies characteristic hazardous wastes in one of six categories according to the following characteristics: ignitability, corrosivity, reactivity, toxicity, acute hazardous or toxic characteristic. These categories are explained below.

An unlisted waste also may be considered a RCRA hazardous waste if it is (1) a by-product derived from the treatment, storage or disposal of a listed hazardous waste;[6] or (2) it is a mixture containing a listed hazardous waste.[7]

Determining whether a waste is hazardous can be a very complex procedure. The company will normally not be enough of an expert to make the determination on its own. In most cases, it is advisable for the company to

---

[5]RCRA Section 1004(27); 42 USC Section 6903(27) contains the definition of solid waste.
[6]This rule is set out in 40 CFR Section 261.3(c)(2)(i).
[7]The components of the mixture rule are set out in 40 CFR Sections 261.3(a)(2)(iv), (b)(2), (c)(1), and (d)(2).

retain a hazardous waste manager or consultant to make the determination. The money spent on consulting and testing fees should be regarded as insurance against the risk of fines and/or liability that may be associated with noncompliance with hazardous waste laws and regulations.

Also remember that some states have more stringent laws than the federal government. The company should obtain specific information about its state regulatory program. The best place to start is with the hazardous waste management office in the state. Appendix 3 contains the names, addresses and phone numbers for the various state agencies. A discussion of the individual state laws is beyond the scope of this book.

3.6 "LISTED" AND "CHARACTERISTIC" HAZARDOUS WASTES

Listed Wastes are wastes that are found on EPA's list of hazardous wastes (40 CFR Part 261)

Characteristic Wastes are wastes that exhibit one or more of the following characteristics:
- It is easily combustible or ignitable, such as paint wastes, spent solvents, or degreasers;
- It dissolves metals or corrodes other materials, or burns skin, such as paint removers or battery acids;
- It is unstable and undergoes a rapid or violent chemical reaction when exposed to water or other materials, such as waste bleaches or other oxidizers;
- It is tested under laboratory analysis using the Toxicity Characteristic Leachate Procedure (TCLP) and is found to be toxic. Such wastes generally contain high concentrations of heavy metals.

3.7 HAZARDOUS WASTE CLASSIFICATION

Once it has been determined that a waste is hazardous, the waste will fall into one or more categories or classes of waste.[8] The EPA classification system is outlined below.

3.8 EPA HAZARD CLASSES

Ignitable Waste (I)

Ignitable waste is a liquid with a flash point less than 60 degrees centigrade, a solid capable of causing fire that burns vigorously and continuously so that it creates a hazard, an ignitable compressed gas or an oxidizer.

Wastes exhibiting the (I) characteristic have the EPA Hazardous Waste Number of D001.

Corrosive Waste (C)

Corrosive waste has a pH less than 2 or greater than 12.5 or corrodes steel at a rate greater than 6.35 mm/year.

Wastes exhibiting the (C) characteristic, and that are not listed, have the EPA Hazardous Waste Number of D002.

Reactive Waste (R)

Reactive waste exhibits one or more of the following characteristics: normally unstable and undergoes violent change without detonating; reacts violently with water; forms explosive mixtures with water; generates toxic gases, vapors, or fumes; is a cyanide or sulfide bearing waste that generates gases, vapors

---

[8] 40 CFR Section 261.30(b) provides a description of the six EPA hazard codes. These codes are used to denote the underlying basis for the listing of the waste as hazardous under RCRA.

or fumes when exposed to a pH between 2 and 12.5; is capable of detonation or explosive reaction; or is a forbidden explosive.

Wastes exhibiting the (R) characteristic have the EPA Hazardous Waste Number of D003.

Toxicity Characteristic Waste (E)

Wastes fall into this class when tested under the Toxicity Characteristic Leachate Procedure (TCLP)[9] and are found to produce an extract that contains specific quantities of any of the following: arsenic, barium, cadmium, chromium, lead, mercury, selenium, silver, endrin, lindane, methoxychlor, toxaphene, 2,4-D or 2,4,5-TP.

Acute Hazardous Waste (H)

Acute hazardous wastes include discarded commercial chemical products, containers, spill residues, manufacturing chemical intermediates, off-specification commercial chemical products, all subject to small quantity exclusions.

Toxic Wastes (T)

Toxic wastes are generally listed wastes and include such wastes as toluene (U220), warfarin (U248) and zinc phosphide (U249).

3.9 COMMON HAZARDOUS WASTES

There are literally thousands of different hazardous wastes. By consulting the various lists found in the Code of Federal Regulations, the company will find the principal wastes that are regulated at the federal level. This section describes some of the

---

[9]The TCLP test replaced the formerly used EP Toxicity test on September 9, 1990.

common wastes that a company may encounter in its business operations.

3.9.1 PETROLEUM PRODUCTS

Every company has some potential of discharging petroleum based products into the environment. Used oil is frequently on hand from changing oil from various pieces of equipment. The used oil may be stored in drums, which may leak, or oil may be inadvertently spilled on the ground. The company may have underground storage tanks containing gasoline that are susceptible to leakage. The company must closely monitor its business practices to maintain containers in good condition and to avoid spillage of any petroleum products on the ground. The hazards associated with discharges of petroleum are most apparent in soil and groundwater contamination. Benzene, a component of gasoline, for instance, is a carcinogen.

3.9.2 CHLORINATED HYDROCARBONS

Chlorinated hydrocarbons are most commonly found in general solvents, paint thinners, paint strippers and degreasers. Most every company uses chlorinated hydrocarbons to some extent. Cleaning solvents also are quite commonly in use at companies.

The liquid components of chlorinated hydrocarbons contaminate soil and groundwater, and the vapor component is damaging to air quality, particularly the ozone layer. Quantities as small as one gallon can easily spread through the ground and water to create a wide column of contamination over time. The company must keep this in mind, use chlorinated hydrocarbon products carefully, and initiate responsible disposal practices for waste by-products. Soil samples near the site of use should be taken on a periodic basis to determine the extent of any ground contamination. Check the federal and state hazardous waste regulations to determine what quantities are considered hazardous.

3.9.3 POLYCHLORINATED BIPHENYLS (PCBS)

PCBs were manufactured in the United States during the period 1929 to 1977. They were most frequently used in such products as electrical transformers, switches, voltage regulators, capacitors, hydraulic fluids, carbonless copy paper, paints, adhesives, caulking compounds, and road coverings. PCBs are toxic in low concentrations and cause birth defects, cancer and other diseases. When burned, PCBs form the toxic chemical dioxin.

Section 6(e) of the Toxic Substances Control Act (TSCA) specifically addresses disposal of polychlorinated biphenyls.[10] TSCA prohibits all manufacture or use of PCBs except for enclosed use, and requires labeling and safe disposal practices. The Food and Drug Administration and EPA have additional PCB regulations governing prohibitions and safeguards on use, and final disposal, such as incineration or in hazardous waste landfills.

The main source of PCB contamination results from electrical fires. The company's best safeguard against contamination, and costly cleanup, would be for a due diligence inspector, or environmental consultant to

---
[10] TSCA Section 6(e), 15 USC Section 2605(e).

review the location of any PCB containing and venting facilities located on the premises. It is also wise to replace any old electrical equipment. Wastes that contain PCBs should be properly disposed of in compliance with EPA regulations.

3.9.4 ASBESTOS

Asbestos fibers pose a hazard to human health. Many products contain asbestos, which remains non-friable unless fibers are released by some type of abrasive action.[11] Brake linings, linoleum, cement pipe, plastics, asphalt, vinyl, and roofing materials are among the products that may containing non-friable asbestos. Of particular concern are the products that contain friable asbestos, such as sprayed and troweled insulation, pipe covering paper and insulating board. Friable asbestos can be released into the air and inhaled, causing asbestos-related diseases.[12]

The health hazard posed by asbestos has caused great alarm in the past decade. Pursuant to the Clean Air Act, EPA has designated asbestos as a hazardous air pollutant. A report to Congress by the EPA estimates that some 501,000 public and commercial buildings contain damaged friable asbestos.[13] EPA has estimated the cost of asbestos cleanup in commercial buildings at

---

[11] "Friable" is the term used for air-borne asbestos fibers; non-friable asbestos is encapsulated or contained sufficiently that fibers are not released into the air.

[12] The known diseases caused by asbestos are: (1) asbestosis, which causes a scarring of lung tissue; (2) lung cancer involving a dysfunction of the lung cells; and (3) mesothelioma, which is cancer of the chest lining.

[13] 18 Envt Rep. (BNA) 2257 (Mar. 4, 1988).

near $4 trillion.[14] Obviously, the problem is great, but often removal of asbestos has caused more harm than benefit. Debate goes on over whether asbestos might be better left alone. As long as the asbestos remains intact, it poses no immediate danger. In recent years, school buildings have been a focus of asbestos abatement. EPA's "Asbestos Rule" targets cleanup in schools with an estimated cost of $3.1 billion over 30 years.[15]

The company must perform an investigation to determine where asbestos exists and whether it is in a hazardous condition. An environmental consultant or asbestos abatement firm can recommend the best course of action: removal, encapsulation, enclosure, or just leave it alone. If asbestos requires removal, it must be properly wetted, labeled and containerized for disposal. Chapter 8 describes the proper procedures for proper shipment of asbestos waste to disposal facilities.

Finally, the Occupational Safety and Health Act[16] contains specific standards for regulating asbestos exposure in the workplace. Sections 7.8 and 7.9 of this book provide a detailed discussion of the OSHA Asbestos Standard and the requirements for worker safety compliance.

3.9.5 SULFURIC ACID

Sulfuric acid is commonly used in automobile batteries and large industrial batteries. A company may

---

[14] See *The American Lawyer*, July/August 1988 at p. 6.
[15] 15 USC Sections 2641-2654; 40 CFR Part 763.
[16] 29 USC Section 651 *et seq*.

be faced with disposal of old or damaged batteries. The acid may be recycled, and in the case of lead acid batteries, the lead may be reclaimed. The state hazardous waste management agency should be consulted for advice on where the company may go and how to properly dispose of used batteries. Chapter 8 of this manual provides some tips for proper shipment of used batteries to a disposal facility.

3.9.6 HEAVY METALS

The presence of heavy metals, such as lead, arsenic, barium, cadmium, chromium, mercury, selenium and silver, in certain quantities in waste material may be considered hazardous by EPA or a state agency. When wastes that are tested using the EPA's Toxicity Characteristic Leachate Procedure are found to contain unacceptable levels of heavy metals, the waste must be disposed of in accordance with safe disposal methods used for hazardous wastes. Used oil is a common waste that contains heavy metals. High concentrations of lead are often found in used oil.

4

Hazardous Waste Generators

4.1 INTRODUCTION

Industry generates billions of pounds of hazardous waste each year, and is therefore, the focus of regulation by many of the federal and state environmental laws. Although other groups, such as households, farmers and federal agencies, generate a considerable volume of hazardous waste, industrial activities are regulated much more stringently than others.[1] Businesses are easier to monitor for legal compliance and usually have "deeper pockets" for contribution toward cleanups.

A company is considered a "generator" of hazardous wastes whenever it generates hazardous waste from products it uses. As a generator, the company's first obligation is to identify its wastes.[2] Next, the company will need to obtain an EPA identification number, and if necessary, a state identification

---

[1] Household waste is by its very nature difficult to regulate. In fact, most household wastes go directly into landfills by means of the weekly trash pickup. Household cleaners, paint products, pesticides, and automotive wastes that are disposed of by every household comprise an extraordinary amount of hazardous wastes that go unchecked into the environment. *See* Ferry, "The Toxic Time Bomb: Municipal Liability for Hazardous Waste," 57 Geo. Wash. L.Rev. 197 (Dec. 1988).
[2] Chapter 3 of this book describes the process for identifying hazardous wastes.

number. To apply for the number, the company will need to submit a Notification of Hazardous Waste Activity.[3]

4.2 HAZARDOUS WASTE GENERATOR CATEGORIES

The company will need to comply with the federal and state hazardous waste regulations that apply to the generator class that the company falls under. The classification will be determined by the volume of hazardous wastes that are generated at the company each month. Each container must be clearly labeled with the words "Hazardous Waste" with the date that the first batch was deposited in the container. Once the container is shipped from the site, additional labeling and marking requirements apply. Chapter 8 explains the requirements for transporting hazardous wastes. If any underground storage tanks are used for storing used oil or petroleum products, additional regulations apply.

GENERATOR CLASSES
* Conditionally Exempt Small Quantity Generator (CESQG)
* Small Quantity Generator (SQG)
* Large Quantity Generator (LQG)

4.2.1 CONDITIONALLY EXEMPT SMALL QUANTITY GENERATOR

A Conditionally Exempt Small Quantity Generator generates no more than 100 kilograms of hazardous waste per month.[4] These generators are exempted from the RCRA

---

[3] This form may be obtained from the EPA regional office in the company area. A sample Notification form is reproduced as Appendix 5 at the end of this book.
[4] 40 CFR Section 261.5 contains special requirements for CESQGs.

requirements provided that they do not exceed the following limits for hazardous waste storage or generation:

CESQG LIMITS	Acutely hazardous waste	Hazardous waste
Generation	1 kg/month	100 kg/month
Waste storage	1 kg	1000 kg
Residue from cleanup of spills/leaks	100 kg	

If the company exceeds the above limits, it is subject to the regulations governing Small Quantity Generators (SQGs) or Large Quantity Generators (LQGs).

4.2.2 SMALL QUANTITY GENERATOR (SQG)

Most companies will fall under the Small Quantity Generator category. The EPA regulations for SQGs[5] are the pertinent regulations unless the state has more stringent regulations.[6] A SQG produces between 100 kilograms (about 220 pounds or one-half of a 55 gallon drum) and 1,000 kilograms (about 2,200 pounds) per calendar month. SQGs are permitted to accumulate hazardous waste on site for up to 180 days without a permit provided that the total amount of accumulated

---
[5] 40 CFR Part 262.
[6] Several states have more stringent requirements for SQGs, and California, Louisiana, and Rhode Island have no exemption at all for SQGs. In these states, the SQG is subject to the same requirements as Large Quantity Generators.

waste is not greater than 6,000 kilograms and an employee is available on site or on call to handle hazardous waste emergencies. If the waste is to be transported more than 200 miles, the company may accumulate waste on site for 270 days. If the company exceeds the 180 (or 270) day limit or stores more than 6,000 kilograms of waste, the company will be considered a storage facility which is subject to additional regulatory requirements.[7] All waste containers must be inspected at least once a week. If a leak is detected, the waste must be transferred to another container.

4.2.3 LARGE QUANTITY GENERATOR (LQG)

Relatively few companies will produce enough hazardous waste per month to be classified as a LQG. If the company produces hazardous waste in amounts exceeding 1,000 kilograms per month, it will, however, become subject to the full set of regulations governing hazardous waste generators. Large quantity generators must file a waste management report with the regional EPA office on March 1st of each even numbered year and maintain an operating log that details how wastes are managed at the company. LQGs must also dispose of their wastes within 90 days.

4.3 DETERMINING THE WASTE GENERATOR CATEGORY

In order for the company to determine the relevant waste generator category, it must monitor the volume of waste produced on a monthly basis. The generator category could change each month depending on the

---

[7] 40 CFR Part 264.

amount of hazardous waste generated during the month. If wastes from different months are accumulated and mixed together, regulations for the most stringent generator category that applied to any single month will be applied to the entire accumulated amount. Of course, it is possible that the company could be a CESQG one month and then be regulated as a LQG the following month if the volume of waste generated reaches the LQG level that month. It is crucial that the company monitor its waste levels each month to determine the appropriate generator category.

In order to accurately determine which waste generator category the company falls under, it must properly identify which wastes will be considered "hazardous" by the EPA and DOT. Chapter 3 outlines the process for determining which wastes are considered hazardous. In some instances, it may be unclear to the company whether a particular waste is considered hazardous under the regulations. When in doubt, it is always best to contact the EPA regional office or the state hazardous waste management agency.

Each company site will count as a separate generator for purposes of the generator status determination. A company that operates more than one business at the same site should request a determination from EPA or the state environmental agency regarding whether each business is considered a separate generator or must be counted as one generator.

5

Hazardous Waste Storage and Disposal

5.1 INTRODUCTION

For effective handling and management of hazardous wastes, the company needs to comply with the various laws and regulations pertaining to storage, disposal, recordkeeping and reporting. After the company has identified its wastes, and submitted its application to EPA and the state environmental authority for an identification number, it will then determine how best to manage the wastes it generates. Wastes should be properly stored before shipment for treatment and disposal. Chapter 3 discussed proper techniques for identifying hazardous wastes. Chapter 4 explained how to determine the appropriate generator classification for the company and this chapter will discuss the regulatory requirements for storing and handling hazardous wastes. Chapter 8 examines the proper means of transporting wastes from the company for treatment and disposal. Chapter 6 will discuss all recordkeeping and reporting requirements.

5.2 HAZARDOUS WASTE STORAGE

The most important aspect of hazardous waste storage relates to minimizing the potential for leaks and spills. Proper labeling, recordkeeping, and emergency preparedness are also essential. Different regulatory standards apply to storage, depending upon which generator class the company falls under. In most cases, it is likely that the company will be classified as a Small Quantity Generator (SQG).[1] Storage requirements for SQGs are found in 40 CFR Part 262,[2] and are explained in this Chapter.

5.3 LABELING

Hazardous waste containers must be labeled with the date that the first amount of waste was deposited in the container and the words "Hazardous Waste" must be clearly marked on every container. The chemical name, the physical state (solid, liquid, gas), the hazardous properties of the waste (flammable, corrosive, etc.), composition, and the name/address of the generator must all be on the label.

5.4 EMERGENCY RESPONSE PLANS

In order to store wastes on site, the company is obligated to prepare a contingency plan that ensures proper maintenance and operation of hazardous waste equipment. In the event of a fire, explosion, or other release, the company must carry out the emergency procedures outlined in the contingency plan. The company must also maintain a testing plan that ensures that all required safety and alarm systems are in proper working order. The contingency plan must include the names and addresses of emergency personnel and

---

[1] Chapter 4 explained how the company can determine its generator category.
[2] SQGs that use aboveground storage tanks must also comply with regulations found at 40 CFR Section 265.201.

contain arrangements made with local police, fire and hospital authorities. The plan must contain an evacuation plan for facility personnel and provide step-by-step procedures for handling any imminent or actual emergency situation.

5.5 RECORDKEEPING AND REPORTING REQUIREMENTS

All generators are required to maintain accurate and up-to-date records of generated wastes, training programs, safety procedures, hazardous waste manifests and other important waste management activities. Chapter 6 provides a detailed discussion of all recordkeeping requirements.

The company must submit a biennial report to EPA that covers hazardous waste activities from the previous calendar year.[3] The report must contain information on what wastes were generated, who transported the wastes and where the wastes were shipped.

CHECKLIST FOR SAFE HAZARDOUS WASTE STORAGE

___ Store all wastes out of the path of traffic, including movement zones for forklift and other company equipment, as well as any foot traffic.

___ Proper ventilation must be maintained to prevent buildup of any gases fumes.

___ Ignitable and reactive wastes should be stored at least fifty feet from all property lines.

---

[3] EPA Form 8700-13A. Call your EPA regional office to obtain a copy of the form and instructions for filling it out.

___ Impermeable linings should be placed beneath all containers to prevent leaking of any contaminants into the ground caused by leaking containers or spilled wastes.

___ Containers must be compatible with the wastes stored and be inspected for corrosion or leaks prior to deposit of any wastes.

___ Local regulatory agencies should be consulted prior to design of the storage area to determine whether specific design criteria are expected.

___ Incompatible wastes should not be stored near to each other. Acids and bases should be separated as should oxidizers and flammables. Safety walls and barriers should separate all incompatible wastes.

___ Drainage systems should be in place to prevent contact of hazardous materials with any container exterior. An overflow containment system must be installed to capture any spilled or leaked waste products.

___ Reused containers should be cleaned and inspected for leaks prior to reuse.

5.6 LAND DISPOSAL RESTRICTIONS

Land disposal has been and is still the most commonly used method of waste disposal. Three classes of land disposal sites are in use in the United States. Class I, "Waste Management Units for Hazardous Waste," are used primarily for hazardous waste treatment residues. After May 1992 only hazardous waste treatment residues will be accepted at Class I sites. Class II, "Waste Management Units for Designated Waste," allow only certain solid hazardous wastes (i.e., asbestos).

Class III, "Landfills for Nonhazardous Wastes," accept common household waste and construction debris. The company should check with the state hazardous waste management authority to determine which Class of site will accept its wastes and where individual sites are located.

5.7 RCRA HAZARDOUS AND SOLID WASTE AMENDMENTS

The 1984 Hazardous and Solid Waste Amendments to RCRA[4] added land disposal restrictions (LDRs) to prohibit disposal of any RCRA hazardous waste on or into land unless the waste meets EPA treatment standards for that particular waste.[5] Pursuant to the 1984 amendments, EPA established a schedule that was implemented over a six year period to restrict land disposal of all listed and characteristic hazardous wastes.[6] The table below contains the schedule for the land disposal restrictions. EPA must issue treatment standards for all newly listed or identified hazardous wastes within six months of listing or identification unless a national capacity variance has been established.[7]

---

[4] Pub. L. No. 98-616, 98 Stat. 3221 (1984).
[5] The restrictions apply to all types of land disposal, such as landfills, surface impoundments, injection wells, underground mines, concrete bunkers, waste piles and land treatment facilities.
[6] See Chapter 3 for a description of how wastes are classified as "listed" or "characteristic" hazardous wastes under RCRA.
[7] If EPA determines that there is an insufficient current capacity to treat a restricted waste by the best demonstrated available technology (BDAT), it issues a two-year national capacity variance from the treatment standard.

SCHEDULE FOR LAND DISPOSAL RESTRICTIONS

Type of Waste	Date of Compliance
Solvents and Dioxins	November 8, 1986
California list[8]	July 8, 1987
First 1/3 of RCRA Listed Wastes	August 8, 1988
Second 1/3 of RCRA Listed Wastes	June 8, 1989
Final 1/3 of RCRA Listed Wastes and All Characteristic Wastes	May 8, 1990

Land Disposal of Hazardous Wastes is Prohibited After these Dates Unless EPA Treatment Standards Have Been Met

In order to determine whether a particular waste meets the EPA treatment standards, the company will need to bear the cost of waste analysis. Use of the Toxicity Characteristic Leachate Procedure (TCLP) test may be necessary. The costs of waste treatment using the Best Demonstrated Available Technology (BDAT) may be double or triple the cost of straight disposal.[9] To defray some of these costs, the company should strongly

---

[8] The "California List" includes RCRA hazardous waste that contains cyanides, RCRA metals, PCBs or wastes that are corrosive, as well as any liquid or non-liquid hazardous waste containing halogenated organic compounds.

[9] Hundreds of treatment technologies are currently available, however, four basic types are generally used. Most organic wastes are treated by incineration or solvent extraction. Most waste metals are treated by precipitation. Activated carbon is used for a variety of liquid and gaseous wastes.

consider recycling and waste minimization whenever possible.

5.8 LDR TREATMENT STANDARDS

Although EPA did not eliminate the availability of land disposal for disposing of hazardous wastes, the company cannot use land disposal unless the company's wastes have first been properly treated according to the EPA standards.[10] For each waste, EPA is in charge of identifying a best demonstrated available technology (BDAT) for treatment prior to land disposal. EPA generally expresses a treatment standard in terms of a concentration level to provide generators with some flexibility in choosing its treatment method. The concentration limits are calculated by means of the BDAT (i.e., incineration) for the particular waste or a similar waste. When EPA specifies a certain type of treatment method as the BDAT treatment standard, the company may request permission to use an alternative treatment method by submitting an Equivalent Treatment Method Petition to the EPA or the authorized state hazardous waste management authority. In the petition, the company must demonstrate that the proposed alternate method is as effective in reducing the toxicity and/or mobility of the waste as the EPA prescribed method of treatment. The company must consult the regulations to determine the EPA treatment

---

[10]*See* 40 CFR Section 268. The EPA has also issued several guidance documents to aid the company in determining the BDAT standard for particular wastes. The company should call the regional office to learn whether a guidance document exists for a particular waste.

method/standard for each waste. In some instances, EPA has designated "no land disposal" as the standard. For those wastes, the company must dispose of the wastes in some other way. Waste generated by CESQGs[11] is not subject to the land disposal restrictions, however, generators in this category must meet additional recordkeeping requirements.[12]

5.9 LDR VARIANCES

Occasionally, the EPA may grant the company a variance from the EPA treatment standards. The company must submit a variance application after the waste has been treated using the BDAT, which states that the waste cannot meet the EPA regulatory standards for the waste. EPA may then grant a variance establishing an alternative treatment standard for the waste prior to land disposal.

5.10 WASTE MIXTURES

In order for the company to dispose of mixed wastes by means of land disposal, the waste mixture must meet all treatment standards applicable to the individual wastes in the mixture. If two standards apply to any single waste constituent, the more restrictive standard will apply.

5.11 CONTAMINATED SOILS

Soils contaminated by a hazardous waste must meet the treatment standards for that waste. It is often

---

[11] See Chapter 4 for a discussion of the waste generator categories.
[12] See Chapter 6 for a full discussion of recordkeeping requirements.

difficult to treat contaminated soils that contain the RCRA regulated waste according to the required EPA standard because the nature of the soil will generally differ significantly from the properties of the contaminant. EPA has issued a two-year capacity variance for soils contaminated from CERCLA response actions and RCRA corrective actions at sites contaminated with solvent and dioxin wastes. EPA is expected to promulgate a regulation governing treatability of soil containing hazardous wastes.

5.12 WASTE MINIMIZATION

To cut down on the high costs of treating and disposing of hazardous wastes, it is beneficial for every company to attempt to minimize as much as possible the amount of wastes generated. Minimization can be as simple as purchasing fewer toxic producing materials or utilizing recycling whenever possible. Moreover, under RCRA, the company is required to take steps to minimize waste generation to the most economically feasible extent possible.[13] On every hazardous waste manifest, the company must sign a statement certifying that it made minimization efforts. Every company must also document its waste minimization efforts in its biennial generator report.

Waste minimization not only permits the company to operate in an environmentally conscious manner, it produces economic and other benefits for the company. First, by instituting an effective waste minimization program, good public relations are developed. With increasing public concern about environmentally sound

---
[13] 42 USC Section 6922(b).

business practices, the company can enhance its public image and breed goodwill with its customers. Besides enhancing its public image, the company can also realize substantial economic savings. The checklist below shows some of the benefits of waste minimization.

WASTE MINIMIZATION BENEFITS
* lower recordkeeping costs
* lower on site handling costs
* lower waste disposal and shipping costs
* reduced size of waste storage area
* less need for expensive pollution control equipment
* reduced waste treatment and testing costs
* reduced risk of potential hazardous waste liability
* reduced risk of exposure to employees

5.13 SAMPLE WASTE MINIMIZATION PROGRAM

Set and Prioritize Objectives:
The company should seek to identify realistically achievable objectives.

Conduct Periodic Waste Stream Reduction Audits:
The company should identify the types, amounts, levels and sources of wastes generated.

Prioritize Which Wastes Streams to Minimize:
Determine which waste streams can be effectively minimized by looking at how the waste is produced and who is producing it.

Obtain Management Support:
Assure that business management is behind the minimization program financially and conceptually.

*Perform On-Site Assessments:
Evaluate the operations at the company and determine where wastes can be reduced.

*Involve and Educate Company Personnel:
Emphasize the value of waste reduction and reward employees who show demonstrable progress in waste minimization.

*Design and Evaluate Action Plans:
Carefully conceived action plans should be developed to target specific company activities that could implement waste reduction.

*Test Selected Action Plans:
Try out the action plan on an isolated activity to determine the feasibility of the plan before proceeding on a larger scale.

*Obtain Financial Commitment to the Plan:
Assure that funding is behind the minimization effort so that time and energy is not wasted in the preparation for implementation.

*Revise Waste Accounting Methods Where Necessary:
Assure that inspection, tracking and accounting systems take the waste minimization goals into account.

*Revise Operational Procedures Where Necessary:
Assure that relevant internal procedures are modified to take the new waste minimization goals into account.

*Implement the Waste Minimization Program:
Once preparatory steps have been carried out, implement the program.

Evaluate the Program and Follow-up:
Institute periodic checks to determine how the program is working and make necessary adjustments to fine tune the success of the program.

6

Hazardous Waste Reporting and Recordkeeping

6.1 INTRODUCTION

The company will need to know what to do in the event of a hazardous waste spill or release. It should have a contingency plan in place to respond effectively in an emergency situation. The federal laws governing hazardous wastes have specific reporting provisions that must be followed in the event of a spill or release. This chapter will provide an overview of the reporting and recordkeeping requirements under the various federal laws. States often have additional requirements that the company must follow. Check with your state hazardous waste management agency to obtain information on state reporting and recordkeeping requirements. The names and addresses found in Appendix 3 at the end of the book will show you whom to contact.

6.2 RCRA HAZARDOUS WASTE SPILL REPORTING

As is generally the case with hazardous waste management, RCRA plays the primary role in the handling of spills and releases. Whenever there is a release or spill of a hazardous waste, the company must notify state and local authorities. A contingency plan should be in place and implemented upon discovery of a release.

A visual assessment may suffice to define the extent of the release. However, the company may be required to confirm that all contamination has been removed from the environment. If so, the company will need to implement a soil sampling and analysis program according to EPA standards, and the company may be required to determine whether the release has impacted groundwater.

Depending on how thoroughly the response contractor handles the spill, the company may need to accumulate materials on site for a time pending removal to an acceptable disposal site. The company will need to keep spill materials in appropriate containers in an appropriate location. As a general rule, the company is not permitted to store the materials longer than 90 days. If it looks like more than 90 days will be required before removal, the company should request a 30 day extension from EPA under 40 CFR Section 262.34(b). In addition, the company may request an emergency permit to treat, store or dispose of hazardous waste under 40 CFR Section 270.61.

HOW TO RESPOND TO A SPILL OR RELEASE
* eliminate the source of the release or stop the release of material into the environment.
* assess the character, amount, and overall extent of the release.
* contain the release to minimize its impact on the environment.
* recover the spilled material, including any contaminated soils.

6.3 UNDERGROUND STORAGE TANK RELEASES

Owners and operators of underground storage tanks must report any of the conditions outlined in the checklist below to the authorized environmental agency within twenty-four hours. Suspected releases other than the last two conditions will require a tank tightness test or site investigation.[1] Inventory discrepancies need not be reported until after a second month's observation confirms the existence of a possible leak.[2]

UST REPORTING

* Any spill or overfill of petroleum resulting in the release into the environment of an amount in excess of 25 gallons (or other reportable quantities of other regulated substances).[3]

* The discovery of a leak or release, including free product or vapor in the soil, sewers, or waters.

* Suspected releases must be investigated within 7 days. An investigation consists of tank tightness tests, a site assessment (soil sampling and analyses), or both, depending on the reason for suspecting a release.[4]

* Instances of malfunction or erratic behavior of UST equipment, unless the owner can confirm that the equipment is defective, but not leaking, and repairs the equipment immediately.

* Monitoring results that indicate a possible release, unless the monitoring device is found to be defective and immediately repaired, replaced, or recalibrated.[5]

---

[1] 40 CFR Sections 280.50-280.53.
[2] 40 CFR Section 280.50(c)(2).
[3] 40 CFR Section 280.53.
[4] 40 CFR Section 280.52.
[5] 40 CFR Section 280.50.

6.4 UST CORRECTIVE ACTIONS

Upon confirmation of a release, the owner or operator must (a) report the release to the authorized agency, (b) take immediate action to prevent further release (for example, close a leaking valve or remove the product from the UST), (c) remove all free product from the environment, and (d) mitigate fire, explosion, and vapor hazards.[6] Within twenty days after confirmation, the company must report to the authorized agency, summarizing the initial response and providing any additional information.

After consultation with the agency, the company must perform a site assessment, develop a response plan, and complete remedial action. The remedial action may include aeration or replacement of contaminated soil and prevention of groundwater contamination.[7] The nature of remedial action depends largely on site-specific factors, such as the type and quantity of substance released, the proximity to surface and groundwater, and the type of soil. A release causing groundwater contamination becomes very expensive and may necessitate long-term groundwater treatment or provision of alternate sources of drinking water for affected residents. A company's liability for contamination of a municipal water supply could be financially devastating.

Most state UST programs do not require removal from the site of soil contaminated with petroleum in every case, but the company must manage or treat it to prevent or minimize environmental risk. The normal

---

[6]40 CFR Sections 280.61-280.63.
[7]40 CFR Sections 280.65-280.66.

procedure for cleaning up small petroleum spills is to (1) dig up the contaminated soil;[8] (2) place the soil on an impermeable sheet to prevent contaminants from returning to the soil; and (3) periodically turn the soil over to allow hydrocarbon contaminants to escape into the air. During aeration, the contaminated soil should also be covered with a top plastic sheet if necessary to prevent rain water from leaching contaminants from the contaminated soil. After aeration has reduced the hydrocarbon levels below applicable levels, the soil may be returned to the excavation. As an alternative, contaminated soil can often be removed and transported to a municipal landfill for aeration and disposal, provided that the landfill is permitted to receive the type and levels of contamination in the soil requiring disposal.

More extensive or severe contamination will probably require more expensive remediation, including disposal of contaminated material in a hazardous waste or special waste landfill, treatment of groundwater, discharge of treated groundwater into the public sewer system, soil venting, incineration of contaminated materials, or biodegradation. Excavation and aeration or disposal of a small leak might cost up to $25,000 to $30,000. The cost of correcting more extensive leaks requiring biodegradation or treatment of groundwater could easily exceed $100,000.

The company should retain an experienced environmental response contractor as soon as there is any suspicion of contamination. A good consultant can

---

[8] The depth is normally to approximately five feet below the level of the leak.

more readily define the scope of the problem and develop a cost-effective solution. Any remediation plan or site assessment must satisfy not only the regulations governing the substantive and technical aspects of the cleanup, but also those governing reimbursement from the state UST trust fund, if applicable. Many state UST trust funds impose conditions on recovery from the fund, including additional reports and approvals.

6.5 CERCLA HAZARDOUS WASTE SPILL REPORTING

Pursuant to the Comprehensive Environmental Response, Compensation and Liability Act (CERCLA), a company is obligated to report any hazardous waste release in any amount equal to or greater than the reportable quantities outlined in the regulations.[9]

---

[9] CERCLA Section 103(a), 42 USC Section 9603(a). *See* 40 CFR Section 302.4 --List of Hazardous Substances and Reportable Quantities. The column headed "Category" lists the code letters "X", "A", "B", "C", and "D", which are associated with reportable quantities of 1, 10, 100, 1000, and 5000 pounds, respectively. The "Pounds (kg)" column provides the reportable quantity for each hazardous substance in pounds and kilograms.
40 CFR Section 302.5 Determination of reportable quantities.
(a) *Listed hazardous substances*. The quantity listed in the column "Final RQ" for each substance in Table 302.4, or in Appendix B to Table 302.4, is the reportable quantity (RQ) for that substance. The RQs in Table 302.4 are in units of pounds based on chemical toxicity, while the RQs in Appendix B to Table 302.4 are in units of curies based on radiation hazard. Whenever the RQs in Table 302.4 and Appendix B to the table are in conflict, the lowest RQ shall apply.
(b) *Unlisted hazardous substances*. Unlisted hazardous substances designated by 40 CFR 302.4(b) have the reportable quantity of 100 pounds, except for those unlisted hazardous wastes which exhibit extraction

The company must notify the National Response Center immediately at (800) 424-8802. Failure to notify the National Response Center immediately[10] upon discovery of the reportable release is punishable by a fine of up to $10,000 and/or up to one year imprisonment.[11] The appropriate state agency will also need to be notified.

Under SARA Title III,[12] the company is required to immediately report accidental spills or releases of hazardous substances to a designated "Community Coordinator."[13] There are two classifications of hazardous substances that must be reported under Title III. The first is when the substance is on the U.S. EPA's published list of approximately 400 "Extremely

---

procedure (EP) toxicity identified in 40 CFR 261.24. Unlisted hazardous wastes which exhibit EP toxicity have the reportable quantities listed in Table 302.4 for the contaminant on which the characteristic of EP toxicity is based. The reportable quantity applies to the waste itself, not merely to the toxic contaminant. If an unlisted hazardous waste exhibits EP toxicity on the basis of more than one contaminant, the reportable quantity for that waste shall be the lowest of the reportable quantities listed in Table 302.4 for those contaminants. If an unlisted hazardous waste exhibits the characteristic of EP toxicity and one or more of the other characteristics referenced in 40 CFR 302.4(b), the reportable quantity for that waste shall be the lowest of the applicable reportable quantities. [51 Fed. Reg. 34547 (Sept. 29, 1986); 54 Feg. Reg. 22538 (May 24, 1989).]

[10]Delays of even one or two hours have resulted in penalties.
[11]CERCLA Section 103(b)(3), 42 USC Section 9603(b)(3).
[12]SARA Title III, 42 USC Sections 11001-11050, was enacted as part of CERCLA in 1986 and is commonly known as the Community Right-to-Know Act. *See* Chapter 7, "Hazard Communication," for a detailed discussion of SARA Title III.
[13]42 USC Section 11004.

Hazardous Substances."[14] The second is when notice of a spill or release is required pursuant to CERCLA.[15]

6.6 CLEAN WATER ACT HAZARDOUS WASTE SPILL REPORTING

Under the Clean Water Act, as with the preceding environmental laws discussed in this chapter, releases of hazardous substances into U.S. waters must be reported.[16] Regulations provide the list of designated hazardous substances[17] and the reportable quantities.[18] Failure to provide notice[19] of a release will result in penalties as high as $250,000.[20]

---

[14] *See* 40 CFR Section 302.4 for the EPA list.
[15] CERCLA Section 103(a), 42 USC Section 9603.
[16] Clean Water Act Section 311(b)(5), 33 USC Section 1321(b)(5) requires notification of discharges into waters of U.S. to appropriate governmental agency or fine of up to $10,000 and/or up to one year imprisonment.
[17] 40 CFR Section 116.4 "Designation of Hazardous Substances," Tables 116.4A and 116.4B contain the list of hazardous substances.
[18] 40 CFR Section 117.3, "Determination of Reportable Quantities," provides the levels at which the release of a listed hazardous substance must be reported.
Many of the listed substances have been assigned reportable quantities under CERCLA regulations. Substances in Table 117.3 that are also listed in Table 302.4, 40 CFR Part 302, are assigned the reportable quantity listed in Table 302.4 for that substance.
[19] Notice must be given in accordance with the procedures outlined in 33 CFR Section 153.203.
[20] 40 CFR Section 117.22, "Penalties," calls for a civil penalty of up to $5,000 per violation or the possibility that a civil action will be commenced under Section 311(b)(6)(B) to impose a penalty up to $50,000 unless the discharge is the result of willful negligence or willful misconduct, in which case the penalty may be as high as $250,000.

6.7 DEVELOPING A CONTINGENCY PLAN

Although only Treatment, Storage and Disposal Facilities (TSDFs) are required to have a contingency plan in place in the event of a spill or release, it is a good practice for the company to have such a plan in case of a sudden spill or release. Company personnel should be advised on what to do in the event of a spill. The contingency plan should be placed in a binder in a location easily accessible to all employees. The checklist below outlines the recommended components for a contingency plan.

CONTINGENCY PLAN COMPONENTS

* actions to be taken by the company personnel in the event of an emergency.
* arrangements or agreements with local or state emergency authorities.
* names, addresses, and telephone numbers of all persons qualified to act as emergency coordinators.
* listing of all emergency equipment including location, physical description, and an outline of capabilities.
* an evacuation plan for company personnel.

6.8 EMERGENCY COORDINATOR

An emergency coordinator should be designated and be on site or on call at all times. The emergency coordinator must be familiar with all aspects of the company operation and emergency procedures and must have the authority to implement the contingency plan. In the event of an emergency, the emergency coordinator must identify the character, source and extent of the release and assess any potential hazard to human health and the environment. The coordinator must immediately notify appropriate local authorities when the hazard exists and promptly arrange for the treatment, storage or disposal of recovered waste, contaminated soil, or surface water. Details of the release should be kept in a log and reported to EPA in the "Toxic Release Inventory" (called Form R).[21]

6.9 HAZARDOUS WASTE RECORDKEEPING

The surest way to effectively manage hazardous wastes is for the company to maintain good records of all hazardous waste activities on the premises. Many recordkeeping requirements are mandatory and fines may be assessed for failure to keep records on file. Recordkeeping is a sound business practice and enables the company to keep track of its hazardous wastes, maintain a safe working environment, and be better prepared for emergencies and inspections.

6.9.1 RCRA RECORDKEEPING REQUIREMENTS

Pursuant to RCRA, EPA has issued regulations specifying the types of records that must be kept by generators of hazardous wastes.[22] As a generator, the company must keep all records relating to hazardous waste management. The table below shows the necessary retention periods and filing frequency for RCRA

---

[21] 42 USC Section 11023. See Chapter 7 for a description of Form R.
[22] RCRA Section 3002(a)(1), 42 USC Section 6922(a)(1); 40 CFR Sections 262.40-262.44.

records. The following list shows some of the records that should be kept to satisfy the RCRA requirements:

* common name description and all applicable EPA hazardous waste numbers.
* physical form of waste (liquid, solid, sludge, contained gas).
* estimated or manifest-reported weight, or volume of waste.
* method and date of storage or disposal of waste.

RECORD RETENTION PERIOD AND FILING FREQUENCY

Type of Record	Retention Period
Copy of each manifest	3 years
Test results	3 years
Waste analysis	3 years
Biennial Report	3 years
Manifest exception report	Active life
Training Records	Until closure or 3 yrs after employee last worked at company
Contingency plan	Active life
EPA ID #	Active life

Type of Report	Frequency
Biennial Report	2 years
Manifest exception	Each event report

6.9.2 UST RECORDKEEPING REQUIREMENTS

There are a series of recordkeeping requirements for underground storage tanks. The checklist below contains a listing of various reports that must be filed with the authorized environmental agency, as well as other records that must be retained.

Records may be kept on-site for immediate inspection by the authorized agency, or may be kept in a readily available alternative site, such as corporate or division headquarters, for inspection upon request by the agency.[23] In some states, decals or annual permits must be affixed to the fill-pipe or displayed at the UST site.

In addition to the required records, the company should keep complete and organized files of all documents relating to its USTs, includng any technical drawings and photographs of all tank installations. By keeping detailed and organized files the cost of subsequent repairs or retro-fits can be reduced. These files also should contain records of warranties and performance claims, any schematics and blueprints, and all diagrams, maps, or surveys of tanks, piping, and monitoring wells. All records should be kept on hand at least until the tank is permanently closed.

UST RECORDKEEPING CHECKLIST
Reports to File With the Environmental Agency
___ Notification of all new UST systems, including certification of proper installation.
___ Reports of releases and suspected releases.
___ Corrective action planned or taken.
___ Notification of permanent closure or change-in-service.

Records to Maintain on File
___ A corrosion expert's analysis if corrosion protection is not used.

---
[23] 40 CFR Section 280.34(c).

___ Documentation that corrosion protection equipment has been tested and is properly operating.
___ Documentation relating to repairs.
___ Compliance with release detection requirements.
___ Site investigations conducted at permanent closure.[24]

6.9.3 CERCLA RECORDKEEPING REQUIREMENTS

Under CERCLA, if a company releases a "reportable" quantity of a hazardous substance, the National Response Center must be notified immediately at (800) 424-8802.[25] Besides communication of spills and releases, regulated companies must also submit two different annual reports to the relevant agencies pursuant to the Emergency Planning and Community Right-to-Know Act,[26] also known as "Title III" of the Superfund Amendments and Reauthorization Act of 1986 (SARA).[27] These two documents are: (1) a Hazardous Chemical Inventory, and (2) a Toxic Release Inventory Form. The hazardous chemical inventory requires each company to submit a Material Safety Data Sheet (MSDS) for each chemical found on its premises in the threshold quantity specified on EPA's published list of approximately 400 "extremely hazardous substances."[28] These MSDSs are identical to those required under OSHA.[29] The chemical inventory must be filed on March

---

[24] 40 CFR Section 280.34(b)(1)-(5).
[25] See 40 CFR Section 302.4 -- List of Hazardous Substances and Reportable Quantities.
[26] 42 USC Sections 11001-11050. See Chapter 7 of this book for a more detailed discussion of Title III.
[27] 42 USC Sections 9601-9675.
[28] See 40 CFR Sections 302, 304 for the EPA list.
[29] Chapter 7 explains OSHA requirements.

1 of each year on either a Tier One or Tier Two form. The Tier One and Tier Two forms are explained in detail in Chapter 7 of this book.

An additional annual report called the "Toxic Release Inventory" (Form R) must be filed on July 1 with the EPA and the relevant state authority.[30] On Form R, the company reports any releases made during the preceding twelve months. Form R must be filed if the business has ten or more employees and the business manufactures, stores, imports or otherwise uses designated toxic chemicals at or above threshold levels.[31] Chemicals and their respective threshold levels are published on a list available from the government printing office.[32]

6.10 HAZARDOUS WASTE RECORDKEEPING CHECKLIST

It is essential that each company maintain accurate records to track wastes, to prepare for inspections and to facilitate good management practices. The checklist below should serve as a means for assuring that all necessary hazardous waste documentation is accounted for.

DOCUMENT INVENTORY
___ copy of hazardous waste management plan
___ copy of hazardous waste disposal plan
___ hazardous waste log book

---
[30] 42 USC Section 11023.
[31] 42 USC Section 11023(b).
[32] Senate Committee on Environment and Public Works, 99th Cong., 1st Sess.,"Toxic Chemicals Subject to Section 313 of the Emergency Planning and Community Right-to-Know Act of 1986" (Comm. Print 1987).

REPORTING AND RECORDKEEPING 69

___ aerial photographs of site
___ company operations flow chart
___ company board minutes
___ internal environmental review reports
___ external environmental review reports
___ special land use permits
___ building permits
___ list of all hazardous wastes generated
___ list of PCB-containing equipment
___ copy of OSHA worker safety monitoring plan
___ list of locations of asbestos
___ any SEC regulation 10-K filings
___ environmental liability insurance policies
___ notices of violation resulting from RCRA inspection
___ notices of violation resulting from OSHA inspection
___ records of previous RCRA compliance audits
___ UST corrective action records
___ OSHA material safety data sheets
___ OSHA hazard communication program
___ OSHA inspection reports
___ record of employee complaints relating to injury from exposure to hazardous wastes
___ Community Right-to-Know notifications regarding "extremely hazardous substances"
___ toxic chemical release inventory forms
___ notifications to the National Response Center regarding release of hazardous waste
___ annual reports filed pursuant to RCRA
___ SARA Title III hazardous chemical inventory
___ SARA Title III toxic release inventory
___ record of any RCRA or CERCLA enforcement actions

___ listing of each UST at company with size, age, contents, leak protection, tank tightness testing
___ records of any UST inspections
___ records of any UST releases
___ records of any UST penalties
___ asbestos disposal records
___ records of asbestos releases
___ record of hazardous waste transporters used

7

Hazard Communication

7.1 INTRODUCTION

Pursuant to the authority of the Occupational Safety and Health Act,[1] the Occupational Safety and Health Administration (OSHA) has established a variety of regulations to provide for the protection of workers from chemical injury. OSHA has promulgated The Hazard Communication Standard,[2] often referred to as the "Worker Right to Know" Rule, which obligates employers to communicate hazards that may result from hazardous substances in the workplace.

Under the OSHA Hazard Communication Standard, each employer must identify operations in work areas where hazardous substances are present and inform workers about them. Employers must also have material safety data sheets (MSDSs) for each hazardous substance available for review and copying by its employees.[3] Employers are required to train workers regarding the hazardous nature of substances found in individual work environments.

Because the company usually lacks expertise in this area, it makes sense to hire an outside consultant to prepare a written hazard communication program, to

---

[1] OSHA, 29 USC Sections 651-678.
[2] 29 CFR Section 1910.1200.
[3] 29 CFR Section 1910.1200(h)(1).

review material safety data sheets and labels, and to train workers. Although consultants may train groups of workers at the same time, they must evaluate different exposure levels for each job and tell each worker the specific hazards and method of protection for his or her particular work environment.

7.2 OSHA HAZARD COMMUNICATION STANDARD REQUIREMENTS

OSHA's Hazard Communication Standard sets up a compliance program to insure that all employees are informed about the presence of hazardous substances in the workplace. It's a comprehensive and complex regulatory system that must be in place at every company where employees might be exposed to hazardous substances under normal working conditions.

The OSHA program requires that each company obtain information about all hazardous chemicals on the premises, label all containers, develop a written hazard communication plan and inventory, explain to all employees the dangers of chemicals present at the company, and train employees to protect themselves from chemical exposure.

7.3 MATERIAL SAFETY DATA SHEETS

All companies must obtain a material safety data sheet (MSDS) from the manufacturer of each hazardous substance found in the workplace. The manufacturer is required to produce these and supply them to its customers.

The MSDS describes the chemical, its properties, the physical or health hazards it represents, the most likely method of exposure, exposure limits, precautions to take for safe handling and use, and first aid procedures in the event of exposure. The company must keep a MSDS for each chemical in a centralized and easily accessible place.[4]

All manufacturers are required to properly label each hazardous substance. Companies are prohibited from removing these labels and are responsible for insuring that they are properly labeled and remain so.[5]

All labels must be in English, visibly displayed, and contain information that identifies the substance and gives warnings about potential hazards. The labels can be words, symbols, pictures, placards or any combination that effectively conveys this basic information to all employees.

7.4 WRITTEN HAZARD COMMUNICATION PLAN

Each company is required to prepare a written hazard communication plan for its business. This also must be kept in an easily accessible location.

While it can be a simple document, the plan must completely explain the procedures established to ensure that all MSDSs are properly obtained and available, that all labeling requirements are met, and that employee training is performed in accordance with the OSHA regulations. The plan must also contain an up-to-date list of all hazardous materials in use at the company.[6]

---

[4] A sample MSDS is reproduced as Appendix 9 at the end of this book.
[5] 29 CFR Section 1910.1200(f). The American National Standard Institute has published an excellent guide to preparing cautionary labels (ANSI Z129.1 - 1988).
[6] 29 CFR Section 1910.1200(e).

7.5 EMPLOYEE TRAINING PROGRAM

All employees and new hires must be thoroughly trained about the hazardous substances in the workplace, and training for all employees must be updated as new substances are introduced to the workplace. There is no specific format required for the program as long as the training material is clearly presented, understandable and covers all of the specific areas required by the regulations. Appendix 8 at the end of the book contains a copy of useful OSHA guidelines for compliance with the training requirement. The checklist below outlines the basic elements necessary for a good training program.

TRAINING PROGRAM COMPONENTS
*Description of the hazard communication regulations and the rights of the employees to know about workplace hazards.
*Specific methods for detecting the presence and release of hazardous substances.
*What to do in the event of exposure or an emergency.
*Health hazards associated with hazardous substances.
*Specific measures workers should take to assure adequate protection from hazardous substances, such as appropriate work-site practices, emergency procedures, and various forms of protective equipment.
*Company's written hazard communication program, including an explanation of how workers can obtain and use the information.[7]

---

[7] 29 CFR Section 1910.1200(h)(2).

7.5.1 EXEMPT SUBSTANCES

Several types of substances are exempt from OSHA's Hazard Communication Standard regulatory requirements. These are: (1) hazardous wastes (which are regulated by EPA); (2) tobacco products; (3) wood or wood products; (4) "articles;" and (5) "consumer products."

Articles are manufactured products that do not release hazardous chemicals under normal conditions of use. Examples would be office products such as pens, typewriter ribbons, and photocopy machines. Consumer products are products that are intended for use by retail consumers and are used in the workplace in the same manner and frequency as that of the average consumer.

7.6 EMERGENCY PLANNING AND COMMUNITY RIGHT-TO-KNOW REQUIREMENTS

While OSHA requires companies to inform employees of the hazardous substances in the workplace, EPA requires companies to inform representatives of the surrounding communities of the existence of hazardous materials on the premises.

The Emergency Planning and Community Right-to-Know Act,[8] also known as "SARA Title III," was enacted as part of the Superfund Amendments and Reauthorization Act (SARA)[9] in 1986. Title III has two distinct objectives: (1) establishes a mechanism whereby communities set up plans for dealing with emergencies created by chemical leaks and spills; and (2) extends

---

[8] 42 USC Sections 11001-11050.
[9] 42 USC Sections 9601-9675.

to communities the same right-to-know provisions guaranteed to workers by OSHA.

A company will be regulated by Title III if it has a substance in a quantity equal to or greater than the "threshold planning quantity" specified on the U.S. EPA's published list of approximately 400 "Extremely Hazardous Substances."[10]

7.7 REPORTING SPILLS OR RELEASES

Title III requires that a business immediately report accidental spills or releases of hazardous substances to a designated "Community Coordinator."[11] There are two classifications of hazardous substances under Title III. The first is when the substance is on the EPA list of "extremely hazardous substances." The second is when notice of a spill or release is required pursuant to CERCLA.[12]

7.7.1 ANNUAL REPORTING

Besides communication of spills and releases, regulated companies must also submit two different annual reports to the relevant agencies: a Hazardous Chemical Inventory and Toxic Chemical Release Inventory Forms.

The hazardous chemical inventory requires each company to submit a MSDS for each chemical found on its premises in the threshold quantity. These MSDSs are identical to those required under OSHA. The Chemical

---

[10]*See* 40 CFR Sections 302, 304 for the EPA list.
[11]42 USC Section 11004.
[12]CERCLA Section 103(a), 42 USC Section 9603. *See* Chapter 6 for a description of the CERCLA spill requirements.

Inventory must be filed on March 1 of each year on either a Tier One or Tier Two form.

An additional annual report called the "Toxic Chemical Release Inventory" (Form R) must be filed on July 1 of each year with the EPA and the relevant state authority.[13] On Form R, the company reports any releases made during the preceding twelve months. Form R must be filed if the business has ten or more full-time employees and the business manufactures, stores, imports or otherwise uses designated toxic chemicals at or above threshold levels.[14] Chemicals and their respective threshold levels are published on a list available from the government printing office.[15]

7.7.2 TIER ONE FORM

All companies who receive MSDSs from their suppliers are required to inform their state environmental authorities, any local emergency response teams, and the local fire department. Notification is done by means of the EPA Emergency and Hazardous Chemical Inventory Tier One form. All hazardous materials at the site must be categorized on this form as to whether they are fire hazards, pressure hazards or are in some manner reactive. Health hazards also

---

[13] 42 USC Section 11023.
[14] 42 USC Section 11023(b). Generally, a company must file an annual toxic chemical release inventory form (Form R) if it manufactures, imports, or processes at least 25,000 pounds of a listed chemical, or if it uses at least 10,000 pounds of a listed chemical during the previous calendar year.
[15] Title III List of Lists: Consolidated List of Chemicals Subject to Reporting Under the Emergency Planning and Community Right-to-Know Act, EPA 560/4-91-011 (Jan. 1991).

must be described. The location and amount of each chemical must be reported.

All reporting on the Tier One form is done in the aggregate and describes total amounts of hazardous substances grouped by their hazardous properties. A sample Tier One form with instruction is included as Appendix 6 at the end of this book.

7.7.3 TIER TWO FORM

A second form, the Tier Two form, is more specific in that it requires quantity, location and hazard information for each specific hazardous material on site. A Tier Two form must be submitted by a company if it is requested by the state or local environmental authorities or by the local fire department.

Some companies may find it easier to complete this form instead of the Tier One since it requires basically the same information that must be ascertained to complete the Tier One form. The company provides the information by individual chemical rather than grouping chemicals by hazardous characteristic. The Tier Two may be submitted in place of the Tier One to the state and local authorities and fire department. Whether the Tier One or the Tier Two is submitted, an MSDS for each chemical must be included along with the inventory form.

A sample Tier Two form with instructions is included as Appendix 7 at the end of this book. For more information about filling out the Tier One and Tier Two forms, call the EPA Title III hotline at (800) 535-0202.

7.7.4 FORM R

With the passage of the Pollution Prevention Act of 1990 (PPA)[16] new requirements were added to the Form R. Section 6607 of the PPA expands and makes mandatory source reduction and recycling information on the EPCRA list of more than 300 chemicals and 20 chemical categories. These requirements have been added by EPA through modifications to sections 6, 7 and 8 of the Form R. Section 6 has been modified so that off-site location and transfer amounts are reported together, including amounts sent off-site for recycling. Section 7 has been modified to include detailed information about on-site recycling activities, as well as changes to the information provided for treatment activities. Section 8 contains the majority of the new source reduction and recycling reporting requirements.

Reporting of the new information commenced on July 1, 1991, therefore, Form Rs filed on July 1, 1992 must contain this additional information.[17] If a company has

---

[16] The Pollution Prevention Act of 1990 (PPA), 42 USC Section 13101 - 13109, was passed to focus industry, government, and public attention on reducing the amount of pollution produced at the source as opposed to the traditional government approach of treating and disposing of wastes. Source reduction is the key term and accorded the highest value under the PPA. The Act defines "source reduction" as any practice that reduces the amount of any hazardous substance, pollutant or contaminant prior to recycling, treatment or disposal. Source reduction refers to in-process recycling whereby a substance is reused or recirculated during the production process to reduce the amount of waste produced. This is in contrast to treatment whereby a substance is sent off-site to be burned and the heat value recovered.

[17] The filing period for the 1992 Form R was extended to September 1, 1992.

reported on Form R in past years, the new Form R will automatically be sent with instructions. For more information about the Form R, call the Emergency Planning and Community Right-to-Know hotline at (800) 535-0202. The checklist below provides a list of the most significant changes to Form R.

7.7.5 REPORTING REQUIREMENTS ADDED TO 1992 FORM R

The following information must now be provided for each toxic chemical for which a Form R is submitted:

*The quantity of the toxic chemical entering any waste stream (or otherwise released into the environment) prior to recycling, treatment, or disposal during the past twelve month period, the percentage change from the previous year, and estimates for the next two years.

*The amount of the toxic chemical being recycled on -site or elsewhere; the percentage change from the previous year; estimates for the next two years; and the recycling processes used.

*The amount of toxic chemical that is treated on-site or elsewhere during the year and the percentage change from the previous year.

*The amount of toxic chemical released into the environment as the result of a catastrophic event, remedial action, or other one-time event and which is not associated with production processes.

*Source reduction practices used with respect to the toxic chemical at the site.

*Techniques used to identify source reduction opportunities, including employee recommendations, external and internal audits, participative team management, and material balance audits.

*A ratio of production in the reporting year to production in the previous year.

7.7.6 EXEMPT SUBSTANCES

Products that are packaged for general consumption are exempt form the Title III reporting requirements. This means that products that are used at the company in the same containers as they are sold to the public, such as lubricating oil and antifreeze are not subject to the Title III regulations. Also, products that are used solely for personal, family, or household purposes are exempt from the regulations. Products used for routine agricultural operations and fertilizers are also exempt.

7.8 OSHA ASBESTOS STANDARD

Probably the most likely encountered and one of the most hazardous substances found at a company is asbestos. Company employees and their customers may be exposed to the release of cancer-causing asbestos fibers during work hours. Asbestos may be carried home on work clothing, exposing workers and their families to possible asbestos-related diseases. It is, therefore, essential that every company know the dangers posed by asbestos, and effectively minimize the risk to its employees.

OSHA issued health standards for exposure to asbestos in 1986. Under these regulations, companies must assure that none of its employees is exposed to more than .2 fibers per cubic centimeter in an eight-hour, time-weighted, average exposure. The asbestos standard also calls for monitoring of asbestos exposure levels at companies that have exposure levels at or above .1 fibers per cubic centimeter during an eight-hour, time-weighted, average exposure. Companies with a .1 threshold level of exposure must also keep medical surveillance records on its employees, and provide a training program for all employees who may be exposed to asbestos.

7.9 MONITORING OF ASBESTOS EXPOSURE

Each company must determine the levels of possible asbestos exposure of its employees. This is accomplished through testing of an air sample. Companies may obtain the names of qualified testing labs by calling EPA at (800) 334-8571, extension 6741. After the initial determination, the air must be tested at least every six months, unless the first test or subsequent tests show that employee exposure to asbestos is below the .1 threshold level. Monitoring must be resumed if company work practices change such that exposure levels could be increased.

7.9.1 EMPLOYEE NOTIFICATION

The company must notify its employees in writing within 15 days of receiving test results. The company may notify employees individually or by posting the results in an easily accessible location. If the test shows asbestos levels above the .2 level, the written notice must state measures being taken by the company to reduce the employee exposure below this level.

7.9.2 RECORD OF EXPOSURE MEASUREMENTS

The company must keep accurate records of all asbestos monitoring activity for 30 years. The record must contain: the date of measurement; the activity causing the asbestos exposure that is being monitored; the sampling and analytical testing method used and evidence of their accuracy; the number, duration and results of any air samples tested; type of respiratory protective equipment worn, if any; and the name, social security number and exposure level of each employee whose exposure is represented by each sample.

7.9.3 REQUIREMENTS FOR COMPANIES EXCEEDING THE .1 THRESHOLD

Every company is required to institute an employee training program for employees exposed to asbestos at levels above the .1 threshold. Training must be given to all existing employees, to all new hires at the time of employment, and training must be given at least annually thereafter. The training program must include discussion of: health effects of asbestos exposure; relationship between smoking and asbestos exposure in producing lung cancer; work practices associated with the employee's job that are designed to reduce asbestos exposure, including regular and emergency cleanup procedures and protective equipment that must be used; purpose and proper usage of respirators and protective equipment whenever appropriate; purpose and description of the medical surveillance program; and description of the OSHA asbestos standard. Companies must make a copy of the OSHA asbestos standard available to all employes upon request and must maintain employee training records for one year beyond the last date of an employee's employment by the company.

7.9.4 MEDICAL SURVEILLANCE

Companies must institute an employee medical surveillance program for all employees whose exposure exceeds the .1 threshold level. The medical surveillance program must include the following:

(1) examination by a medical doctor at the company's expense for every employee assigned to a job involving asbestos exposure. A complete physical examination of the respiratory system, the cardiovascular system and the digestive tract must be given, along with a complete medical history, a chest X-ray and pulmonary function tests. The OSHA respiratory disease questionnaire must be completed for each employee examined.

(2) following the examination outlined in (1), annual medical examinations must be made available to employees. The scope of the annual exams is the same as in (1) except that chest X-rays are not given each year, and a shortened OSHA questionnaire is filled out.

(3) if an employee leaves employment at the company, the company must provide a termination of employment examination within 30 days. No examination is required if the employee has been examined within the last year of employment.

(4) the company must provide the examining physician with: a copy of the OSHA asbestos standard and OSHA appendices D and E; a description of the employee's duties as they relate to asbestos exposure; the

employee's measure or expected exposure level; a description of any protective or respiratory equipment used or to be used; and any information from previous medical examinations of the employee that is not otherwise available to the examining physician.

(5) the company must obtain a written and signed opinion from the examining physician that includes: the physician's opinion about whether the employee has any medical conditions that could place the employee at an increased risk of health impairment from asbestos exposure; any limitations on the employee or upon the use of protective equipment; and a statement that the employee was informed by the physician of the results of the examination. The company must instruct the physician not to reveal any findings or diagnoses to the employee that are unrelated to asbestos exposure. A copy of the physician's written statement must be provided to the employee.

7.9.5 MEDICAL SURVEILLANCE RECORDS

The company must keep medical surveillance records on each employee for 30 years after employment is terminated. The records must include: the employee's name and social security number; all physicians' written statements; any employee medical complaints related to asbestos exposure; a copy of the information provided to the physician before each physical examination.

8

Transporting Hazardous Wastes

8.1 INTRODUCTION

Because the company is more apt to be involved with hazardous wastes as opposed to hazardous materials, this chapter focuses primarily on the regulatory requirements for shipment of hazardous wastes. It is crucial that the company understand and comply with the detailed body of law regulating the transportation of hazardous wastes. Shipment of hazardous wastes and materials are primarily regulated by the EPA and the Department of Transportation (DOT) pursuant to authorization in the Resource Conservation and Recovery Act (RCRA), and Hazardous Materials Transportation Act (HMTA).[1] The DOT uses the broad term "hazardous materials" in its regulations and the EPA uses the broad term "hazardous substances" to refer to both hazardous materials and hazardous wastes. Hazardous wastes are considered a subset of both hazardous materials and hazardous substances.[2]

---

[1] Resource Conservation and Recovery Act, 42 USC Section 6901 *et seq.*; Hazardous Materials Transportation Act, 49 USC Sections 1801-1812.
[2] *See* Chapter 3 for an explanation of the distinction between the two terms.

8.2 HAZARDOUS MATERIALS TRANSPORTATION ACT

The Hazardous Materials Transportation Act (HMTA) is the primary federal statute regulating shipment of hazardous materials. It authorizes the Secretary of Transportation to adopt measures deemed necessary to protect the public from the dangers associated with hazardous material and waste shipments.[3] Sections 105 and 106 of the HMTA empower the Secretary of Transportation to adopt regulations covering all aspects of safe transportation and handling of hazardous materials and wastes, including packaging, labeling, marking and routing.[4] Section 110 of the Act provides for civil penalties of up to $10,000 per violation, and criminal penalties of up to $25,000 and five years imprisonment for willful violation of the statute and its regulations.[5] The main purpose behind passage of the Hazardous Materials Transportation Act was to insure that oversight of hazardous materials transportation was vested in a single federal authority in order to preclude a multiplicity of conflicting state and local regulations.[6] Still, although the federal government retains primary jurisdiction over the regulation of hazardous waste shipments, the state and local governments may impose regulations as long as

---

[3]Specifically, Section 104 of the Act vests authority in the Secretary to designate materials as hazardous to the public health and safety based on their quantity and form. 49 USC Section 1803.
[4]49 USC Sections 1804, 1805.
[5]49 USC Section 1809.
[6]S. Rep. No. 1192, 93rd Congress, 2d Sess. (Sept. 30, 1974).

they are not inconsistent or less stringent than the federal scheme.[7]

8.3 RESOURCE CONSERVATION AND RECOVERY ACT

As described in Chapter 2, the Resource Conservation and Recovery Act (RCRA)[8] was enacted to provide a "cradle-to-grave" scheme for regulating hazardous wastes. The statute vests primary authority in the EPA to oversee implementation of the RCRA program. Associated with this oversight function EPA is empowered to adopt regulations for carrying out the goals of the statute. If the company generates hazardous waste,[9] it must first obtain an EPA identification number, and, if required, a state identification number. To apply for a number, the company must provide EPA or the state with a Notification of Hazardous Waste Activity (EPA Form

---

[7]49 USC Section 1811(a) provides:
  Except as provided in subSection (b) of this Section, any requirement, of a State or political subdivision thereof, which is inconsistent with any requirement set forth in this chapter, or in a regulation issued under this chapter, is preempted.
  49 USC Section 1811(b) provides, in relevant part:
  Any requirement, of a State or political subdivision thereof, which is not consistent with any requirement set forth in this chapter, or in a regulation issued under this chapter, is not preempted if, upon the application of an appropriate State agency, the Secretary determines, in accordance with procedures to be prescribed by regulation, that such requirement (1) affords an equal or greater level of protection to the public than is afforded by the requirements of this chapter or of regulations issued under this chapter and (2) does not unreasonably burden commerce.
[8]42 USC Section 6901 *et seq*.
[9]*See* Chapter 4 for a description of the categories of generators.

8700-12), which is available from the EPA Regional Office in the company's area. The company cannot treat, store, ship or dispose of waste off-site without this identification number. A sample EPA Notification of Hazardous Waste Activity is reproduced as Appendix 5 at the end of the book.

8.4 HAZARDOUS WASTE SHIPMENTS PREPARATION

After identifying any wastes generated by the company and determining whether they are considered hazardous under either the EPA or DOT regulations,[10] EPA must be notified that hazardous wastes are present and an EPA I.D. number must be obtained. The company will also need to verify whether the transporter and the disposal facility have EPA I.D. numbers.

The company must carry out the following steps to prepare the wastes for shipment:
* Determine the hazard class and DOT number corresponding to the waste.
* Prepare the waste shipment according to the DOT labeling, marking, placarding, and packaging regulations.
* Choose a reputable transporter.
* Properly fill out a hazardous waste manifest to accompany the shipment.

---

[10] *See* Chapter 3 for a detailed explanation on how to make this determination. *See, e.g.,* 40 CFR Sections 261 and 262.11; and 49 CFR Section 172.101.

8.5 HAZARD CLASS

The DOT regulations categorize all hazardous materials into twenty-two hazard classes. The six most important classes for the company are: Flammable Liquid; Combustible Liquid; Corrosive Material; ORM-A; ORM-C; and ORM-E. "ORM" stands for "Other Regulated Material." The DOT classes are essential to proper completion of the shipping manifest. The EPA classes described in Chapter 3 are used for completion of the EPA hazardous waste identification number application form but are not used on the shipping manifest.

8.6 DOT SHIPMENT NUMBER

The appropriate DOT number must be determined for the manifest. The number will correspond to the waste being shipped. The DOT regulations contain a listing of each substance considered hazardous with a corresponding shipment number.[11] The regulations are much too lengthy to reproduce here. The company should obtain a copy of the DOT regulations from the government printing office by calling (202) 783-3238. Ask for a copy of Title 49, Code of Federal Regulations.

8.7 PACKAGING REQUIREMENTS

Proper packaging of hazardous waste shipments is not only essential but is also a prudent way to minimize the risk of costly spills. Both EPA and DOT have specific regulatory requirements relating to

---

[11] This list is found at 49 CFR Section 172.101.

packaging.[12] A wide variety of packaging materials and containers are available for hazardous waste shipments. Generally speaking, the packaging materials have inert qualities and must not react chemically with the waste products. The packaging must possess absorbent qualities that will consume moisture. The packaging must also withstand impacts whenever a container is bumped or dropped. The packaging may be reused.

Only DOT approved containers may be used for hazardous waste shipments. Various national companies distribute containers, labels and markings that have been approved for use by the DOT, such as HazMat and Labelmaster.[13] Usually these companies can provide some advice regarding proper containers and packaging for a hazardous waste shipment, however, when in doubt it is always better to contact the DOT.

The most widely used container for hazardous waste is a 55 gallon steel drum. These drums must be 90% full and have no visible liquid. They must contain a sealing gasket and ring that will not deteriorate from the drum contents.

8.8 USED BATTERIES

Special packaging requirements must be observed when disposing of used batteries. The DOT regulations impose different packaging requirements depending on whether "wet" or "dry" batteries are being transported. Wet batteries (which still contain the electrolyte

---

[12] See 40 CFR Section 262.32 for the EPA requirements and 49 CFR Sections 172 and 173 for the DOT requirements.
[13] American Labelmark Co. 5724 North Pulaski Road Chicago, IL 60646 (800) 621-5808

fluid) with asphaltum composition, impregnated rubber, plastic, steel or wooden casings may be shipped in strong fiberboard containers or wooden boxes depending on the weight of the shipment. For batteries weighing up to 25 pounds, 3 batteries are allowed per box ; for batteries weighing up to 15 pounds, 4 batteries are allowed per box; for batteries weighing up to 10 pounds, 5 batteries are allowed per box.

The company may also use pallets to ship large quantities of used batteries. This packaging method demands that all batteries be firmly attached to the pallet and that the pallet and batteries weigh at least 300 pounds. The height of the complete pallet unit must not exceed one and one-half times the pallet width. Each battery package must be carefully cushioned and packed to prevent any short circuits.

8.9 LABELS AND PLACARDS

Every company must assure that every container of waste is properly labeled according to DOT specifications. A four inch square label is used to indicate what is in the container in the event of a spill or accident during shipment. Labels also alert anyone who handles the container about the degree of caution that should be observed.

Generally, if hazardous wastes are shipped in portable tanks, cargo tanks or tank cars, no labeling is required. Shipments of combustible liquids and cylinders of compressed gases shipped via company vehicle are also exempt from the labeling requirements. With most other containers, a four inch square label will need to be affixed to each hazardous waste container. If more than one hazard is present, labels must be affixed for each hazard.

Placards are color-coded, diamond-shaped signs that must be posted on both sides and both ends of the transport vehicle containing hazardous materials. The placard will identify the specific hazard posed by the hazardous material shipment. Small quantity and ORM-classified shipments are not placarded, nor are hazardous **waste** shipments.

Only specified colors, numbers and symbols are used on labels and placards. The Table below shows the color and class number system used on labels and placards.

LABEL/PLACARD COLORS AND CLASS NUMBERS
* Explosives - orange - #1
* Compressed gases, non-flammable - green - #2
* Compressed gases, flammable - red - #2
* Flammable liquids - red - #3
* Flammable solids - red and white stripes - #4
* Oxidizers and organic peroxides - yellow - #5
* Poisons - white - #6
* Etiologic agents - rectangular red and white - #6
* Radioactive materials - white and yellow - #7
* Corrosive materials - white and black - #8
* Miscellaneous - #9

8.10 MARKINGS

In addition to the necessary labeling and placarding, other identification markings with messages, names and numbers may be required. EPA requires that hazardous waste containers holding less than 110 gallons of waste be marked with a statement that federal law prohibits improper disposal of hazardous waste. This marking must also display the generator's name, address, and the hazardous waste manifest document number. Stick-on labels bearing this marking can be purchased and must be affixed to each container requiring the marking. These containers must also be marked with the DOT shipping name and identification number. ORM-hazard class materials must be marked with the proper ORM-A, ORM-C or ORM-E designation.

If a company uses portable tanks or cargo tank trucks for hazardous waste shipment, each tank or cargo tank must also bear an orange rectangular panel on each side and end of the tank that shows a DOT identification number for the tank contents. As an alternative, the portable tanks may display placards that have this identification number printed across the center of the placard.

8.11 CHOOSING A REPUTABLE TRANSPORTER

It is incumbent upon the company to exercise care in selecting the transporter of its hazardous wastes. If the transporter is not licensed or does not have an EPA identification number and a spill occurs, the company may be found liable for cleanup and other damages caused by the spill. The risk of this sort of liability can be prevented by assuring that an experienced and reputable transporter is chosen. The company should take the following steps to verify the qualifications of the transporter:

*Verify that the transporter has an EPA identification number.
*Make sure that the transporter has the necessary state licenses and permits to transport hazardous wastes.
*Call the state agency to find out whether the transporter has ever been cited for any violations.
*Ask an attorney to determine whether the transporter has ever been or is currently involved in hazardous waste litigation.
*Call the Better Business Bureau and the Chamber of Commerce to find out whether any complaints have been lodged against the transporter for unsatisfactory performance or poor business practices.
*Check transporter's knowledge of the DOT regulations by asking questions you know the answers to.
*Ask for references and call to verify satisfactory past performance.

By carrying out this simple investigation, the company will greatly minimize the potential for a mishap during shipment and acquire some peace of mind while the shipment is on its way to the disposal site.

8.12 HAZARDOUS WASTE MANIFEST REQUIREMENTS

The transporter is prohibited from accepting any shipment of hazardous materials or wastes without properly completed shipping papers. For hazardous materials, a bill of lading is required.[14] For hazardous wastes, a hazardous waste manifest is needed.

---

[14] EPA has no jurisdiction over the transportation of hazardous *materials*. DOT has sole jurisdiction.

The normal manifest procedure involves preparation of a separate manifest for each generator, each site, and each shipment of hazardous waste. The manifest consists of multiple copies and is prepared by the generator and given to the transporter with the waste shipment. The transporter signs and gives a copy back to the generator. The generator must sign and send a copy to the authorized regulatory agency, and the transporter and disposal facility each retain a copy. After the disposal facility has accepted the waste, it signs and sends a copy to the regulatory agency, which matches the copy with the one sent by the generator. The disposal facility also signs and sends a copy back to the generator, which also matches with the one on file. The company must keep a copy of the manifest on file for at least three years.

Almost all hazardous waste shipments by Large or Small Quantity Generators must be accompanied by the EPA hazardous waste manifest. Conditionally Exempt Generators do not need to use the manifest for their hazardous waste shipments unless the state agency requires one.[15] Most states have their own manifest form modeled after the federal form. The company should check with the state agency to determine whether a state or federal form is required for the hazardous waste shipment. The procedure for determining which form to use is outlined below.

---

[15]See Chapter 4 for a description of the categories of generators.

> **HOW TO DECIDE WHICH MANIFEST FORM TO USE**
>
> **First** If the destination state for the waste requires completion of a state form, the company must fill out that state's manifest form.
>
> **Second** If the destination state does not require a state manifest form but the company's state does, the company must fill out a state form for the company state.
>
> **Third** If neither the destination state nor the company state require a state form, the federal manifest form must be used.

8.13 MANIFEST EXCEPTION REPORTS

If the company fails to receive a copy of the manifest from the disposal facility within 60 days from the date of shipment, the company must file an Exception Report with the EPA Regional Office in the company's area. For Small Quantity Generators, the Exception Report can be in the form of a photocopy of the manifest and a note indicating that the confirmation copy of the manifest was never received from the disposal facility. Many states also require that Exception Reports be filed with the authorized state agency.[16]

8.14 MANIFEST EXEMPTIONS

EPA allows an exemption from the manifest requirement when Small Quantity Generators ship wastes

---

[16] Some states require filing of the Exception Report if the disposal facility has not sent the confirmation copy within 30 days.

for recycling pursuant to a contractual recycling agreement with a recycling firm. EPA does not require a manifest as long as (1) the agreement specifies the frequency of recycling shipments; (2) the wastes are transported to the recycling facility in a vehicle that is owned and operated by the recycling facility; and (3) the company retains a copy of the recycling agreement on file for at least three years following termination of the agreement. Some states do not recognize this exemption, therefore, the company should check with the state environmental agency to find out whether the exemption is available. Used oil (with hazardous characteristics), spent solvents, and used batteries are common examples of hazardous wastes that may be recycled by the company.

8.15 TRANSPORTING HAZARDOUS WASTES IN COMPANY VEHICLES

Ordinarily, the company should hire a reputable transporter to ship wastes to the disposal facility. Professional transporters will have special training and experience in shipping the hazardous waste that the company usually lacks.

Federal law, however, does permit the company to transport its waste to the disposal facility in a company vehicle provided that the company complies with the same requirements that other hazardous waste transporters must observe. The company will need an EPA transporter identification number, and will need to comply with the extensive DOT regulations applying to transporters.[17] The high costs of liability insurance

---

[17] *See* 49 CFR Section 177.

alone makes it seldom cost-effective for the company to transport its own wastes.

9

Hazardous Waste Liability Insurance*

9.1 INTRODUCTION

Enactment of hazardous waste laws has exposed companies to increased risk of liability for pollution generating activities. The greatest potential for hazardous waste liability arises under RCRA and CERCLA. In some cases, companies have been hit with cleanup and response costs of potentially bankrupting proportions.

Companies need to evaluate their business practices to assure compliance with the various hazardous waste laws to minimize the risk of costly hazardous waste liability. This book is intended as a useful guide to RCRA compliance. Companies must also assure compliance with CERCLA and other pollution control laws mentioned in this book.

No matter how careful a company may be in managing its hazardous wastes, accidents may occur. If a hazardous waste spill or release happens, the company must be prepared to handle it. An emergency plan must be in place to minimize the damage caused by any release. Environmental insurance coverage can absorb expensive hazardous waste cleanup and response costs.

---

*This chapter is adapted from an article written by the author, which appeared in the September 1992 issue of Environmental Protection magazine. Dennison, Mark, "Environmental Insurance: Battling the Ghosts of Pollution Past," in Environmental Protection (Sept. 1992). © Stevens Publishing Corporation. Reprinted by permission of Stevens Publishing Corporation, Waco, Texas.

9.2 TYPES OF COVERAGE

Currently, businesses are carefully monitoring their operations to assure environmental compliance and taking all available steps to minimize liability. Environmental insurance is being purchased in one form or another, be it environmental impairment insurance, pollution legal liability insurance, or pollution cleanup coverage insurance. Some environmental laws, such as RCRA, even require environmental liability insurance coverage.[1] Companies need to assess their individual situations, consult with their legal counsel and insurance broker to determine what type of coverage is best for the company.

Insurance is generally written on an "occurrence" or "claims-made" basis, depending on the policy. The difference is that occurrence policies provide coverage on a per occurrence basis, meaning that losses arising from each occurrence during the policy period are covered. A claim can be made years after the policy expires as long as it is shown that the occurrence took place while the policy was in effect. Claims-made policies provide coverage as long as the claim is made during the time period that the policy is in effect. The table below describes the most common types of environmental liability insurance on the market today.

---

[1]*See, e.g.,* 40 CFR Section 280.97 (insurance requirement for RCRA underground storage tanks).

COMMON TYPES OF ENVIRONMENTAL LIABILITY INSURANCE

Pollution Legal Liability Insurance - This type of policy expressly provides claims-made coverage during the policy period for claims arising from pollution generating activities. It is generally issued only after an inspection by engineers and the cost of the inspection, which may run $5,000 or more, is borne by the insured regardless of whether a policy is issued. Premiums can run $10,000 or more per year.

Pollution Cleanup Coverage - This type of policy is specifically written to provide coverage for the costs of environmental cleanup on the insured's own property. It is intended to complement other policies such as the Pollution Legal Liability Insurance or CGL coverage, which do not cover losses for environmental damage to the insured's own property. Premiums are at least $10,000 per year and coverage is excluded for known pollution on the property at the time the policy is issued. Coverage is generally limited to $1 million.

Environmental Impairment Liability (EIL) Insurance - This type of policy is another type of claims-made coverage for losses resulting from environmental damage to property. Environmental damage to the insured's own property may be covered. CERCLA sites are excluded from coverage. Commerce & Industry Insurance Company, a subsidiary of AIG (American International Group) offers the greatest amount of coverage - $40 million per claim with a $40 million annual aggregate limit in most cases. Special purpose EIL policies are available for underground storage tanks, cleanup contractors and consultants.

Underground Storage Tank Liability Insurance - Under 40 CFR Section 280.97, owners of USTs are required to purchase insurance to protect against leaks and spills. Special policies are available to cover pollution liability for USTs. Coverage is usually limited to $1-2 million per claim. Premiums run around $1,500 per year.

9.3 THE CURRENT PROBLEM

In earlier years, before RCRA and CERCLA were laws, most companies did not have such a strong incentive to purchase environmental liability insurance. Previously, the only insurance most companies had to cover environmental liability was the Comprehensive General Liability Insurance (CGL) policy.

Although different types of environmental liability insurance are currently available, albeit with very high premiums, high deductibles and low limits of liability, the major problem faced by companies today is coverage for past pollution activities, which are only now being discovered. Courts are looking to the CGL policy to determine who is responsible for paying environmental damages, cleanup, and response costs. Several issues are commonly litigated in environmental insurance coverage disputes, which are discussed in turn below.

9.4 STANDARD CGL POLICY

All CGL policies have an insuring clause containing the insurance carrier's promise to defend and indemnify the insured. The standard language of the policy provides that:

> "The Company will pay on behalf of the insured all sums which the insured shall become legally obligated to pay as damages because of bodily injury or property damage to which this insurance applies, caused by an occurrence, and the Company shall have the right and duty to defend any suit against the insured seeking damages on account of such bodily injury or

property damage, even if any of the allegations of the suit are groundless, false or fraudulent ...."

9.5 DUTY TO DEFEND

CGL policies impose the duty on the insurer to defend any suit against the insured seeking damages for bodily injury or property damage. The duty to defend has been broadly interpreted by the courts to cover defense of suits by private parties alleging bodily injury or property damage resulting from the insured's release of hazardous substances into the environment, as well as suits brought against the insured pursuant to RCRA, CERCLA and other environmental laws. The duty to defend does not apply, however, in instances where no covered "occurrence" exists during the insurer's policy period, or where "property damage" is not found within the meaning of the policy or where some exclusion is found to preclude coverage.

The most widely debated issue in regard to the duty to defend is whether a PRP letter[2] or a state compliance order constitutes a "suit" triggering the duty to defend. The controversy centers on whether the term "suit" should be construed to mean a lawsuit in the traditional sense or whether the term should include all claims that may result in liability against the insured. The courts are about evenly split on this issue.

---

[2] "Potentially Responsible Party" letter from EPA notifying a party that it may be liable for CERCLA liability.

9.6 TYPES OF DAMAGES COVERED BY THE CGL POLICY

The standard CGL policy covers claims for bodily injury and property damage. Bodily injury is defined in the CGL policy as "bodily injury, sickness or disease sustained by any person which occurs during the policy period, including death at any time resulting therefrom." Bodily injury is usually easily proven. As a general rule, it does not include purely emotional or mental distress and suffering unless accompanied by physical injury.

In the environmental liability context, "property damage" is the more common claim for damages under the CGL policy and is the subject of tremendous legal interpretation and struggle. The main issue is whether the CGL policy is intended to cover claims for environmental cleanup costs or governmental response costs. Unfortunately, this issue is far from settled. In most cases, an insured who has been called upon by the government to reimburse the response costs associated with an environmental cleanup can claim those costs as damages,[3] however, a few courts have held that response costs are merely an economic loss and not "property damages."[4] The courts are fairly evenly split on whether cleanup costs are property

---
[3]See, e.g., Intel Corp. v. Hartford Accident & Indemnity Co., 692 F. Supp. 1171 (N.D. Cal. 1988); Village of Morrisville Water & Light Dept. v. United States Fidelity & Guar. Co., 775 F. Supp 718 (D. Vt. 1991); Boeing Co. v. Aetna Cas. & Sur. Co., 113 Wash. 2d 869, 784 P.2d 507 (1990); Avondale Indus., Inc. v. Travelers Indem. Co., 887 F.2d 1200 (2d Cir. 1989), cert. denied, 110 S.Ct. 2588 (1990).

[4]See, e.g., Mraz v. Canadian Universal Insurance Co., 804 F.2d 1345 (4th Cir. 1986); Travelers Indemnity v. Allied Signal, Inc., 718 F. Supp. 1252 (D. Md. 1989).

damages under the CGL policy. Insurers argue that the term property damage is intended only to mean "legal damages" and not environmental cleanup costs, which are generally considered to be a form of equitable relief. The First, Fourth and Eighth Circuit Courts of Appeals take the insurers' position.[5] Insureds contend that the term property damage should be given its every day ordinary meaning to include any damage to property. This interpretation makes no distinction between whether losses are incurred on a legal or equitable basis. The Second, Third, Ninth and D.C. Circuit Courts of Appeals follow the insureds' position.[6] On the whole insureds have faired better in state courts, which have more consistently taken the position that cleanup costs are property damages under the CGL policy.[7]

---

[5] *See, e.g.,* Maryland Casualty v. Armco, Inc., 822 F.2d 1348 (4th Cir. 1987), *cert. denied,* 484 U.S. 1008 (1988); Continental Insurance Cos. v. Northeastern Pharmaceutical & Chemical Co., 842 F.2d 977 (8th Cir.), *cert. denied,* 109 S.Ct. 88 (1988); A. Johnson & Co. v. Aetna Casualty and Surety Co., 933 F.2d 66 (1st Cir. 1991).

[6] *See, e.g.,* Avondale Indus., Inc. v. Travelers Indem. Co., 887 F.2d 1200 (2d Cir. 1989), *cert. denied,* 110 S.Ct. 2588 (1990); Independent Petrochemical Corp. v. Aetna Casualty and Sur. Co., 944 F.2d 940 (D.C. Cir. 1991); Aetna Casualty and Sur. Co. v. Pintlar Corp., 948 F.2d 1507 (9th Cir. 1991); New Castle County v. Hartford Accident & Indem. Co., 933 F.2d 1162 (3d Cir. 1991).

[7] In fact, the highest courts of the states of California, Washington, Massachusetts, Wisconsin, Wyoming, and Iowa have ruled that cleanup costs are recoverable under the CGL policy, whereas the highest courts of the states of Maine and New Hampshire take a minority position in holding that cleanup costs are not recoverable.

9.7 DEFINITION OF OCCURRENCE

In order for coverage to be triggered, it is necessary for an "occurrence" to take place during the policy period. The common CGL definition of occurrence states that:

> "'Occurrence' means an accident, including continuous or repeated exposure to conditions, which results in personal injury or property damage neither expected nor intended from the standpoint of the insured."

In determining whether there has been an occurrence, there must be an accident (which courts have construed to mean some fortuitous event) that was not expected nor intended by the insured. Some courts have stated that this "neither expected nor intended" language means that the insured must not have willfully intended to damage property.[8] Other courts construe this language to mean that a reasonable person in the position of the insured could not have anticipated the resulting property damage.[9]

9.8 TRIGGER OF COVERAGE

In the environmental damage context, pollutants may have been released into the environment many years before they cause injury or before the harm is discovered. This circumstance causes a real problem in trying to determine when an occurrence happened, especially when pollution is ongoing over time. Insurers will acknowledge that an occurrence took

---
[8] This view is commonly referred to as the subjective test.
[9] This view is commonly referred to as the objective test.

place, however, will refuse coverage by claiming that the time of the occurrence was not during their policy period. Determining the exact time of occurrence has become the most elusive aspect of environmental liability coverage disputes. Often multiple carriers may become involved in a suit because the insured changed insurance carriers several times over a period of years when the property damage began, continued, and was finally discovered.

The courts have been as much at odds as the plaintiffs and defendants in these suits. No single method has emerged to determine the trigger of coverage. Instead, disparate courts have developed different theories to deal with this issue. The four tests are commonly referred to as the "exposure," "manifestation," "injury in fact," and "continuous trigger" theories of coverage.

WHAT IS THE "TRIGGER" OF COVERAGE?
Exposure Trigger - Under this theory, coverage is triggered at the time that a hazardous substance is released into the environment, which results in bodily injury or property damage. The insurer whose policy is in effect at the time of exposure is obligated to defend and indemnify the insured.[10]
Manifestation Trigger - Under this theory, the time when bodily injury or property damage caused by a hazardous substance first manifests itself, triggers coverage under the policy in effect at that time.[11]

---

[10] see, e.g., Insurance Co. of North America v. 40-8 Installations, 633 F.2d 1212 (6th Cir. 1980), *aff'd on rehearing,* 657 F.2d 814 (6th Cir. 1981), *cert. denied,* 545 U.S. 1009 (1989).
[11] In an asbestos bodily injury case, Eagle-Picher Industries, Inc. v. Liberty Mutual Insurance Co., 682 F.2d 12 (1st Cir. 1982), *cert. denied,* 460 U.S. 1028

> *Injury In Fact* - Under this theory, the court looks to when damage actually occurred and not to the time of initial exposure or when injury first manifested itself. This trigger falls somewhere between manifestation and exposure. The policy in effect at the time that actual damage resulted from exposure to hazardous substances determines which insurer is liable to defend and indemnify the insured.[12]

> *Triple or Continuous Trigger of Coverage* - Under this theory, courts hold that there is no temporal limitation on the term "injury" in the CGL policy and reason that an insurer's risk should cover the period from the time of initial exposure to the time that injury manifests itself. Thus, all insurers whose policies were in effect during the time that covered persons or property were exposed, injured in fact, or when the injury was manifested, are each obligated to defend and indemnify the insured. The continuous trigger has been applied in cases of bodily injury and in property damage cases.[13]

9.9 "SUDDEN AND ACCIDENTAL" POLLUTION EXCLUSION

Beginning in the early 1970's, insurers sought to limit their liability for damages stemming from environment harm by adding pollution exclusion clauses

---

(1983), the court found that the time that a claimant exhibited symptoms was the time of manifestation.

[12] This theory was developed in another asbestos bodily injury case in which the Court found fault with both the exposure and manifestation concepts, since exposure did not always cause injury and injury often occurred "in fact" before it manifested itself. *See* American Home Products v. Liberty Mutual Insurance Co., 565 F. Supp. 1485 (S.D.N.Y. 1983), *aff'd as modified*, 748 F.2d 760 (2d Cir. 1984).

[13] *See, e.g.,* Keene Corp. v. Insurance Co. of North America, 667 F.2d 1034 (D.C. Cir. 1981), *cert. denied*, 455 U.S. 1007 (1982), *rehearing denied*, 456 U.S. 951 (1982); New Castle County v. Continental Casualty Co., 725 F. Supp. 800 (D. Del. 1989) ("New Castle III"), *aff'd in part, rev'd in part and remanded*, 933 F.2d 1162 (3rd Cir. 1991).

to the standard CGL policy. The standard CGL policy as revised by the Insurance Service Office (ISO) in 1973 added the following provision:

> [This policy does not apply] "to bodily injury or property damage arising out of the discharge, dispersal, release or escape of smoke, vapors, soot, fumes, acids, alkalis, toxic chemicals, liquids or gases, waste materials or other irritants, contaminants or pollutants into or upon land, the atmosphere or any water course or body of water; *but this exclusion does not apply if such discharge, dispersal, release or escape is sudden and accidental."* (emphasis added)

The last phrase caused the clause to be commonly known as the "sudden and accidental" pollution exclusion. This sudden and accidental language has been the subject of tremendous litigation. Most courts have found that the pollution exclusion does not bar coverage just because the discharge or release of pollutants was not "sudden." These courts consider the language of the exclusion (1) to be ambiguous and therefore should be construed in a manner most favorable to the insured;[14] or (2) to be a restatement of the definition of the term "occurrence" and therefore coverage should not be excluded where the damage to the environment was "neither expected nor intended from the standpoint of the insured."[15] Some

---

[14]*See, e.g.,* Aetna Casualty & Surety Co. v. General Dynamics Corp., No. 88-2220C (A), 783 F. Supp. 1199 (E.D. Mo. 1991).
[15]*See, e.g.,* Hecla Mining Co. v. New Hampshire Insurance Co., 811 P.2d 1083 (Colo. 1991); Claussen v. Aetna Casualty & Sur. Co., 754 F. Supp. 1576 (S.D. Ga. 1990). *But see* Bentz v. Mutual Fire, Marine & Inland

recent cases, have found that long-term hazardous waste disposal cannot meet the definition of "sudden and accidental."[16] These courts look to the actual intent of the insured rather than the nature of the damages to determine whether the coverage should be excluded.[17] Where the insured is found to have disposed of hazardous waste over time knowing that injury would result, the disposal cannot be considered sudden and accidental.[18]

9.10 THE "ABSOLUTE POLLUTION EXCLUSION"

In response to court decisions finding the "sudden and accidental" pollution exclusion ambiguous and construed in favor of insureds, the insurance industry modified the wording of the exclusion in an effort to bar recovery for damages due to pollution. The new pollution exclusion was included in policies issued in the late 70's and early 80's and is known as the "absolute pollution exclusion." This exclusion excludes coverage whether or not the discharge or release was sudden and accidental. One version of the absolute pollution exclusion provides:

Ins. Co., 83 Md. App. 524, 575 A.2d 795 (Md. Ct. of Spec. App. June 29, 1990).
[16] *See, e.g.,* Outboard Marine Corp. v. Liberty Mutual Ins. Co.. 212 Ill. App. 3d 231, 570 N.E.2d 1154 (App. Ct.), *appeal granted,* 139 Ill.2d 598, 575 N.E.2d 917 (1991); Fireman's Fund Ins. Cos. v. Ex-Cell-O Corp., 702 F. Supp. 1317 (E.D. Mich. 1988), *reconsideration denied,* 720 F. Supp. 597 (E.D. Mich 1989).
[17] *See. e.g.,* Liberty Mutual Ins. Co. v. Triangle Indus., 765 F. Supp. 881 (N.D. W. Va. 1991).
[18] *See, e.g.,* Fireman's Fund Inc., Cos. v. Meenan Oil Co., 755 F. Supp. 547 (E.D.N.Y. 1991); Borg-Warner Corp. v. Liberty Mutual Ins. Co., No. 88-539 (N.Y. Sup. Ct. Jan. 24, 1991).

"The Company shall have no obligation under this policy (1) to settle or defend any claim or suit against any insured alleging actual or threatened injury or damage of any nature or kind to persons or property which arises out of or would not have occurred but for the pollution hazard; (2) to pay any damages, judgments, settlements, losses, costs or expenses of any kind or nature that may be awarded or incurred by reason of any such claim or suit or any such actual or threatened injury or damage; (3) for any losses, costs or expenses arising out of any obligation, order, direction or request of or upon any insured, including but not limited to any government obligation, order, direction or request, to test for, monitor, clean up, remove, contain, treat, detoxify or neutralize irritants, contaminants or pollutants.

'Pollution hazard' means an actual exposure or threat of exposure to the corrosive, toxic or other harmful properties of any solid, liquid, gaseous or thermal pollutants, contaminants, irritants or toxic substances, including smoke, vapors, soot, fumes, acids or alkalis, and waste materials consisting of or containing any of the foregoing arising out of the discharge, dispersal or release or escape of any of the aforementioned irritants, contaminants or pollutants into or upon land, the atmosphere or any water course or body of water. Waste material includes any materials which are intended to be or have been recycled, reconditioned or reclaimed."

Although the newer CGL policies containing this exclusion have not yet been the subject of very much litigation, a growing number of courts have considered the applicability of the absolute pollution exclusion. Most courts have found the exclusion unambiguous, and accordingly precluded coverage.[19] Very few courts have refused to apply an absolute pollution exclusion, and usually do where factual issues remained concerning policy language.[20]

9.11 THE "OWNED PROPERTY" EXCLUSION

The standard CGL policy excludes coverage for damage to "(1) property owned or occupied by or rented to the insured, (2) property used by the insured; or (3) property in the care, custody or control of the insured or as to which the insured is for any purpose exercising physical control." Insurers usually take the position that this language precludes coverage for the costs of cleanup on property owned by the insured.[21] Still, a significant number of courts have refused to apply this exclusion on the basis that the insured did

---

[19] *See, e.g.,* Ascon Properties v. Illinois Union Insurance Co., 908 F.2d 976 (9th Cir. 1990); New Castle County v. Hartford Accident & Indemnity Co., 673 F. Supp. 1359 (D. Del. 1987), *aff'd in part, rev'd in part and remanded on pollution exclusion grounds,* 933 F. 2d 1162 (3rd Cir. 1991).
[20] *See, e.g.,* In re Hub Recycling Inc., 106 B.R. 372 (D. N.J. 1989); Titan Holdings Syndicate, Inc. v. City of Keene, 898 F.2d 265 (1st Cir. 1990).
[21] *See, e.g.,* Diamond Shamrock Chem. Co. v. Aetna Casualty and Surety Co., 231 N.J. Super. 1, 554 A.2d 1342 (1989).

not "own" the groundwater or the public's "natural resources" which may have been, or were threatened to be, contaminated.[22]

9.12 INSURING AGAINST ENVIRONMENTAL RISKS

Presently, companies that continue to produce pollutants in everyday business operations, especially hazardous wastes, should purchase some form of environmental liability insurance. Coverage is expensive; however, like all forms of insurance, the premiums are set according to risk. The price of insurance is meager in comparison to the enormous liability that is covered. The type, amount of coverage and deductible sought, must be decided according to the degree of involvement a company has with hazardous waste and other pollution generating activities. Today's CGL policies have effectively excluded coverage for damage caused by pollutants. Therefore, in order to secure some form of coverage, each company should look to the sound advice of its insurance broker and legal counsel. The broker can explain what insurance options are available and legal counsel can review particular policies to assure that sufficient coverage is supplied to cover the risk of environmental liability.

---

[22]*See, e.g.*, State of New York v. New York Central Mutual Fire Ins. Co., 147 A.D.2d 77, 542 N.Y.S.2d 402 (1989); Claussen v. Aetna Casualty & Surety Co., 754 F. Supp. 1576 (S.D. Ga. 1990).

10

Hazardous Waste Penalties, Costs and Compliance Audits

10.1 INTRODUCTION

Foremost in the mind of the company should be the cost of noncompliance with the hazardous waste laws and regulations. Although every company will wish to operate a safe and legal program for handling and disposing of hazardous wastes, hazardous waste penalties are an added incentive to regulatory compliance at the federal, state and local levels. Naturally, willful and knowing violations of the laws and regulations carry the heaviest penalties, whereas negligent or inadvertent violations carry lesser penalties. Still, ignorance of the law is never an excuse, so it is incumbent on every company to know the laws and how they may impact the company. Moreover, fines and penalties may only be a small portion of the cost of noncompliance. The expense associated with liabilities stemming from faulty and illegal hazardous waste management practices could be staggering. It is always in the best interest of the company to maintain regulatory compliance.

10.2 HAZARDOUS WASTE PENALTIES

The table below outlines some of the possible penalties associated with violations of hazardous waste laws. The penalties are per violation amounts for noncompliance with federal statutes. Individual state laws have their own penalty provisions as well. Both civil and criminal penalties may be assessed depending on the nature of the violation and whether or not a violation is intentional.

PENALTIES FOR NONCOMPLIANCE		
Federal Statute	Civil Penalty	Criminal Penalty
RCRA	up to $25,000 per day[1]	up to $1 million and/or 15 yrs imprisonment[2]
RCRA Underground Storage Tanks		up to $10,000 per day[3]
CERCLA	liability for all cleanup and damage costs[4]	punitive damages up to three times cost of cleanup[5]
Clean Water Act	$10,000 per day[6]	$2,500 - $25,000 per day and/or 1 yr imprisonment[7]
TSCA	up to $25,000 per day[8]	$25,000/day and/or 1 yr imprisonment[9]

[1] Resource Conservation and Recovery Act Section 3008(a)(3), 42 USC Section 6928(a)(3).
[2] RCRA Section 3008(d), (e), 42 USC Section 6928(d), (e).
[3] 42 USC Section 6991e(d).
[4] Comprehensive Environmental Response, Compensation and Liability Act Section 107(c)(2), 42 USC Section 9607(c)(2).
[5] CERCLA Section 107(c)(3), 42 USC Section 9607(c)(3).
[6] Federal Water Pollution Control Act Section 309(d), 33 USC Section 1319(d).
[7] Federal Water Pollution Control Act Section 309(c), 33 USC Section 1319(c).
[8] Toxic Substances Control Act Section 16(a), 17 USC Section 2615(a).
[9] TSCA Section 16(b), 17 USC Section 2615(b).

Federal Statute	Civil Penalty	Criminal Penalty
OSHA	up to $7,000 per violation[10]	up to $70,000 per violation and/or 6 mos imprisonment[11]
Clean Air Act		up to $25,000/day and/or 1 yr imprisonment[12]

10.3 HAZARDOUS WASTE MANAGEMENT COSTS

Proper management of hazardous wastes is hardly inexpensive. Still, it should always be remembered that the costs of improper or illegal hazardous waste practices can be enormous, especially when a company finds itself part of a costly hazardous waste liability litigation.

The table below presents a list of common costs associated with hazardous waste management. These costs can certainly add up, but must be considered part of the expense of sound business practice. No amount of money saved in avoiding these costs can outweigh the potential exposure to liability for improper hazardous waste management. These costs must be budgeted like all the other ordinary expenses of a business. Many are also deductible expenses, and the company should consult with its tax adviser about claiming some of these expenses as business deductions on tax returns.

---

[10] 29 USC Section 666(b).
[11] 29 USC Section 666(e).
[12] Clean Air Act Section 113(c), 42 USC Section 7413(c).

> **COST OF HAZARDOUS WASTE MANAGEMENT**
> * signs and placards
> * UST permit fees
> * construction modification costs
> * monitoring costs
> * testing costs
> * special equipment costs
> * protective clothing costs
> * employee training costs
> * cost of hazardous waste tracking
> * administrative recordkeeping costs
> * hazardous waste disposal costs
> * hazardous waste shipping costs
> * cost of special containers and packaging
> * cost of bad publicity
> * hazardous waste cleanup costs
> * cost for consultants/legal advice
> * cost of environmental liability insurance

10.4 HAZARDOUS WASTE AUDITS

The focus of this manual has been on effective hazardous waste management techniques for the company. The preceding chapters have outlined the best means for understanding and complying with the vast body of laws and regulations concerning hazardous wastes. This section focuses on steps the company should take to assure that the company's business remains in compliance with all applicable hazardous waste laws and regulations. Particular emphasis is placed on conducting periodic hazardous waste assessments and audits at the company site.

It is important to perform routine checks of hazardous waste activities at the company to assure continued compliance with the laws, and continued maintenance of safe hazardous waste practices in on-going operations at the company site. An operational site analysis or audit is a systematic means of assessing compliance. It is the process of determining whether all or selected levels of the organization are in compliance with the environmental regulations, internal policies and accepted business practices. It is a check of the environmental status of the business. The level of analysis at the company will depend upon the degree of hazardous waste activity at the site.

LEVELS OF ASSESSMENT

Known Contaminant Impact Analysis - if a problem is known to exist or suspected at the site, an in-depth analysis should be undertaken to determine the type of hazardous waste, quantity and the extent of any possible or actual contamination.

Regulatory Compliance Analysis - this involves an assessment of regulatory compliance with all environmental laws and regulations at the site.

Corporate Policy Compliance - this involves an assessment of corporate compliance with internal policies, such as a written hazardous waste management plan for the site.

Best Management Practice Compliance - this involves assessing whether the hazardous waste management practices at the company are in line with generally accepted industry standards. This reflects what the rest of the industry is doing about hazardous waste.

10.5 PREPARING FOR AN AUDIT

Several guidelines should be followed when carrying out a hazardous waste assessment. It should be remembered that whenever any problems are uncovered through an audit, they must be reported to the relevant authorities and steps must be taken to remedy any problems discovered. Care should be exercised in systematically performing the assessment. Ordinarily, the company will hire an outside consultant to carry out an audit. A reputable consultant must be chosen who has expertise in completing hazardous waste audits. Only when hazardous waste activities at the company are minimal in nature should the company consider performing its own assessment. Still, even when the scope of hazardous waste activity is limited, the money invested in using an outside consultant will far outweigh the potential liability associated with hazardous wastes.

The company should always meet with the consultant to review the hazardous waste activity at the site and to explain the hazardous waste management program and procedures followed at the company. It is important for the consultant to understand how the business operates to fully determine where hazardous waste problems may be found. The company should also go over the process that the consultant will utilize in conducting an audit.

10.6 PERFORMING AN AUDIT

The scope of the actual assessment will vary depending on the degree of hazardous waste activity at the company site. The checklist below shows the general steps taken in performing an audit:

(1) *Pre-audit* - Select the consultant, schedule the audit, and gather background information for the consultant.

(2) *Site Analysis* - The consultant meets with company managers to go over the basic routine of the audit. The consultant gathers written records from the company files, conducts interviews with company employees, makes a visual review of the company site, reviews procedures for the handling, storage and disposal of wastes, reviews emergency planning and preparedness procedures, reviews recordkeeping and reporting procedures, does necessary testing, notes findings on an audit checklist, and then reviews the initial findings with company managers.

(3) *Post-audit* - The consultant analyzes and interprets the data from the site analysis. The consultant then recommends an action plan for specific actions to be completed by specific dates. The company should designate responsible employees and periodically review the action plan to make sure that each component of the plan is fully implemented.

The audit is designed to alert the company to any problems concerning hazardous waste management at the company. A wide array of problems are common and will vary from site to site. Actual hazardous waste discharges may be uncovered. Improper labeling may be present. Inadequate training of employees may be discovered. Hazardous waste minimization may need improvement by the company. Each problem area must be addressed and corrected in a timely manner. Violations of laws and regulations must be remedied as quickly as possible to avoid fines, penalties or liabilities.

The checklist on the next page may be used as a quick reference to assess company compliance with hazardous waste laws and regulations. For more detailed analysis of particular issues, consult appropriate portions of this book.

HAZARDOUS WASTE COMPLIANCE CHECK

___ Hazardous wastes identified
___ Hazardous waste samples tested
___ Log book kept for all hazardous wastes
___ List of state regulatory offices with phone numbers
___ List of EPA offices with phone numbers
___ Emergency response plan in place
___ Proper hazardous waste storage maintained
___ Proper labeling of containers
___ Biennial reports filed
___ EPA ID number obtained
___ Containers properly packaged for transport
___ Containers properly labeled/marked for transport
___ Hazardous waste manifests properly filled out
___ Hazardous waste manifests signed/dated by transporter
___ Signed manifest received from disposal site
___ Compliance with UST registration requirements
___ Compliance with UST system performance standards
___ Compliance with UST operation & maintenance requirements
___ Compliance with UST closure requirements
___ Compliance with UST financial responsibility requirements
___ OSHA employee training carried out
___ OSHA hazard communication plan in place
___ Material safety data sheets on file
___ Contingency plan for hazardous waste releases
___ Asbestos disposal procedures followed
___ Used batteries properly disposed of
___ Check of recordkeeping practices

Appendix 1

U.S. EPA Regional Offices

EPA Region I
*Connecticut, Massachusetts, Maine, New Hampshire,
Rhode Island, Vermont*

State Waste Programs Branch
JFK Federal Building
Boston, MA 02203
(617) 223-3468

EPA Region II
New Jersey, New York, Puerto Rico, Virgin Islands

Air and Waste Management Division
26 Federal Plaza
New York, NY 10278
(212) 264-5175

EPA Region III
*Delaware, Maryland, Pennsylvania, Virginia,
West Virginia, District of Columbia*

Waste Management Branch
841 Chestnut Street
Philadelphia, PA 19107
(215) 597-0980

EPA Region IV
*Alabama, Florida, Georgia, Kentucky, Mississippi,
North Carolina, South Carolina, Tennessee*

Hazardous Waste Management Division
345 Courtland Street, N.E.
Atlanta, GA 30365
(404) 347-3016

EPA Region V
Illinois, Indiana, Michigan, Minnesota, Ohio, Wisconsin

RCRA Activities
230 South Dearborn Street
Chicago, IL 60604
(312) 353-2000

EPA Region VI
Arkansas, Louisiana, New Mexico, Oklahoma, Texas

Air and Hazardous Materials Division
1201 Elm Street
Dallas, TX 75270
(214) 767-2600

EPA Region VII
Iowa, Kansas, Missouri, Nebraska

RCRA Branch
726 Minnesota Avenue
Kansas City, KS 66101
(913) 236-2800

EPA Region VIII
Colorado, Montana, North Dakota, South Dakota, Utah, Wyoming

Waste Management Division
One Denver Place
999 18th Street, Suite 1300
Denver, CO 80202
(303) 293-1720

EPA Region IX
Arizona, California, Hawaii, Nevada, American Samoa, Guam, Trust Territories of the Pacific

Toxics and Waste Management Division
214 Fremont Street
San Francisco, CA 94105
(415) 974-7472

EPA Region X
Alaska, Idaho, Oregon, Washington

Waste Management Branch
1200 6th Avenue
Seattle, WA 98101
(206) 442-2777

Appendix 2

Federal Hazardous Waste Regulatory Offices

<u>Disposal Sites - Landfills</u>

Environmental Protection Agency
401 M Street, S.W.
Washington, DC 20460

Office of Solid Waste
Municipal Solid Waste Programs Division
(202) -382-3346
* *information on potential contamination from current and former municipal disposal sites*

Office of Solid Waste Characterization & Assessment Division
(202) 382-4770
* *information on land disposal restrictions*

Office of Waste Programs Enforcement
RCRA Guidance and Evaluation
(202) 382-5392
* *technical assistance for site evaluation*

Office of Emergency & Remedial Response (Superfund)
Hazardous Site Evaluation Division
(202) 475-8103

<u>Hazardous Substance Management - Use, Storage and Disposal</u>

Environmental Protection Agency
401 M Street, S.W.
Washington, DC 20460

Office of Solid Waste Characterization & Assessment Division
(202) 382-4761
* *information on types of solid and hazardous wastes*

Chemical Emergency Preparedness and Prevention Office
(202) 475-9361
* *chemical reporting per SARA Title III*

Office of Emergency and Remedial Response
Superfund Hazardous Site Evaluation Division
(202) 475-8103
* *information on sites scored for Superfund evaluation*

Office of Water Regulations and Standards
Analysis and Evaluation Division
(202) 382-5392
* *information on NPDES permits for wastewater*

Office of Waste Programs Enforcement
RCRA Enforcement: Technical Assistance
(202) 475-8544
* *hazardous waste compliance history, required permits etc.*

Asbestos

Environmental Protection Agency
401 M Street, S.W.
Washington, DC 20460

Office of Toxic Substances - Division of Environmental Assistance
(202) 382-3949
* *standards for asbestos in schools*

Department of Labor
Asbestos Program, OSHA
200 Constitution Avenue, N.W. - Room 3469N
Washington, DC 20210
(202) 523-8036
* *standards for worker protection*

Aboveground Storage Tanks (ASTs)

Environmental Protection Agency
401 M Street, S.W.
Washington, DC 20460

Office of Aboveground Storage Tanks
(202) 382-4130

Underground Storage Tanks (USTs)

Environmental Protection Agency
401 M Street, S.W.
Washington, DC 20460
Office of Underground Storage Tanks
(202) 382-4756

APPENDIX 2—FEDERAL REGULATORY OFFICES

Polychlorinated Biphenyls (PCBs)

Environmental Protection Agency
401 M Street, S.W.
Washington, DC 20460

Office of Toxic Substances
Exposure Evaluation Division
(202) 382-3569

Pesticides

Environmental Protection Agency
401 M Street, S.W.
Washington, DC 20460

Office of Pesticide Programs
Registration Division
(202) 557-7410

Wetlands

U.S. Army Corps of Engineers
20 Massachusetts Avenue, N.W.
Casimir Pulaski Building
Washington, D.C. 20314

Regulatory Branch
(202) 272-0201

Environmental Protection Agency
401 M Street, S.W.
Washington, DC 20460

Office of Wetlands Protection
(202) 475-7799

U.S. Fish and Wildlife Service
Department of Interior
1849 C Street, N.W.
Washington, D.C. 20240

Division of Habitat
(703) 358-2201

Coastal Dunes/Beaches

Federal Emergency Management Agency (FEMA)
Federal Center Plaza
500 C Street, S.W.
Washington, D.C. 20472

Federal Insurance Administration
(202) 646-2774

U.S. Fish and Wildlife Service
Department of Interior
1849 C Street, N.W.
Washington, D.C. 20240

Coastal Barriers
(703) 358-2161

National Park Service
Department of Interior
1849 C Street, N.W.
Washington, D.C. 20240

Office of Land Resources
(202) 208-5881

National Oceanic & Atmospheric Administration
Department of Commerce
Universal Building South
1825 Connecticut Avenue, N.W.
Washington, D.C. 20235

Ocean and Coastal Resource Management Office
(202) 673-5138

Groundwater Protection

Environmental Protection Agency
401 M Street, S.W.
Washington, DC 20460

Office of Groundwater Protection
(202) 382-7077

APPENDIX 2—FEDERAL REGULATORY OFFICES

U.S. ARMY CORPS OF ENGINEERS DIVISION OFFICES*

Please note that some states are within the jurisdiction of more than one divisional office because Corps' divisions are organized by watershed area and not by state boundary.

New England Division

424 Trapelo Road
Waltham, MA 02254-9149
(617) 647-8778

Maine
Vermont
Connecticut
New Hampshire
Massachusetts
Rhode Island

North Atlantic Division

90 Church Street
New York, NY 10007-2979
(212) 264-7500

New York
Pennsylvania
Virginia
Maryland
Delaware
West Virginia
Vermont

South Atlantic Division

Room 313
77 Forsyth Street, S.W.
Atlanta, GA 30335-8801
(404) 331-6715

Virginia
North Carolina
South Carolina
Georgia
Florida
Alabama

Mississippi
Tennessee
Puerto Rico
U.S. Virgin Islands

Ohio River Division

P.O. Box 1159
Cincinnati, OH 45201-1159
(513) 684-3010

Pennsylvania
Virginia
West Virginia
Kentucky
Tennessee
Ohio
Indiana
Illinois
Alabama
North Carolina
Georgia

North Central Division

536 South Clark Street
Chicago, IL 60605-6319
(312) 353-6319

North Dakota
Minnesota
South Dakota
Iowa
Missouri
Illinois
Wisconsin
Michigan
Indiana
Ohio

Lower Mississippi Valley Division

P.O. Box 80
Vicksburg, MS 39180-0080
(601) 631-5052

Louisiana
Mississippi

APPENDIX 2—FEDERAL REGULATORY OFFICES

Tennessee
Arkansas
Missouri
Illinois
Kentucky

Missouri River Division

P.O. Box 103
Downtown Station
Omaha, NE 68101-0103
(402) 221-7208

Montana
North Dakota
South Dakota
Wyoming
Nebraska
Iowa
Colorado
Kansas
Missouri

Southwestern Division

1114 Commerce Street
Dallas, TX 75242-0216
(214) 767-2510

Colorado
Kansas
Missouri
New Mexico
Oklahoma
Arkansas
Texas
Louisiana

North Pacific Division

P.O. Box 2870
Portland, OR 97208-2870
(503) 326-3768

Washington
Oregon
Idaho

Montana
Wyoming
Nevada
Alaska

South Pacific Division

630 Sansome Street
Room 720
San Francisco, CA 94111-2206
(415) 705-2405

Oregon
California
Arizona
New Mexico
Colorado
Utah
Wyoming
Idaho
Nevada

Pacific Ocean Division

Building 230
Fort Shafter, HI 96858-5440
(808) 438-9258

Hawaii

Appendix 3

State Hazardous and Solid Waste Regulatory Offices (HW/SW)*

* *Please note that (HW) preceding the address indicates a hazardous waste contact and (SW) preceding the address indicates a solid waste contact. Some states have only one office for both.*

Alabama (HW/SW)

Land Division
Alabama Dept. of Environmental Management
1751 Federal Drive
Montgomery, AL 36130
(205) 271-7730

Alaska

(HW)
U.S. EPA Region X
Waste Management Branch
MS HW-112
1200 Sixth Avenue
Seattle, WA 98101
(206) 442-0151

(SW)
Air and Solid Waste Management
Dept. of Envt'l Conservation
Pouch O
Juneau, AK 99801
(907) 465-2666

American Samoa (HW/SW)

Environmental Quality Commission
Government of American Samoa
Pago Pago, American Samoa 96799
(684) 663-2304

Arizona (HW/SW)

Office of Waste & Water Quality Management
Arizona Department of Environmental Quality
2005 No. Central Ave., Room 304
Phoenix, AZ 85004
(602) 257-2305

Arkansas (HW/SW)

Arkansas Dept. of Pollution Control & Ecology
P.O. Box 9583
Little Rock, AK 72219
(501) 562-7444

California

(HW)
California Dept. of Health Services
Toxic Substances Control Division
P.O. Box 942732 - 400 P Street
Sacramento, CA 95814
(916) 323-2913

(SW)
State Water Resources Control Bd.
P.O. Box 100
Sacramento, CA 95801
(916) 445-1553

Colorado (HW/SW)

Hazardous Material & Waste Management Division
Colorado Dept. of Health
4210 East 11th Avenue
Denver, CO 80220
(303) 331-8844

Connecticut

(HW)
Waste Management Bureau
Dept. of Envt'l Protection
State Office Building
Hartford, CT 06106
(203) 566-8844

APPENDIX 3—STATE REGULATORY OFFICES

(SW)
Connecticut Resource Recovery Authority
179 Allyn Street - Suite 603
Professional Building
Hartford, CT 06103
(203) 549-6390

Delaware (HW/SW)

Delaware Dept. of Natural Resources and Envt'l Control
Div. of Air and Waste Mgmt.
Hazardous Waste Mgmt. Bureau
P.O. Box 1401 - 89 Kings Highway
Dover, DE 19903
(302) 736-3689

District of Columbia (HW/SW)

Dept. of Consumer and Regulatory Affairs
Environmental Control Division
Pesticides & Hazardous Waste Bureau
2100 Martin Luther King Jr. Avenue, S.W. - Room 204
Washington, DC 20020
(202) 783-3194

Florida (HW/SW)

Hazardous Waste Section
Dept. of Envt'l Regulations
Twin Tower Office Building
2600 Blair Stone Road
Tallahassee, FL 32399-2400
(904) 488-0300

Georgia (HW/SW)

Land Protection Branch
Industrial and Hazardous Waste Management Program
Floyd Towers East
205 Butler Street, S.E.
Atlanta, GA 30334
(404) 656-2833

Guam (HW/SW)

Guam Envt'l Protection Agency
IT&E
Harmon Plaza Complex, Unit D-107
130 Rojas Street
Harmon, Guam 96911
(671) 646-7579

Hawaii (HW/SW)

Department of Health
Hazardous Waste Program
P.O. Box 3378
Honolulu, HI 96801
(808) 548-2270

Idaho (HW/SW)

Idaho Dept. of Health & Welfare
Tower Building - 3rd FL
450 West State Street
Boise, ID 83720
(208) 334-5879

Illinois

(HW)
U.S. EPA Region V
RCRA Activities
Waste Management Division
P.O. Box A3597
Chicago, IL 60690
(312) 886-4001

(SW)
Div. of Land Pollution Control
Environmental Protection Agency
2200 Churchill Road
Springfield, IL 62706
(217) 782-6760

Indiana (HW/SW)

Indiana Dept. of Envt'l Mgmt.
105 S. Meridian St. - P.O. Box 6015
Indianapolis, IN 46225
(317) 232-3210

APPENDIX 3—STATE REGULATORY OFFICES

Iowa (HW/SW)

U.S. EPA Region VII
RCRA Branch
726 Minnesota Avenue
Kansas City, KS 66101
(913) 236-2852

Kansas

(HW)
Bureau of Air & Waste Mgmt.
Dept. of Health & Environment
Forbes Field - Building 740
Topeka, KS 66620
(913) 296-1600

(SW)
Bureau of Waste Management
Dept. of Health & Environment
Forbes Field - Building 321
Topeka, KS 66620
(913) 862-9360

Kentucky (HW/SW)

Div. of Waste Management
Dept. of Envt'l Protection
Cabinet for Natural Resources & Enviromental Protection
Fort Boone Plaza, Building #2
Frankfort, KY 40601
(502) 564-6716

Louisiana (HW/SW)

Louisiana Dept. of Envt'l Quality
Dept. of Solid & Hazardous Waste
P.O. Box 44307
Baton Rouge, LA 70804
(504) 342-1354

Maine (HW/SW)

Bureau of Oil and Hazardous Materials Control
Dept. of Environmental Protection
Ray Building - Station #17
Augusta, ME 04333
(207) 289-2651

Maryland (HW/SW)

Maryland Dept. of the Environment
Waste Mgmt. Administration
2500 Broening Highway
Baltimore, MD 21224
(301) 631-3304

Massachusetts (HW/SW)

Division of Hazardous Waste
Dept. of Environmental Protection
One Winter Street - 5th FL
Boston, MA 02108
(617) 292-5851

Michigan (HW/SW)

Waste Management Division
Environmental Protection Bureau
Dept. of Natural Resources
Box 30038
Lansing, MI 48909
(517) 373-2730

Minnesota (HZ/SW)

Solid and Hazardous Waste Div.
Pollution Control Agency
520 Lafayette Road, North
St. Paul, MN 55155
(612) 296-7282

Mississippi (HW/SW)

Hazardous Waste Division
Bureau of Pollution Control
Dept. of Environmental Quality
P.O. Box 10385
Jackson, MI 39289-0385
(601) 961-5062

APPENDIX 3—STATE REGULATORY OFFICES

Missouri (HW/SW)

Waste Management Program
Dept. of Natural Resources
Jefferson Building
205 Jefferson Street - 13/14 FL
P.O. Box 176
Jefferson City, MO 65102
(314) 751-3176

Montana (HW/SW)

Solid & Hazardous Waste Bureau
Dept. of Health & Environmental Sciences
Cogswell Building - Room B-201
Helena, MT 59620
(406) 444-1430

Nebraska (HW/SW)

Hazardous Waste Mgmt. Section
Dept. of Environmental Control
State House Station
P.O. 98922
Lincoln, NE 68509-8922
(402) 471-4215

Nevada (HW/SW)

Waste Management Bureau
Div. of Environmental Protection
Dept. of Conservation and Natural Resources
Capitol Complex
123 West Nye Lane
Carson City, NV 89710
(702) 687-5872

New Hampshire (HW/SW)

Dept. of Environmental Services
Waste Management Division
6 Hazen Drive
Concord, NH 03301
(603) 271-2900

New Jersey (HW/SW)

New Jersey Department of Envt'l Protection
Div. of Waste Management
Bureau of Hazardous Waste Classification and Manifests
401 East State Street - CN-028
Trenton, NJ 08625
(609) 292-8341

New Mexico (HW/SW)

New Mexico Health & Environment Department
Hazardous Waste Bureau
1190 St. Francis Drive
Santa Fe, NM 87503
(505) 827-2929

New York (HW/SW)

New York Dept. of Environmental Conservation
Div. of Hazardous Waste Substances Regulation
P.O. Box 12820
Albany, NY 12212
(518) 457-0530

North Carolina (HW/SW)

Hazardous Waste Section
Div. of Solid Waste Management
Dept. of Environment, Health and Natural Resources
P.O. Box 27687
Raleigh, NC 27611-7687
(919) 733-2178

North Dakota (HW/SW)

Div. of Waste Management
Dept. of Health and Consolidate Laboratories
1200 Missouri Avenue
P.O. Box 5520
Bismarck, ND 58502-5520
(701) 221-5150

APPENDIX 3—STATE REGULATORY OFFICES

Northern Mariana Islands (HW/SW)

Dept. of Public Health and Environmental Quality
Div. of Environmental Quality
Dr. Torres Hospital - P.O. Box 1304
Saipan, Mariana Islands 96950
(676) 234-6984

Ohio

(HW)
U.S. EPA Region V
RCRA Activities
Waste Management Division
P.O. Box A3587
Chicago, IL 60690
(312) 886-4001

(SW)
Division of Solid & Hazardous Waste Management
Ohio Envt'l Protection Agency
361 East Broad Street
Columbus, OH 43215
(614) 466-7220

Oklahoma (HW/SW)

Oklahoma State Dept. of Health
Industrial Waste Division
1000 Northeast 10th Street
Oklahoma City, OK 73152
(405) 271-5338

Oregon (HW/SW)

Oregon Dept. of Envt'l Quality
Hazardous Waste Operations
811 Southwest 6th Avenue
Portland, OR 97204
(503) 229-5913

Pennsylvania (HW/SW)

Pennsylvania Dept. of Environmental Resources
Bureau of Waste Management
P.O. Box 2063
Harrisburg, PA 17120
(717) 787-9870

Puerto Rico (HW/SW)

Puerto Rico Envt'l Quality Bd.
Land Pollution Control Area
Inspection, Monitoring & Surveillance
P.O. Box 11488
Santurce, PR 00910-1488
(809) 722-0439

Rhode Island (HW/SW)

Div. of Air & Hazardous Materials
Dept. of Environmental Mgmt.
291 Promenade Street
Providence, RI 02908-5767
(401) 277-2808

South Carolina (HW/SW)

Bureau of Solid Waste Management
Hazardous Waste Management
Dept. of Health & Envt'l Control
2600 Bull Street
Columbia, SC 29201
(803) 734-2500

South Dakota (HW/SW)

Office of Waste Management
Dept. of Water & Natural Resources
Joe Foss Building
523 East Capitol Street
Pierre, SD 57501-3181
(605) 773-3153

Tennessee (HW/SW)

Div. of Solid Waste Management
Tennessee Dept. of Health & Environment
701 Broadway - Customs House, 4th FL
Nashville, TN 37247-3530
(615) 741-3424

Texas

(HW)
Texas Water Commission
Compliance Assistance Unit
Hazardous & Solid Waste Div.
P.O. Box 13087 - Capitol Station
Austin, TX 78711-3087
(512) 463-8175

(SW)
Division of Solid Waste
Texas Department of Health
1100 West 49th St., T-610A
Austin, TX 78756-3199
(512) 458-7271

Utah (HW/SW)

Bureau of Solid and Hazardous Waste Management
Department of Health
288 North 1460 West - P.O. Box 16690
Salt Lake City, UT 84116-0690
(801) 538-6170

Vermont (HW/SW)

Hazardous Materials Mgmt. Div.
Dept. of Envt'l Conservation
103 South Main Street
Waterbury, VT 95676
(802) 244-8702

Virgin Islands (HW/SW)

Virgin Islands Dept. of Planning and Natural Resources
Div. of Environmental Protection
179 Altona and Welgunst
St. Thomas, VI 00801
(809) 774-3320

Virginia (HW/SW)

Virginia Dept. of Waste Mgmt.
Monroe Building - 11th FL
101 North 14th Street
Richmond, VA 23219
(804) 225-2667

Washington (HW/SW)

Solid & Hazardous Waste Mgmt Div.
Dept. of Ecology
Mail Stop PV-11
Olympia, WA 98504
(206) 459-6369

West Virginia (HW/SW)

West Virginia Div. of Natural Resources
Waste Management Section
1356 Hansford Street
Charleston, WV 25301
(304) 358-5393

Wisconsin (HW/SW)

Bureau of Solid Waste
Dept. of Natural Resources
P.O. Box 7921
Madison, WI 53707
(608) 266-1327

Wyoming

(HW)
U.S. EPA Region VIII
Hazardous Waste Management Division (8HWM-ON)
999 18th Street - Suite 500
Denver, CO 80202-2405
(303) 293-1795

(SW)
Solid Waste Management Program
State of Wyoming
Dept. of Environmental Quality
122 West 25th Street
Herschler Building
Cheyenne, WY 82002
(307) 777-7752

Appendix 4

RCRA Hazardous Wastes

Environmental Protection Agency § 261.31

TABLE 1—MAXIMUM CONCENTRATION OF CONTAMINANTS FOR THE TOXICITY CHARACTERISTIC—Continued

EPA HW No.[1]	Contaminant	CAS No.[2]	Regulatory Level (mg/L)
D043	Vinyl chloride	75-01-4	0.2

[1] Hazardous waste number.
[2] Chemical abstracts service number.
[3] Quantitation limit is greater than the calculated regulatory level. The quantitation limit therefore becomes the regulatory level.
[4] If o-, m-, and p-Cresol concentrations cannot be differentiated, the total cresol (D026) concentration is used. The regulatory level of total cresol is 200 mg/l.

[55 FR 11862, Mar. 29, 1990, as amended at 55 FR 22684, June 1, 1990; 55 FR 26987, June 29, 1990]

Subpart D—Lists of Hazardous Wastes

§ 261.30 General.

(a) A solid waste is a hazardous waste if it is listed in this subpart, unless it has been excluded from this list under §§ 260.20 and 260.22.

(b) The Administrator will indicate his basis for listing the classes or types of wastes listed in this subpart by employing one or more of the following Hazard Codes:

Ignitable Waste .... (I)
Corrosive Waste .... (C)
Reactive Waste .... (R)
Toxicity Characteristic Waste ..... (E)
Acute Hazardous Waste .... (H)
Toxic Waste .... (T)

Appendix VII identifies the constituent which caused the Administrator to list the waste as a Toxicity Characteristic Waste (E) or Toxic Waste (T) in §§ 261.31 and 261.32.

(c) Each hazardous waste listed in this subpart is assigned an EPA Hazardous Waste Number which precedes the name of the waste. This number must be used in complying with the notification requirements of Section 3010 of the Act and certain recordkeeping and reporting requirements under parts 262 through 265, 268, and part 270 of this chapter.

(d) The following hazardous wastes listed in § 261.31 or § 261.32 are subject to the exclusion limits for acutely hazardous wastes established in § 261.5: EPA Hazardous Wastes Nos. F020, F021, F022, F023, F026, and F027.

[45 FR 33119, May 19, 1980, as amended at 48 FR 14294, Apr. 1, 1983; 50 FR 2000, Jan. 14, 1985; 51 FR 40636, Nov. 7, 1986; 55 FR 11863, Mar. 29, 1990]

§ 261.31 Hazardous wastes from non-specific sources.

(a) The following solid wastes are listed hazardous wastes from non-specific sources unless they are excluded under §§ 260.20 and 260.22 and listed in appendix IX.

Industry and EPA hazardous waste No.	Hazardous waste	Hazard code
Generic:		
F001	The following spent halogenated solvents used in degreasing: Tetrachloroethylene, trichloroethylene, methylene chloride, 1,1,1-trichloroethane, carbon tetrachloride, and chlorinated fluorocarbons; all spent solvent mixtures/blends used in degreasing containing, before use, a total of ten percent or more (by volume) of one or more of the above halogenated solvents or those solvents listed in F002, F004, and F005; and still bottoms from the recovery of these spent solvents and spent solvent mixtures.	(T)
F002	The following spent halogenated solvents: Tetrachloroethylene, methylene chloride, trichloroethylene, 1,1,1-trichloroethane, chlorobenzene, 1,1,2-trichloro-1,2,2-trifluoroethane, ortho-dichlorobenzene, trichlorofluoromethane, and 1,1,2-trichloroethane; all spent solvent mixtures/blends containing, before use, a total of ten percent or more (by volume) of one or more of the above halogenated solvents or those listed in F001, F004, or F005; and still bottoms from the recovery of these spent solvents and spent solvent mixtures.	(T)

§ 261.31   40 CFR Ch. I (7-1-91 Edition)

Industry and EPA hazardous waste No.	Hazardous waste	Hazard code
F003	The following spent non-halogenated solvents: Xylene, acetone, ethyl acetate, ethyl benzene, ethyl ether, methyl isobutyl ketone, n-butyl alcohol, cyclohexanone, and methanol; all spent solvent mixtures/blends containing, before use, only the above spent non-halogenated solvents; and all spent solvent mixtures/blends containing, before use, one or more of the above non-halogenated solvents, and, a total of ten percent or more (by volume) of one or more of those solvents listed in F001, F002, F004, and F005; and still bottoms from the recovery of these spent solvents and spent solvent mixtures.	(I)*
F004	The following spent non-halogenated solvents: Cresols and cresylic acid, and nitrobenzene; all spent solvent mixtures/blends containing, before use, a total of ten percent or more (by volume) of one or more of the above non-halogenated solvents or those solvents listed in F001, F002, and F005; and still bottoms from the recovery of these spent solvents and spent solvent mixtures.	(T)
F005	The following spent non-halogenated solvents: Toluene, methyl ethyl ketone, carbon disulfide, isobutanol, pyridine, benzene, 2-ethoxyethanol, and 2-nitropropane; all spent solvent mixtures/blends containing, before use, a total of ten percent or more (by volume) of one or more of the above non-halogenated solvents or those solvents listed in F001, F002, or F004; and still bottoms from the recovery of these spent solvents and spent solvent mixtures.	(I,T)
F006	Wastewater treatment sludges from electroplating operations except from the following processes: (1) Sulfuric acid anodizing of aluminum; (2) tin plating on carbon steel; (3) zinc plating (segregated basis) on carbon steel; (4) aluminum or zinc-aluminum plating on carbon steel; (5) cleaning/stripping associated with tin, zinc and aluminum plating on carbon steel; and (6) chemical etching and milling of aluminum.	(T)
F007	Spent cyanide plating bath solutions from electroplating operations.	(R, T)
F008	Plating bath residues from the bottom of plating baths from electroplating operations where cyanides are used in the process.	(R, T)
F009	Spent stripping and cleaning bath solutions from electroplating operations where cyanides are used in the process.	(R, T)
F010	Quenching bath residues from oil baths from metal heat treating operations where cyanides are used in the process.	(R, T)
F011	Spent cyanide solutions from salt bath pot cleaning from metal heat treating operations.	(R, T)
F012	Quenching waste water treatment sludges from metal heat treating operations where cyanides are used in the process.	(T)
F019	Wastewater treatment sludges from the chemical conversion coating of aluminum except from zirconium phosphating in aluminum can washing when such phosphating is an exclusive conversion coating process.	(T)
F020	Wastes (except wastewater and spent carbon from hydrogen chloride purification) from the production or manufacturing use (as a reactant, chemical intermediate, or component in a formulating process) of tri- or tetrachlorophenol, or of intermediates used to produce their pesticide derivatives. (This listing does not include wastes from the production of Hexachlorophene from highly purified 2,4,5-trichlorophenol.).	(H)
F021	Wastes (except wastewater and spent carbon from hydrogen chloride purification) from the production or manufacturing use (as a reactant, chemical intermediate, or component in a formulating process) of pentachlorophenol, or of intermediates used to produce its derivatives.	(H)
F022	Wastes (except wastewater and spent carbon from hydrogen chloride purification) from the manufacturing use (as a reactant, chemical intermediate, or component in a formulating process) of tetra-, penta-, or hexachlorobenzenes under alkaline conditions.	(H)
F023	Wastes (except wastewater and spent carbon from hydrogen chloride purification) from the production of materials on equipment previously used for the production or manufacturing use (as a reactant, chemical intermediate, or component in a formulating process) of tri- and tetrachlorophenols. (This listing does not include wastes from equipment used only for the production or use of Hexachlorophene from highly purified 2,4,5-trichlorophenol.).	(H)
F024	Process wastes, including but not limited to, distillation residues, heavy ends, tars, and reactor clean-out wastes, from the production of certain chlorinated aliphatic hydrocarbons by free radical catalyzed processes. These chlorinated aliphatic hydrocarbons are those having carbon chain lengths ranging from one to and including five, with varying amounts and positions of chlorine substitution. (This listing does not include wastewaters, wastewater treatment sludges, spent catalysts, and wastes listed in § 261.31 or § 261.32.).	(T)
F025	Condensed light ends, spent filters and filter aids, and spent desiccant wastes from the production of certain chlorinated aliphatic hydrocarbons, by free radical catalyzed processes. These chlorinated aliphatic hydrocarbons are those having carbon chain lengths ranging from one to and including five, with varying amounts and positions of chlorine substitution.	(T)

Environmental Protection Agency § 261.31

Industry and EPA hazardous waste No.	Hazardous waste	Hazard code
F026	Wastes (except wastewater and spent carbon from hydrogen chloride purification) from the production of materials on equipment previously used for the manufacturing use (as a reactant, chemical intermediate, or component in a formulating process) of tetra-, penta-, or hexachlorobenzene under alkaline conditions.	(H)
F027	Discarded unused formulations containing tri-, tetra-, or pentachlorophenol or discarded unused formulations containing compounds derived from these chlorophenols. (This listing does not include formulations containing Hexachlorophene sythesized from prepurified 2,4,5-trichlorophenol as the sole component.).	(H)
F028	Residues resulting from the incineration or thermal treatment of soil contaminated with EPA Hazardous Waste Nos. F020, F021, F022, F023, F026, and F027.	(T)
F032 [1]	Wastewaters, process residuals, preservative drippage, and spent formulations from wood preserving processes generated at plants that currently use or have previously used chlorophenolic formulations (except potentially cross-contaminated wastes that have had the F032 waste code deleted in accordance with § 261.35 of this chapter and where the generator does not resume or initiate use of chlorophenolic formulations). This listing does not include K001 bottom sediment sludge from the treatment of wastewater from wood preserving processes that use creosote and/or pentachlorophenol. (NOTE: The listing of wastewaters that have not come into contact with process contaminants is stayed administratively. The listing for plants that have previously used chlorophenolic formulations is administratively stayed whenever these wastes are covered by the F034 or F035 listings. These stays will remain in effect until further administrative action is taken.).	(T)
F034 [1]	Wastewaters, process residuals, preservative drippage, and spent formulations from wood preserving process generated at plants that use creosote formulations. This listing does not include K001 bottom sediment sludge from the treatment of wastewater from wood preserving processes that use creosote and/or pentachlorophenol. (NOTE: The listing of wastewaters that have not come into contact with process contaminants is stayed administratively. The stay will remain in effect until further administrative action is taken.).	(T)
F035 [1]	Wastewaters, process residuals, preservative drippage, and spent formulations from wood preserving process generated at plants that use inorganic preservatives containing arsenic or chromium. This listing does not include K001 bottom sediment sludge from the treatment of wastewater from wood preserving processes that use creosote and/or pentachlorophenol. (NOTE: The listing of wastewaters that have not come into contact with process contaminants is stayed administratively. The stay will remain in effect until further administrative action is taken.).	(T)
F037	Petroleum refinery primary oil/water/solids separation sludge—Any sludge generated from the gravitational separation of oil/water/solids during the storage or treatment of process wastewaters and oily cooling wastewaters from petroleum refineries. Such sludges include, but are not limited to, those generated in: oil/water/solids separators; tanks and impoundments; ditches and other conveyances; sumps; and stormwater units receiving dry weather flow. Sludge generated in stormwater units that do not receive dry weather flow, sludges generated from non-contact once-through cooling waters segregated for treatment from other process or oily cooling waters, sludges generated in aggressive biological treatment units as defined in § 261.31(b)(2) (including sludges generated in one or more additional units after wastewaters have been treated in aggressive biological treatment units) and K051 wastes are not included in this listing.	(T)
F038	Petroleum refinery secondary (emulsified) oil/water/solids separation sludge—Any sludge and/or float generated from the physical and/or chemical separation of oil/water/solids in process wastewaters and oily cooling wastewaters from petroleum refineries. Such wastes include, but are not limited to, all sludges and floats generated in: induced air flotation (IAF) units, tanks and impoundments, and all sludges generated in DAF units. Sludges generated in stormwater units that do not receive dry weather flow, sludges generated from non-contact once-through cooling waters segregated for treatment from other process or oily cooling waters, sludges and floats generated in aggressive biological treatment units as defined in § 261.31(b)(2) (including sludges and floats generated in one or more additional units after wastewaters have been treated in aggressive biological treatment units) and F037, K048, and K051 wastes are not included in this listing.	(T)
F039	Leachate (liquids that have percolated through land disposed wastes) resulting from the disposal of more than one restricted waste classified as hazardous under subpart D of this part. (Leachate resulting from the disposal of one or more of the following EPA Hazardous Wastes and no other Hazardous Wastes retains its EPA Hazardous Waste Number(s): F020, F021, F022, F026, F027, and/or F028.).	(T)

[1] The F032, F034, and F305 listings are administratively stayed with respect to the process area receiving drippage of these wastes provided persons desiring to continue operating notify EPA by August 6, 1991 of their intent to upgrade or install drip pads, and by November 6, 1991 provide evidence to EPA that they have adequate financing to pay for drip pad upgrades or installation, as provided in the administrative stay. The stay of the listings will remain in effect until February 6, 1992 for existing drip pads and until May 6, 1992 for new drip pads.

*(I,T) should be used to specify mixtures containing ignitable and toxic constituents.

§ 261.32

(b) Listing Specific Definitions: (1) For the purposes of the F037 and F038 listings, oil/water/solids is defined as oil and/or water and/or solids.

(2) (i) For the purposes of the F037 and F038 listings, aggressive biological treatment units are defined as units which employ one of the following four treatment methods: activated sludge; trickling filter; rotating biological contactor for the continuous accelerated biological oxidation of wastewaters; or high-rate aeration. High-rate aeration is a system of surface impoundments or tanks, in which intense mechanical aeration is used to completely mix the wastes, enhance biological activity, and (A) the units employs a minimum of 6 hp per million gallons of treatment volume; and either (B) the hydraulic retention time of the unit is no longer than 5 days; or (C) the hydraulic retention time is no longer than 30 days and the unit does not generate a sludge that is a hazardous waste by the Toxicity Characteristic.

(ii) Generators and treatment, storage and disposal facilities have the burden of proving that their sludges are exempt from listing as F037 and F038 wastes under this definition. Generators and treatment, storage and disposal facilities must maintain, in their operating or other onsite records, documents and data sufficient to prove that: (A) the unit is an aggressive biological treatment unit as defined in this subsection; and (B) the sludges sought to be exempted from the definitions of F037 and/or F038 were actually generated in the aggressive biological treatment unit.

(3) (i) For the purposes of the F037 listing, sludges are considered to be generated at the moment of deposition in the unit, where deposition is defined as at least a temporary cessation of lateral particle movement.

(ii) For the purposes of the F038 listing,

(A) sludges are considered to be generated at the moment of deposition in the unit, where deposition is defined as at least a temporary cessation of lateral particle movement and

(B) floats are considered to be generated at the moment they are formed in the top of the unit.

[46 FR 4617, Jan. 16, 1981]

EDITORIAL NOTE: For FEDERAL REGISTER citations affecting § 261.31, see the List of CFR Sections Affected in the Finding Aids section of this volume.

§ 261.32 Hazardous wastes from specific sources.

The following solid wastes are listed hazardous wastes from specific sources unless they are excluded under §§ 260.20 and 260.22 and listed in appendix IX.

Industry and EPA hazardous waste No.	Hazardous waste	Hazard code
Wood preservation: K001	Bottom sediment sludge from the treatment of wastewaters from wood preserving processes that use creosote and/or pentachlorophenol.	(T)
Inorganic pigments:		
K002	Wastewater treatment sludge from the production of chrome yellow and orange pigments.	(T)
K003	Wastewater treatment sludge from the production of molybdate orange pigments	(T)
K004	Wastewater treatment sludge from the production of zinc yellow pigments	(T)
K005	Wastewater treatment sludge from the production of chrome green pigments	(T)
K006	Wastewater treatment sludge from the production of chrome oxide green pigments (anhydrous and hydrated).	(T)
K007	Wastewater treatment sludge from the production of iron blue pigments	(T)
K008	Oven residue from the production of chrome oxide green pigments	(T)
Organic chemicals:		
K009	Distillation bottoms from the production of acetaldehyde from ethylene	(T)
K010	Distillation side cuts from the production of acetaldehyde from ethylene	(T)
K011	Bottom stream from the wastewater stripper in the production of acrylonitrile	(R, T)
K013	Bottom stream from the acetonitrile column in the production of acrylonitrile	(R, T)
K014	Bottoms from the acetonitrile purification column in the production of acrylonitrile	(T)
K015	Still bottoms from the distillation of benzyl chloride	(T)
K016	Heavy ends or distillation residues from the production of carbon tetrachloride	(T)
K017	Heavy ends (still bottoms) from the purification column in the production of epichlorohydrin.	(T)

Environmental Protection Agency § 261.32

Industry and EPA hazardous waste No.	Hazardous waste	Hazard code
K018	Heavy ends from the fractionation column in ethyl chloride production	(T)
K019	Heavy ends from the distillation of ethylene dichloride in ethylene dichloride production.	(T)
K020	Heavy ends from the distillation of vinyl chloride in vinyl chloride monomer production.	(T)
K021	Aqueous spent antimony catalyst waste from fluoromethanes production	(T)
K022	Distillation bottom tars from the production of phenol/acetone from cumene	(T)
K023	Distillation light ends from the production of phthalic anhydride from naphthalene	(T)
K024	Distillation bottoms from the production of phthalic anhydride from naphthalene	(T)
K025	Distillation bottoms from the production of nitrobenzene by the nitration of benzene	(T)
K026	Stripping still tails from the production of methy ethyl pyridines	(T)
K027	Centrifuge and distillation residues from toluene diisocyanate production	(R, T)
K028	Spent catalyst from the hydrochlorinator reactor in the production of 1,1,1-trichloroethane.	(T)
K029	Waste from the product steam stripper in the production of 1,1,1-trichloroethane	(T)
K030	Column bottoms or heavy ends from the combined production of trichloroethylene and perchloroethylene.	(T)
K083	Distillation bottoms from aniline production	(T)
K085	Distillation or fractionation column bottoms from the production of chlorobenzenes	(T)
K093	Distillation light ends from the production of phthalic anhydride from ortho-xylene	(T)
K094	Distillation bottoms from the production of phthalic anhydride from ortho-xylene	(T)
K095	Distillation bottoms from the production of 1,1,1-trichloroethane	(T)
K096	Heavy ends from the heavy ends column from the production of 1,1,1-trichloroethane.	(T)
K103	Process residues from aniline extraction from the production of aniline	(T)
K104	Combined wastewater streams generated from nitrobenzene/aniline production	(T)
K105	Separated aqueous stream from the reactor product washing step in the production of chlorobenzenes.	(T)
K107	Column bottoms from product separation from the production of 1,1-dimethylhydrazine (UDMH) from carboxylic acid hydrazines.	(C,T)
K108	Condensed column overheads from product separation and condensed reactor vent gases from the production of 1,1-dimethylhydrazine (UDMH) from carboxylic acid hydrazides.	(I,T)
K109	Spent filter cartridges from product purification from the production of 1,1-dimethylhydrazine (UDMH) from carboxylic acid hydrazides.	(T)
K110	Condensed column overheads from intermediate separation from the production of 1,1-dimethylhydrazine (UDMH) from carboxylic acid hydrazides.	(T)
K111	Product washwaters from the production of dinitrotoluene via nitration of toluene	(C,T)
K112	Reaction by-product water from the drying column in the production of toluenediamine via hydrogenation of dinitrotoluene.	(T)
K113	Condensed liquid light ends from the purification of toluenediamine in the production of toluenediamine via hydrogenation of dinitrotoluene.	(T)
K114	Vicinals from the purification of toluenediamine in the production of toluenediamine via hydrogenation of dinitrotoluene.	(T)
K115	Heavy ends from the purification of toluenediamine in the production of toluenediamine via hydrogenation of dinitrotoluene.	(T)
K116	Organic condensate from the solvent recovery column in the production of toluene diisocyanate via phosgenation of toluenediamine.	(T)
K117	Wastewater from the reactor vent gas scrubber in the production of ethylene dibromide via bromination of ethene.	(T)
K118	Spent adsorbent solids from purification of ethylene dibromide in the production of ethylene dibromide via bromination of ethene.	(T)
K136	Still bottoms from the purification of ethylene dibromide in the production of ethylene dibromide via bromination of ethene.	(T)
Inorganic chemicals:		
K071	Brine purification muds from the mercury cell process in chlorine production, where separately prepurified brine is not used.	(T)
K073	Chlorinated hydrocarbon waste from the purification step of the diaphragm cell process using graphite anodes in chlorine production.	(T)
K106	Wastewater treatment sludge from the mercury cell process in chlorine production	(T)
Pesticides:		
K031	By-product salts generated in the production of MSMA and cacodylic acid	(T)
K032	Wastewater treatment sludge from the production of chlordane	(T)
K033	Wastewater and scrub water from the chlorination of cyclopentadiene in the production of chlordane.	(T)
K034	Filter solids from the filtration of hexachlorocyclopentadiene in the production of chlordane.	(T)
K035	Wastewater treatment sludges generated in the production of creosote	(T)
K036	Still bottoms from toluene reclamation distillation in the production of disulfoton	(T)
K037	Wastewater treatment sludges from the production of disulfoton	(T)
K038	Wastewater from the washing and stripping of phorate production	(T)

APPENDIX 4—RCRA HAZARDOUS WASTES

§ 261.32 — 40 CFR Ch. I (7-1-91 Edition)

Industry and EPA hazardous waste No.	Hazardous waste	Hazard code
K039	Filter cake from the filtration of diethylphosphorodithioic acid in the production of phorate.	(T)
K040	Wastewater treatment sludge from the production of phorate	(T)
K041	Wastewater treatment sludge from the production of toxaphene	(T)
K042	Heavy ends or distillation residues from the distillation of tetrachlorobenzene in the production of 2,4,5-T.	(T)
K043	2,6-Dichlorophenol waste from the production of 2,4-D	(T)
K097	Vacuum stripper discharge from the chlordane chlorinator in the production of chlordane.	(T)
K098	Untreated process wastewater from the production of toxaphene	(T)
K099	Untreated wastewater from the production of 2,4-D	(T)
K123	Process wastewater (including supernates, filtrates, and washwaters) from the production of ethylenebisdithiocarbamic acid and its salt.	(T)
K124	Reactor vent scrubber water from the production of ethylenebisdithiocarbamic acid and its salts.	(C, T)
K125	Filtration, evaporation, and centrifugation solids from the production of ethylenebisdithiocarbamic acid and its salts.	(T)
K126	Baghouse dust and floor sweepings in milling and packaging operations from the production or formulation of ethylenebisdithiocarbamic acid and its salts.	(T)
K131	Wastewater from the reactor and spent sulfuric acid from the acid dryer from the production of methyl bromide.	(C, T)
K132	Spent absorbent and wastewater separator solids from the production of methyl bromide.	(T)
Explosives:		
K044	Wastewater treatment sludges from the manufacturing and processing of explosives	(H)
K045	Spent carbon from the treatment of wastewater containing explosives	(R)
K046	Wastewater treatment sludges from the manufacturing, formulation and loading of lead-based initiating compounds.	(T)
K047	Pink/red water from TNT operations	(R)
Petroleum refining:		
K048	Dissolved air flotation (DAF) float from the petroleum refining industry	(T)
K049	Slop oil emulsion solids from the petroleum refining industry	(T)
K050	Heat exchanger bundle cleaning sludge from the petroleum refining industry	(T)
K051	API separator sludge from the petroleum refining industry	(T)
K052	Tank bottoms (leaded) from the petroleum refining industry	(T)
Iron and steel:		
K061	Emission control dust/sludge from the primary production of steel in electric furnaces.	(T)
K062	Spent pickle liquor generated by steel finishing operations of facilities within the iron and steel industry (SIC Codes 331 and 332).	(C,T)
Primary copper:		
K064	Acid plant blowdown slurry/sludge resulting from the thickening of blowdown slurry from primary copper production.	(T)
Primary lead:		
K065	Surface impoundment solids contained in and dredged from surface impoundments at primary lead smelting facilities.	(T)
Primary zinc:		
K066	Sludge from treatment of process wastewater and/or acid plant blowdown from primary zinc production.	(T)
Primary aluminum:		
K088	Spent potliners from primary aluminum reduction	(T)
Ferroalloys:		
K090	Emission control dust or sludge from ferrochromiumsilicon production	(T)
K091	Emission control dust or sludge from ferrochromium production	(T)
Secondary lead:		
K069	Emission control dust/sludge from secondary lead smelting. (NOTE: This listing is stayed administratively for sludge generated from secondary acid scrubber systems. The stay will remain in effect until further administrative action is taken. If EPA takes further action effecting this stay, EPA will publish a notice of the action in the **Federal Register**.	(T)
K100	Waste leaching solution from acid leaching of emission control dust/sludge from secondary lead smelting.	(T)
Veterinary pharmaceuticals:		
K084	Wastewater treatment sludges generated during the production of veterinary pharmaceuticals from arsenic or organo-arsenic compounds.	(T)
K101	Distillation tar residues from the distillation of aniline-based compounds in the production of veterinary pharmaceuticals from arsenic or organo-arsenic compounds.	(T)
K102	Residue from the use of activated carbon for decolorization in the production of veterinary pharmaceuticals from arsenic or organo-arsenic compounds.	(T)

Environmental Protection Agency § 261.33

Industry and EPA hazardous waste No.	Hazardous waste	Hazard code
Ink formulation:		
K086	Solvent washes and sludges, caustic washes and sludges, or water washes and sludges from cleaning tubs and equipment used in the formulation of ink from pigments, driers, soaps, and stabilizers containing chromium and lead.	(T)
Coking:		
K060	Ammonia still lime sludge from coking operations	(T)
K087	Decanter tank tar sludge from coking operations	(T)

[46 FR 4618, Jan. 16, 1981]

EDITORIAL NOTE: For FEDERAL REGISTER citations affecting § 261.32, see the List of CFR Sections Affected in the Finding Aids section of this volume.

§ 261.33 Discarded commercial chemical products, off-specification species, container residues, and spill residues thereof.

The following materials or items are hazardous wastes if and when they are discarded or intended to be discarded as described in § 261.2(a)(2)(i), when they are mixed with waste oil or used oil or other material and applied to the land for dust suppression or road treatment, when they are otherwise applied to the land in lieu of their original intended use or when they are contained in products that are applied to the land in lieu of their original intended use, or when, in lieu of their original intended use, they are produced for use as (or as a component of) a fuel, distributed for use as a fuel, or burned as a fuel.

(a) Any commercial chemical product, or manufacturing chemical intermediate having the generic name listed in paragraph (e) or (f) of this section.

(b) Any off-specification commercial chemical product or manufacturing chemical intermediate which, if it met specifications, would have the generic name listed in paragraph (e) or (f) of this section.

(c) Any residue remaining in a container or in an inner liner removed from a container that has held any commercial chemical product or manufacturing chemical intermediate having the generic name listed in paragraphs (e) or (f) of this section, unless the container is empty as defined in § 261.7(b) of this chapter.

[Comment: Unless the residue is being beneficially used or reused, or legitimately recycled or reclaimed; or being accumulated, stored, transported or treated prior to such use, re-use, recycling or reclamation, EPA considers the residue to be intended for discard, and thus, a hazardous waste. An example of a legitimate re-use of the residue would be where the residue remains in the container and the container is used to hold the same commercial chemical product or manufacturing chemical intermediate it previously held. An example of the discard of the residue would be where the drum is sent to a drum reconditioner who reconditions the drum but discards the residue.]

(d) Any residue or contaminated soil, water or other debris resulting from the cleanup of a spill into or on any land or water of any commercial chemical product or manufacturing chemical intermediate having the generic name listed in paragraph (e) or (f) of this section, or any residue or contaminated soil, water or other debris resulting from the cleanup of a spill, into or on any land or water, of any off-specification chemical product and manufacturing chemical intermediate which, if it met specifications, would have the generic name listed in paragraph (e) or (f) of this section.

[Comment: The phrase "commercial chemical product or manufacturing chemical intermediate having the generic name listed in . . ." refers to a chemical substance which is manufactured or formulated for commercial or manufacturing use which consists of the commercially pure grade of the chemical, any technical grades of the chemical that are produced or marketed, and all formulations in which the chemical is the sole active ingredient. It does not refer to a material, such as a manufacturing process waste, that contains any of the substances listed in paragraph (e) or (f). Where a manufacturing process waste is deemed to be a hazardous waste because it contains a substance listed in paragraph (e) or (f), such waste will be listed in either § 261.31 or

§ 261.33

[§ 261.32 or will be identified as a hazardous waste by the characteristics set forth in subpart C of this part.]

(e) The commercial chemical products, manufacturing chemical intermediates or off-specification commercial chemical products or manufacturing chemical intermediates referred to in paragraphs (a) through (d) of this section, are identified as acute hazardous wastes (H) and are subject to be the small quantity exclusion defined in § 261.5(e).

[*Comment:* For the convenience of the regulated community the primary hazardous properties of these materials have been indicated by the letters T (Toxicity), and R (Reactivity). Absence of a letter indicates that the compound only is listed for acute toxicity.]

These wastes and their corresponding EPA Hazardous Waste Numbers are:

Hazardous waste No.	Chemical abstracts No.	Substance
P023	107-20-0	Acetaldehyde, chloro-
P002	591-08-2	Acetamide, N-(aminothioxomethyl)-
P057	640-19-7	Acetamide, 2-fluoro-
P058	62-74-8	Acetic acid, fluoro-, sodium salt
P002	591-08-2	1-Acetyl-2-thiourea
P003	107-02-8	Acrolein
P070	116-06-3	Aldicarb
P004	309-00-2	Aldrin
P005	107-18-6	Allyl alcohol
P006	20859-73-8	Aluminum phosphide (R,T)
P007	2763-96-4	5-(Aminomethyl)-3-isoxazolol
P008	504-24-5	4-Aminopyridine
P009	131-74-8	Ammonium picrate (R)
P119	7803-55-6	Ammonium vanadate
P099	506-61-6	Argentate(1-), bis(cyano-C)-, potassium
P010	7778-39-4	Arsenic acid H_3AsO_4
P012	1327-53-3	Arsenic oxide As_2O_3
P011	1303-28-2	Arsenic oxide As_2O_5
P011	1303-28-2	Arsenic pentoxide
P012	1327-53-3	Arsenic trioxide
P038	692-42-2	Arsine, diethyl-
P036	696-28-6	Arsonous dichloride, phenyl-
P054	151-56-4	Aziridine
P067	75-55-8	Aziridine, 2-methyl-
P013	542-62-1	Barium cyanide
P024	106-47-8	Benzenamine, 4-chloro-
P077	100-01-6	Benzenamine, 4-nitro-
P028	100-44-7	Benzene, (chloromethyl)-
P042	51-43-4	1,2-Benzenediol, 4-[1-hydroxy-2-(methylamino)ethyl]-, (R)-
P046	122-09-8	Benzeneethanamine, alpha,alpha-dimethyl-
P014	108-98-5	Benzenethiol
P001	[1] 81-81-2	2H-1-Benzopyran-2-one, 4-hydroxy-3-(3-oxo-1-phenylbutyl)-, & salts, when present at concentrations greater than 0.3%
P028	100-44-7	Benzyl chloride
P015	7440-41-7	Beryllium
P017	598-31-2	Bromoacetone
P018	357-57-3	Brucine
P045	39196-18-4	2-Butanone, 3,3-dimethyl-1-(methylthio)-, O-[methylamino)carbonyl] oxime
P021	592-01-8	Calcium cyanide
P021	592-01-8	Calcium cyanide Ca(CN)$_2$
P022	75-15-0	Carbon disulfide
P095	75-44-5	Carbonic dichloride
P023	107-20-0	Chloroacetaldehyde
P024	106-47-8	p-Chloroaniline
P026	5344-82-1	1-(o-Chlorophenyl)thiourea
P027	542-76-7	3-Chloropropionitrile
P029	544-92-3	Copper cyanide
P029	544-92-3	Copper cyanide Cu(CN)
P030	....	Cyanides (soluble cyanide salts), not otherwise specified
P031	460-19-5	Cyanogen
P033	506-77-4	Cyanogen chloride

Environmental Protection Agency § 261.33

Hazardous waste No.	Chemical abstracts No.	Substance
P033	506-77-4	Cyanogen chloride (CN)Cl
P034	131-89-5	2-Cyclohexyl-4,6-dinitrophenol
P016	542-88-1	Dichloromethyl ether
P036	696-28-6	Dichlorophenylarsine
P037	60-57-1	Dieldrin
P038	692-42-2	Diethylarsine
P041	311-45-5	Diethyl-p-nitrophenyl phosphate
P040	297-97-2	O,O-Diethyl O-pyrazinyl phosphorothioate
P043	55-91-4	Diisopropylfluorophosphate (DFP)
P004	309-00-2	1,4,5,8-Dimethanonaphthalene, 1,2,3,4,10,10-hexa- chloro-1,4,4a,5,8,8a,-hexahydro-, (1alpha,4alpha,4abeta,5alpha,8alpha,8abeta)-
P060	465-73-6	1,4,5,8-Dimethanonaphthalene, 1,2,3,4,10,10-hexa- chloro-1,4,4a,5,8,8a-hexahydro-, (1alpha,4alpha,4abeta,5beta,8beta,8abeta)-
P037	60-57-1	2,7:3,6-Dimethanonaphth[2,3-b]oxirene, 3,4,5,6,9,9-hexachloro-1a,2,2a,3,6,6a,7,7a-octahydro-, (1aalpha,2beta,2aalpha,3beta,6beta,6aalpha,7beta, 7aalpha)-
P051	[1] 72-20-8	2,7:3,6-Dimethanonaphth [2,3-b]oxirene, 3,4,5,6,9,9-hexachloro-1a,2,2a,3,6,6a,7,7a-octahydro-, (1aalpha,2beta,2abeta,3alpha,6alpha,6abeta,7beta, 7aalpha)-, & metabolites
P044	60-51-5	Dimethoate
P046	122-09-8	alpha,alpha-Dimethylphenethylamine
P047	[1] 534-52-1	4,6-Dinitro-o-cresol, & salts
P048	51-28-5	2,4-Dinitrophenol
P020	88-85-7	Dinoseb
P085	152-16-9	Diphosphoramide, octamethyl-
P111	107-49-3	Diphosphoric acid, tetraethyl ester
P039	298-04-4	Disulfoton
P049	541-53-7	Dithiobiuret
P050	115-29-7	Endosulfan
P088	145-73-3	Endothall
P051	72-20-8	Endrin
P051	72-20-8	Endrin, & metabolites
P042	51-43-4	Epinephrine
P031	460-19-5	Ethanedinitrile
P066	16752-77-5	Ethanimidothioic acid, N-[[(methylamino)carbonyl]oxy]-, methyl ester
P101	107-12-0	Ethyl cyanide
P054	151-56-4	Ethyleneimine
P097	52-85-7	Famphur
P056	7782-41-4	Fluorine
P057	640-19-7	Fluoroacetamide
P058	62-74-8	Fluoroacetic acid, sodium salt
P065	628-86-4	Fulminic acid, mercury(2+) salt (R,T)
P059	76-44-8	Heptachlor
P062	757-58-4	Hexaethyl tetraphosphate
P116	79-19-6	Hydrazinecarbothioamide
P068	60-34-4	Hydrazine, methyl-
P063	74-90-8	Hydrocyanic acid
P063	74-90-8	Hydrogen cyanide
P096	7803-51-2	Hydrogen phosphide
P060	465-73-6	Isodrin
P007	2763-96-4	3(2H)-Isoxazolone, 5-(aminomethyl)-
P092	62-38-4	Mercury, (acetato-O)phenyl-
P065	628-86-4	Mercury fulminate (R,T)
P082	62-75-9	Methanamine, N-methyl-N-nitroso-
P064	624-83-9	Methane, isocyanato-
P016	542-88-1	Methane, oxybis[chloro-
P112	509-14-8	Methane, tetranitro- (R)
P118	75-70-7	Methanethiol, trichloro-
P050	115-29-7	6,9-Methano-2,4,3-benzodioxathiepin, 6,7,8,9,10,10-hexachloro-1,5,5a,6,9,9a-hexahydro-, 3-oxide
P059	76-44-8	4,7-Methano-1H-indene, 1,4,5,6,7,8,8-heptachloro-3a,4,7,7a-tetrahydro-
P066	16752-77-5	Methomyl
P068	60-34-4	Methyl hydrazine
P064	624-83-9	Methyl isocyanate
P069	75-86-5	2-Methyllactonitrile
P071	298-00-0	Methyl parathion
P072	86-88-4	alpha-Naphthylthiourea
P073	13463-39-3	Nickel carbonyl
P073	13463-39-3	Nickel carbonyl Ni(CO)$_4$, (T-4)-

§ 261.33   40 CFR Ch. I (7-1-91 Edition)

Hazardous waste No.	Chemical abstracts No.	Substance
P074	557-19-7	Nickel cyanide
P074	557-19-7	Nickel cyanide Ni(CN)$_2$
P075	[1] 54-11-5	Nicotine,. & salts
P076	10102-43-9	Nitric oxide
P077	100-01-6	p-Nitroaniline
P078	10102-44-0	Nitrogen dioxide
P076	10102-43-9	Nitrogen oxide NO
P078	10102-44-0	Nitrogen oxide NO$_2$
P081	55-63-0	Nitroglycerine (R)
P082	62-75-9	N-Nitrosodimethylamine
P084	4549-40-0	N-Nitrosomethylvinylamine
P085	152-16-9	Octamethylpyrophosphoramide
P087	20816-12-0	Osmium oxide OsO$_4$, (T-4)-
P087	20816-12-0	Osmium tetroxide
P088	145-73-3	7-Oxabicyclo[2.2.1]heptane-2,3-dicarboxylic acid
P089	56-38-2	Parathion
P034	131-89-5	Phenol, 2-cyclohexyl-4,6-dinitro-
P048	51-28-5	Phenol, 2,4-dinitro-
P047	[1] 534-52-1	Phenol, 2-methyl-4,6-dinitro-, & salts
P020	88-85-7	Phenol, 2-(1-methylpropyl)-4,6-dinitro-
P009	131-74-8	Phenol, 2,4,6-trinitro-, ammonium salt (R)
P092	62-38-4	Phenylmercury acetate
P093	103-85-5	Phenylthiourea
P094	298-02-2	Phorate
P095	75-44-5	Phosgene
P096	7803-51-2	Phosphine
P041	311-45-5	Phosphoric acid, diethyl 4-nitrophenyl ester
P039	298-04-4	Phosphorodithioic acid, O,O-diethyl S-[2-(ethylthio)ethyl] ester
P094	298-02-2	Phosphorodithioic acid, O,O-diethyl S-[(ethylthio)methyl] ester
P044	60-51-5	Phosphorodithioic acid, O,O-dimethyl S-[2-(methylamino)-2-oxoethyl] ester
P043	55-91-4	Phosphorofluoridic acid, bis(1-methylethyl) ester
P089	56-38-2	Phosphorothioic acid, O,O-diethyl O-(4-nitrophenyl) ester
P040	297-97-2	Phosphorothioic acid, O,O-diethyl O-pyrazinyl ester
P097	52-85-7	Phosphorothioic acid, O-[4-[(dimethylamino)sulfonyl]phenyl] O,O-dimethyl ester
P071	298-00-0	Phosphorothioic acid, O,O,-dimethyl O-(4-nitrophenyl) ester
P110	78-00-2	Plumbane, tetraethyl-
P098	151-50-8	Potassium cyanide
P098	151-50-8	Potassium cyanide K(CN)
P099	506-61-6	Potassium silver cyanide
P070	116-06-3	Propanal, 2-methyl-2-(methylthio)-, O-[(methylamino)carbonyl]oxime
P101	107-12-0	Propanenitrile
P027	542-76-7	Propanenitrile, 3-chloro-
P069	75-86-5	Propanenitrile, 2-hydroxy-2-methyl-
P081	55-63-0	1,2,3-Propanetriol, trinitrate (R)
P017	598-31-2	2-Propanone, 1-bromo-
P102	107-19-7	Propargyl alcohol
P003	107-02-8	2-Propenal
P005	107-18-6	2-Propen-1-ol
P067	75-55-8	1,2-Propylenimine
P102	107-19-7	2-Propyn-1-ol
P008	504-24-5	4-Pyridinamine
P075	[1] 54-11-5	Pyridine, 3-(1-methyl-2-pyrrolidinyl)-, (S)-, & salts
P114	12039-52-0	Selenious acid, dithallium(1+) salt
P103	630-10-4	Selenourea
P104	506-64-9	Silver cyanide
P104	506-64-9	Silver cyanide Ag(CN)
P105	26628-22-8	Sodium azide
P106	143-33-9	Sodium cyanide
P106	143-33-9	Sodium cyanide Na(CN)
P108	[1] 57-24-9	Strychnidin-10-one, & salts
P018	357-57-3	Strychnidin-10-one, 2,3-dimethoxy-
P108	[1] 57-24-9	Strychnine, & salts
P115	7446-18-6	Sulfuric acid, dithallium(1+) salt
P109	3689-24-5	Tetraethyldithiopyrophosphate
P110	78-00-2	Tetraethyl lead
P111	107-49-3	Tetraethyl pyrophosphate

Environmental Protection Agency § 261.33

Hazardous waste No.	Chemical abstracts No.	Substance
P112	509-14-8	Tetranitromethane (R)
P062	757-58-4	Tetraphosphoric acid, hexaethyl ester
P113	1314-32-5	Thallic oxide
P113	1314-32-5	Thallium oxide Tl_2O_3
P114	12039-52-0	Thallium(I) selenite
P115	7446-18-6	Thallium(I) sulfate
P109	3689-24-5	Thiodiphosphoric acid, tetraethyl ester
P045	39196-18-4	Thiofanox
P049	541-53-7	Thioimidodicarbonic diamide [$(H_2N)C(S)$]$_2$NH
P014	108-98-5	Thiophenol
P116	79-19-6	Thiosemicarbazide
P026	5344-82-1	Thiourea, (2-chlorophenyl)-
P072	86-88-4	Thiourea, 1-naphthalenyl-
P093	103-85-5	Thiourea, phenyl-
P123	8001-35-2	Toxaphene
P118	75-70-7	Trichloromethanethiol
P119	7803-55-6	Vanadic acid, ammonium salt
P120	1314-62-1	Vanadium oxide V_2O_5
P120	1314-62-1	Vanadium pentoxide
P084	4549-40-0	Vinylamine, N-methyl-N-nitroso-
P001	[1] 81-81-2	Warfarin, & salts, when present at concentrations greater than 0.3%
P121	557-21-1	Zinc cyanide
P121	557-21-1	Zinc cyanide $Zn(CN)_2$
P122	1314-84-7	Zinc phosphide Zn_3P_2, when present at concentrations greater than 10% (R,T)

[1] CAS Number given for parent compound only.

(f) The commercial chemical products, manfacturing chemical intermediates, or off-specification commercial chemical products referred to in paragraphs (a) through (d) of this section, are identified as toxic wastes (T), unless otherwise designated and are subject to the small quantity generator exclusion defined in § 261.5 (a) and (g).

[*Comment:* For the convenience of the regulated community, the primary hazardous properties of these materials have been indicated by the letters T (Toxicity), R (Reactivity), I (Ignitability) and C (Corrosivity). Absence of a letter indicates that the compound is only listed for toxicity.]

These wastes and their corresponding EPA Hazardous Waste Numbers are:

Hazardous waste No.	Chemical abstracts No.	Substance
U001	75-07-0	Acetaldehyde (I)
U034	75-87-6	Acetaldehyde, trichloro-
U187	62-44-2	Acetamide, N-(4-ethoxyphenyl)-
U005	53-96-3	Acetamide, N-9H-fluoren-2-yl-
U240	[1] 94-75-7	Acetic acid, (2,4-dichlorophenoxy)-, salts & esters
U112	141-78-6	Acetic acid ethyl ester (I)
U144	301-04-2	Acetic acid, lead(2+) salt
U214	563-68-8	Acetic acid, thallium(1+) salt
see F027	93-76-5	Acetic acid, (2,4,5-trichlorophenoxy)-
U002	67-64-1	Acetone (I)
U003	75-05-8	Acetonitrile (I,T)
U004	98-86-2	Acetophenone
U005	53-96-3	2-Acetylaminofluorene
U006	75-36-5	Acetyl chloride (C,R,T)
U007	79-06-1	Acrylamide
U008	79-10-7	Acrylic acid (I)
U009	107-13-1	Acrylonitrile
U011	61-82-5	Amitrole
U012	62-53-3	Aniline (I,T)
U136	75-60-5	Arsinic acid, dimethyl-
U014	492-80-8	Auramine

§ 261.33   40 CFR Ch. I (7-1-91 Edition)

Hazardous waste No.	Chemical abstracts No.	Substance
U015	115-02-6	Azaserine
U010	50-07-7	Azirino[2',3':3,4]pyrrolo[1,2-a]indole-4,7-dione, 6-amino-8-[[(aminocarbonyl)oxy]methyl]-1,1a,2,8,8a,8b-hexahydro-8a-methoxy-5-methyl-, [1aS-(1aalpha, 8beta,8aalpha,8balpha)]-
U157	56-49-5	Benz[j]aceanthrylene, 1,2-dihydro-3-methyl-
U016	225-51-4	Benz[c]acridine
U017	98-87-3	Benzal chloride
U192	23950-58-5	Benzamide, 3,5-dichloro-N-(1,1-dimethyl-2-propynyl)-
U018	56-55-3	Benz[a]anthracene
U094	57-97-6	Benz[a]anthracene, 7,12-dimethyl-
U012	62-53-3	Benzenamine (I,T)
U014	492-80-8	Benzenamine, 4,4'-carbonimidoylbis[N,N-dimethyl-
U049	3165-93-3	Benzenamine, 4-chloro-2-methyl-, hydrochloride
U093	60-11-7	Benzenamine, N,N-dimethyl-4-(phenylazo)-
U328	95-53-4	Benzenamine, 2-methyl-
U353	106-49-0	Benzenamine, 4-methyl-
U158	101-14-4	Benzenamine, 4,4'-methylenebis[2-chloro-
U222	636-21-5	Benzenamine, 2-methyl-, hydrochloride
U181	99-55-8	Benzenamine, 2-methyl-5-nitro-
U019	71-43-2	Benzene (I,T)
U038	510-15-6	Benzeneacetic acid, 4-chloro-alpha-(4-chlorophenyl)-alpha-hydroxy-, ethyl ester
U030	101-55-3	Benzene, 1-bromo-4-phenoxy-
U035	305-03-3	Benzenebutanoic acid, 4-[bis(2-chloroethyl)amino]-
U037	108-90-7	Benzene, chloro-
U221	25376-45-8	Benzenediamine, ar-methyl-
U028	117-81-7	1,2-Benzenedicarboxylic acid, bis(2-ethylhexyl) ester
U069	84-74-2	1,2-Benzenedicarboxylic acid, dibutyl ester
U088	84-66-2	1,2-Benzenedicarboxylic acid, diethyl ester
U102	131-11-3	1,2-Benzenedicarboxylic acid, dimethyl ester
U107	117-84-0	1,2-Benzenedicarboxylic acid, dioctyl ester
U070	95-50-1	Benzene, 1,2-dichloro-
U071	541-73-1	Benzene, 1,3-dichloro-
U072	106-46-7	Benzene, 1,4-dichloro-
U060	72-54-8	Benzene, 1,1'-(2,2-dichloroethylidene)bis[4-chloro-
U017	98-87-3	Benzene, (dichloromethyl)-
U223	26471-62-5	Benzene, 1,3-diisocyanatomethyl- (R,T)
U239	1330-20-7	Benzene, dimethyl- (I,T)
U201	108-46-3	1,3-Benzenediol
U127	118-74-1	Benzene, hexachloro-
U056	110-82-7	Benzene, hexahydro- (I)
U220	108-88-3	Benzene, methyl-
U105	121-14-2	Benzene, 1-methyl-2,4-dinitro-
U106	606-20-2	Benzene, 2-methyl-1,3-dinitro-
U055	98-82-8	Benzene, (1-methylethyl)- (I)
U169	98-95-3	Benzene, nitro-
U183	608-93-5	Benzene, pentachloro-
U185	82-68-8	Benzene, pentachloronitro-
U020	98-09-9	Benzenesulfonic acid chloride (C,R)
U020	98-09-9	Benzenesulfonyl chloride (C,R)
U207	95-94-3	Benzene, 1,2,4,5-tetrachloro-
U061	50-29-3	Benzene, 1,1'-(2,2,2-trichloroethylidene)bis[4-chloro-
U247	72-43-5	Benzene, 1,1'-(2,2,2-trichloroethylidene)bis[4- methoxy-
U023	98-07-7	Benzene, (trichloromethyl)-
U234	99-35-4	Benzene, 1,3,5-trinitro-
U021	92-87-5	Benzidine
U202	[1] 81-07-2	1,2-Benzisothiazol-3(2H)-one, 1,1-dioxide, & salts
U203	94-59-7	1,3-Benzodioxole, 5-(2-propenyl)-
U141	120-58-1	1,3-Benzodioxcle, 5-(1-propenyl)-
U090	94-58-6	1,3-Benzodioxole, 5-propyl-
U064	189-55-9	Benzo[rst]pentaphene
U248	[1] 81-81-2	2H-1-Benzopyran-2-one, 4-hydroxy-3-(3-oxo-1-phenyl-butyl)-, & salts, when present at concentrations of 0.3% or less
U022	50-32-8	Benzo[a]pyrene
U197	106-51-4	p-Benzoquinone
U023	98-07-7	Benzotrichloride (C,R,T)
U085	1464-53-5	2,2'-Bioxirane
U021	92-87-5	[1,1'-Biphenyl]-4,4'-diamine
U073	91-94-1	[1,1'-Biphenyl]-4,4'-diamine, 3,3'-dichloro-
U091	119-90-4	[1,1'-Biphenyl]-4,4'-diamine, 3,3'-dimethoxy-
U095	119-93-7	[1,1'-Biphenyl]-4,4'-diamine, 3,3'-dimethyl-
U225	75-25-2	Bromoform

Environmental Protection Agency § 261.33

Hazardous waste No.	Chemical abstracts No.	Substance
U030	101-55-3	4-Bromophenyl phenyl ether
U128	87-68-3	1,3-Butadiene, 1,1,2,3,4,4-hexachloro-
U172	924-16-3	1-Butanamine, N-butyl-N-nitroso-
U031	71-36-3	1-Butanol (I)
U159	78-93-3	2-Butanone (I,T)
U160	1338-23-4	2-Butanone, peroxide (R,T)
U053	4170-30-3	2-Butenal
U074	764-41-0	2-Butene, 1,4-dichloro- (I,T)
U143	303-34-4	2-Butenoic acid, 2-methyl-, 7-[[2,3-dihydroxy-2-(1-methoxyethyl)-3-methyl-1-oxobutoxy]methyl]-2,3,5,7a-tetrahydro-1H-pyrrolizin-1-yl ester, [1S-[1alpha(Z),7(2S*,3R*),7aalpha]]-
U031	71-36-3	n-Butyl alcohol (I)
U136	75-60-5	Cacodylic acid
U032	13765-19-0	Calcium chromate
U238	51-79-6	Carbamic acid, ethyl ester
U178	615-53-2	Carbamic acid, methylnitroso-, ethyl ester
U097	79-44-7	Carbamic chloride, dimethyl-
U114	[1] 111-54-6	Carbamodithioic acid, 1,2-ethanediylbis-, salts & esters
U062	2303-16-4	Carbamothioic acid, bis(1-methylethyl)-, S-(2,3-dichloro-2-propenyl) ester
U215	6533-73-9	Carbonic acid, dithallium(1+) salt
U033	353-50-4	Carbonic difluoride
U156	79-22-1	Carbonochloridic acid, methyl ester (I,T)
U033	353-50-4	Carbon oxyfluoride (R,T)
U211	56-23-5	Carbon tetrachloride
U034	75-87-6	Chloral
U035	305-03-3	Chlorambucil
U036	57-74-9	Chlordane, alpha & gamma isomers
U026	494-03-1	Chlornaphazin
U037	108-90-7	Chlorobenzene
U038	510-15-6	Chlorobenzilate
U039	59-50-7	p-Chloro-m-cresol
U042	110-75-8	2-Chloroethyl vinyl ether
U044	67-66-3	Chloroform
U046	107-30-2	Chloromethyl methyl ether
U047	91-58-7	beta-Chloronaphthalene
U048	95-57-8	o-Chlorophenol
U049	3165-93-3	4-Chloro-o-toluidine, hydrochloride
U032	13765-19-0	Chromic acid H_2CrO_4, calcium salt
U050	218-01-9	Chrysene
U051		Creosote
U052	1319-77-3	Cresol (Cresylic acid)
U053	4170-30-3	Crotonaldehyde
U055	98-82-8	Cumene (I)
U246	506-68-3	Cyanogen bromide (CN)Br
U197	106-51-4	2,5-Cyclohexadiene-1,4-dione
U056	110-82-7	Cyclohexane (I)
U129	58-89-9	Cyclohexane, 1,2,3,4,5,6-hexachloro-, (1alpha,2alpha,3beta,4alpha,5alpha,6beta)-
U057	108-94-1	Cyclohexanone (I)
U130	77-47-4	1,3-Cyclopentadiene, 1,2,3,4,5,5-hexachloro-
U058	50-18-0	Cyclophosphamide
U240	[1] 94-75-7	2,4-D, salts & esters
U059	20830-81-3	Daunomycin
U060	72-54-8	DDD
U061	50-29-3	DDT
U062	2303-16-4	Diallate
U063	53-70-3	Dibenz[a,h]anthracene
U064	189-55-9	Dibenzo[a,i]pyrene
U066	96-12-8	1,2-Dibromo-3-chloropropane
U069	84-74-2	Dibutyl phthalate
U070	95-50-1	o-Dichlorobenzene
U071	541-73-1	m-Dichlorobenzene
U072	106-46-7	p-Dichlorobenzene
U073	91-94-1	3,3'-Dichlorobenzidine
U074	764-41-0	1,4-Dichloro-2-butene (I,T)
U075	75-71-8	Dichlorodifluoromethane
U078	75-35-4	1,1-Dichloroethylene
U079	156-60-5	1,2-Dichloroethylene

§ 261.33

40 CFR Ch. I (7-1-91 Edition)

Hazardous waste No.	Chemical abstracts No.	Substance
U025	111-44-4	Dichloroethyl ether
U027	108-60-1	Dichloroisopropyl ether
U024	111-91-1	Dichloromethoxy ethane
U081	120-83-2	2,4-Dichlorophenol
U082	87-65-0	2,6-Dichlorophenol
U084	542-75-6	1,3-Dichloropropene
U085	1464-53-5	1,2:3,4-Diepoxybutane (I,T)
U108	123-91-1	1,4-Diethyleneoxide
U028	117-81-7	Diethylhexyl phthalate
U086	1615-80-1	N,N'-Diethylhydrazine
U087	3288-58-2	O,O-Diethyl S-methyl dithiophosphate
U088	84-66-2	Diethyl phthalate
U089	56-53-1	Diethylstilbesterol
U090	94-58-6	Dihydrosafrole
U091	119-90-4	3,3'-Dimethoxybenzidine
U092	124-40-3	Dimethylamine (I)
U093	60-11-7	p-Dimethylaminoazobenzene
U094	57-97-6	7,12-Dimethylbenz[a]anthracene
U095	119-93-7	3,3'-Dimethylbenzidine
U096	80-15-9	alpha,alpha-Dimethylbenzylhydroperoxide (R)
U097	79-44-7	Dimethylcarbamoyl chloride
U098	57-14-7	1,1-Dimethylhydrazine
U099	540-73-8	1,2-Dimethylhydrazine
U101	105-67-9	2,4-Dimethylphenol
U102	131-11-3	Dimethyl phthalate
U103	77-78-1	Dimethyl sulfate
U105	121-14-2	2,4-Dinitrotoluene
U106	606-20-2	2,6-Dinitrotoluene
U107	117-84-0	Di-n-octyl phthalate
U108	123-91-1	1,4-Dioxane
U109	122-66-7	1,2-Diphenylhydrazine
U110	142-84-7	Dipropylamine (I)
U111	621-64-7	Di-n-propylnitrosamine
U041	106-89-8	Epichlorohydrin
U001	75-07-0	Ethanal (I)
U174	55-18-5	Ethanamine, N-ethyl-N-nitroso-
U155	91-80-5	1,2-Ethanediamine, N,N-dimethyl-N'-2-pyridinyl-N'-(2-thienylmethyl)-
U067	106-93-4	Ethane, 1,2-dibromo-
U076	75-34-3	Ethane, 1,1-dichloro-
U077	107-06-2	Ethane, 1,2-dichloro-
U131	67-72-1	Ethane, hexachloro-
U024	111-91-1	Ethane, 1,1'-[methylenebis(oxy)]bis[2-chloro-
U117	60-29-7	Ethane, 1,1'-oxybis-(I)
U025	111-44-4	Ethane, 1,1'-oxybis[2-chloro-
U184	76-01-7	Ethane, pentachloro-
U208	630-20-6	Ethane, 1,1,1,2-tetrachloro-
U209	79-34-5	Ethane, 1,1,2,2-tetrachloro-
U218	62-55-5	Ethanethioamide
U226	71-55-6	Ethane, 1,1,1-trichloro-
U227	79-00-5	Ethane, 1,1,2-trichloro-
U359	110-80-5	Ethanol, 2-ethoxy-
U173	1116-54-7	Ethanol, 2,2'-(nitrosoimino)bis-
U004	98-86-2	Ethanone, 1-phenyl-
U043	75-01-4	Ethene, chloro-
U042	110-75-8	Ethene, (2-chloroethoxy)-
U078	75-35-4	Ethene, 1,1-dichloro-
U079	156-60-5	Ethene, 1,2-dichloro-, (E)-
U210	127-18-4	Ethene, tetrachloro-
U228	79-01-6	Ethene, trichloro-
U112	141-78-6	Ethyl acetate (I)
U113	140-88-5	Ethyl acrylate (I)
U238	51-79-6	Ethyl carbamate (urethane)
U117	60-29-7	Ethyl ether (I)
U114	[1] 111-54-6	Ethylenebisdithiocarbamic acid, salts & esters
U067	106-93-4	Ethylene dibromide
U077	107-06-2	Ethylene dichloride
U359	110-80-5	Ethylene glycol monoethyl ether
U115	75-21-8	Ethylene oxide (I,T)
U116	96-45-7	Ethylenethiourea
U076	75-34-3	Ethylidene dichloride

Environmental Protection Agency § 261.33

Hazardous waste No.	Chemical abstracts No.	Substance
U118	97-63-2	Ethyl methacrylate
U119	62-50-0	Ethyl methanesulfonate
U120	206-44-0	Fluoranthene
U122	50-00-0	Formaldehyde
U123	64-18-6	Formic acid (C,T)
U124	110-00-9	Furan (I)
U125	98-01-1	2-Furancarboxaldehyde (I)
U147	108-31-6	2,5-Furandione
U213	109-99-9	Furan, tetrahydro-(I)
U125	98-01-1	Furfural (I)
U124	110-00-9	Furfuran (I)
U206	18883-66-4	Glucopyranose, 2-deoxy-2-(3-methyl-3-nitrosoureido)-, D-
U206	18883-66-4	D-Glucose, 2-deoxy-2-[[(methylnitrosoamino)carbonyl]amino]-
U126	765-34-4	Glycidylaldehyde
U163	70-25-7	Guanidine, N-methyl-N'-nitro-N-nitroso-
U127	118-74-1	Hexachlorobenzene
U128	87-68-3	Hexachlorobutadiene
U130	77-47-4	Hexachlorocyclopentadiene
U131	67-72-1	Hexachloroethane
U132	70-30-4	Hexachlorophene
U243	1888-71-7	Hexachloropropene
U133	302-01-2	Hydrazine (R,T)
U086	1615-80-1	Hydrazine, 1,2-diethyl-
U098	57-14-7	Hydrazine, 1,1-dimethyl-
U099	540-73-8	Hydrazine, 1,2-dimethyl-
U109	122-66-7	Hydrazine, 1,2-diphenyl-
U134	7664-39-3	Hydrofluoric acid (C,T)
U134	7664-39-3	Hydrogen fluoride (C,T)
U135	7783-06-4	Hydrogen sulfide
U135	7783-06-4	Hydrogen sulfide H_2S
U096	80-15-9	Hydroperoxide, 1-methyl-1-phenylethyl- (R)
U116	96-45-7	2-Imidazolidinethione
U137	193-39-5	Indeno[1,2,3-cd]pyrene
U190	85-44-9	1,3-Isobenzofurandione
U140	78-83-1	Isobutyl alcohol (I,T)
U141	120-58-1	Isosafrole
U142	143-50-0	Kepone
U143	303-34-4	Lasiocarpine
U144	301-04-2	Lead acetate
U146	1335-32-6	Lead, bis(acetato-O)tetrahydroxytri-
U145	7446-27-7	Lead phosphate
U146	1335-32-6	Lead subacetate
U129	58-89-9	Lindane
U163	70-25-7	MNNG
U147	108-31-6	Maleic anhydride
U148	123-33-1	Maleic hydrazide
U149	109-77-3	Malononitrile
U150	148-82-3	Melphalan
U151	7439-97-6	Mercury
U152	126-98-7	Methacrylonitrile (I, T)
U092	124-40-3	Methanamine, N-methyl- (I)
U029	74-83-9	Methane, bromo-
U045	74-87-3	Methane, chloro- (I, T)
U046	107-30-2	Methane, chloromethoxy-
U068	74-95-3	Methane, dibromo-
U080	75-09-2	Methane, dichloro-
U075	75-71-8	Methane, dichlorodifluoro-
U138	74-88-4	Methane, iodo-
U119	62-50-0	Methanesulfonic acid, ethyl ester
U211	56-23-5	Methane, tetrachloro-
U153	74-93-1	Methanethiol (I, T)
U225	75-25-2	Methane, tribromo-
U044	67-66-3	Methane, trichloro-
U121	75-69-4	Methane, trichlorofluoro-
U036	57-74-9	4,7-Methano-1H-indene, 1,2,4,5,6,7,8,8-octachloro-2,3,3a,4,7,7a-hexahydro-
U154	67-56-1	Methanol (I)
U155	91-80-5	Methapyrilene
U142	143-50-0	1,3,4-Metheno-2H-cyclobuta[cd]pentalen-2-one, 1,1a,3,3a,4,5,5,5a,5b,6-decachlorooctahydro-
U247	72-43-5	Methoxychlor

§ 261.33   40 CFR Ch. I (7-1-91 Edition)

Hazardous waste No.	Chemical abstracts No.	Substance
U154	67-56-1	Methyl alcohol (I)
U029	74-83-9	Methyl bromide
U186	504-60-9	1-Methylbutadiene (I)
U045	74-87-3	Methyl chloride (I,T)
U156	79-22-1	Methyl chlorocarbonate (I,T)
U226	71-55-6	Methyl chloroform
U157	56-49-5	3-Methylcholanthrene
U158	101-14-4	4,4'-Methylenebis(2-chloroaniline)
U068	74-95-3	Methylene bromide
U080	75-09-2	Methylene chloride
U159	78-93-3	Methyl ethyl ketone (MEK) (I,T)
U160	1338-23-4	Methyl ethyl ketone peroxide (R,T)
U138	74-88-4	Methyl iodide
U161	108-10-1	Methyl isobutyl ketone (I)
U162	80-62-6	Methyl methacrylate (I,T)
U161	108-10-1	4-Methyl-2-pentanone (I)
U164	56-04-2	Methylthiouracil
U010	50-07-7	Mitomycin C
U059	20830-81-3	5,12-Naphthacenedione, 8-acetyl-10-[(3-amino-2,3,6-trideoxy)-alpha-L-lyxo-hexopyranosyl)oxy]-7,8,9,10-tetrahydro-6,8,11-trihydroxy-1-methoxy-, (8S-cis)-
U167	134-32-7	1-Naphthalenamine
U168	91-59-8	2-Naphthalenamine
U026	494-03-1	Naphthalenamine, N,N'-bis(2-chloroethyl)-
U165	91-20-3	Naphthalene
U047	91-58-7	Naphthalene, 2-chloro-
U166	130-15-4	1,4-Naphthalenedione
U236	72-57-1	2,7-Naphthalenedisulfonic acid, 3,3'-[(3,3'-dimethyl[1,1'-biphenyl]-4,4'-diyl)bis(azo)bis[5-amino-4-hydroxy]-, tetrasodium salt
U166	130-15-4	1,4-Naphthoquinone
U167	134-32-7	alpha-Naphthylamine
U168	91-59-8	beta-Naphthylamine
U217	10102-45-1	Nitric acid, thallium(1+) salt
U169	98-95-3	Nitrobenzene (I,T)
U170	100-02-7	p-Nitrophenol
U171	79-46-9	2-Nitropropane (I,T)
U172	924-16-3	N-Nitrosodi-n-butylamine
U173	1116-54-7	N-Nitrosodiethanolamine
U174	55-18-5	N-Nitrosodiethylamine
U176	759-73-9	N-Nitroso-N-ethylurea
U177	684-93-5	N-Nitroso-N-methylurea
U178	615-53-2	N-Nitroso-N-methylurethane
U179	100-75-4	N-Nitrosopiperidine
U180	930-55-2	N-Nitrosopyrrolidine
U181	99-55-8	5-Nitro-o-toluidine
U193	1120-71-4	1,2-Oxathiolane, 2,2-dioxide
U058	50-18-0	2H-1,3,2-Oxazaphosphorin-2-amine, N,N-bis(2-chloroethyl)tetrahydro-, 2-oxide
U115	75-21-8	Oxirane (I,T)
U126	765-34-4	Oxiranecarboxyaldehyde
U041	106-89-8	Oxirane, (chloromethyl)-
U182	123-63-7	Paraldehyde
U183	608-93-5	Pentachlorobenzene
U184	76-01-7	Pentachloroethane
U185	82-68-8	Pentachloronitrobenzene (PCNB)
See F027	87-86-5	Pentachlorophenol
U161	108-10-1	Pentanol, 4-methyl-
U186	504-60-9	1,3-Pentadiene (I)
U187	62-44-2	Phenacetin
U188	108-95-2	Phenol
U048	95-57-8	Phenol, 2-chloro-
U039	59-50-7	Phenol, 4-chloro-3-methyl-
U081	120-83-2	Phenol, 2,4-dichloro-
U082	87-65-0	Phenol, 2,6-dichloro-
U089	56-53-1	Phenol, 4,4'-(1,2-diethyl-1,2-ethenediyl)bis-, (E)-
U101	105-67-9	Phenol, 2,4-dimethyl-
U052	1319-77-3	Phenol, methyl-
U132	70-30-4	Phenol, 2,2'-methylenebis[3,4,6-trichloro-
U170	100-02-7	Phenol, 4-nitro-

Environmental Protection Agency § 261.33

Hazardous waste No.	Chemical abstracts No.	Substance
See F027	87-86-5	Phenol, pentachloro-
See F027	58-90-2	Phenol, 2,3,4,6-tetrachloro-
See F027	95-95-4	Phenol, 2,4,5-trichloro-
See F027	88-06-2	Phenol, 2,4,6-trichloro-
U150	148-82-3	L-Phenylalanine, 4-[bis(2-chloroethyl)amino]-
U145	7446-27-7	Phosphoric acid, lead(2+) salt (2:3)
U087	3288-58-2	Phosphorodithioic acid, O,O-diethyl S-methyl ester
U189	1314-80-3	Phosphorus sulfide (R)
U190	85-44-9	Phthalic anhydride
U191	109-06-8	2-Picoline
U179	100-75-4	Piperidine, 1-nitroso-
U192	23950-58-5	Pronamide
U194	107-10-8	1-Propanamine (I,T)
U111	621-64-7	1-Propanamine, N-nitroso-N-propyl-
U110	142-84-7	1-Propanamine, N-propyl- (I)
U066	96-12-8	Propane, 1,2-dibromo-3-chloro-
U083	78-87-5	Propane, 1,2-dichloro-
U149	109-77-3	Propanedinitrile
U171	79-46-9	Propane, 2-nitro- (I,T)
U027	108-60-1	Propane, 2,2'-oxybis[2-chloro-
U193	1120-71-4	1,3-Propane sultone
See F027	93-72-1	Propanoic acid, 2-(2,4,5-trichlorophenoxy)-
U235	126-72-7	1-Propanol, 2,3-dibromo-, phosphate (3:1)
U140	78-83-1	1-Propanol, 2-methyl- (I,T)
U002	67-64-1	2-Propanone (I)
U007	79-06-1	2-Propenamide
U084	542-75-6	1-Propene, 1,3-dichloro-
U243	1888-71-7	1-Propene, 1,1,2,3,3,3-hexachloro-
U009	107-13-1	2-Propenenitrile
U152	126-98-7	2-Propenenitrile, 2-methyl- (I,T)
U008	79-10-7	2-Propenoic acid (I)
U113	140-88-5	2-Propenoic acid, ethyl ester (I)
U118	97-63-2	2-Propenoic acid, 2-methyl-, ethyl ester
U162	80-62-6	2-Propenoic acid, 2-methyl-, methyl ester (I,T)
U194	107-10-8	n-Propylamine (I,T)
U083	78-87-5	Propylene dichloride
U148	123-33-1	3,6-Pyridazinedione, 1,2-dihydro-
U196	110-86-1	Pyridine
U191	109-06-8	Pyridine, 2-methyl-
U237	66-75-1	2,4-(1H,3H)-Pyrimidinedione, 5-[bis(2-chloroethyl)amino]-
U164	56-04-2	4(1H)-Pyrimidinone, 2,3-dihydro-6-methyl-2-thioxo-
U180	930-55-2	Pyrrolidine, 1-nitroso-
U200	50-55-5	Reserpine
U201	108-46-3	Resorcinol
U202	[1] 81-07-2	Saccharin, & salts
U203	94-59-7	Safrole
U204	7783-00-8	Selenious acid
U204	7783-00-8	Selenium dioxide
U205	7488-56-4	Selenium sulfide
U205	7488-56-4	Selenium sulfide SeS_2 (R,T)
U015	115-02-6	L-Serine, diazoacetate (ester)
See F027	93-72-1	Silvex (2,4,5-TP)
U206	18883-66-4	Streptozotocin
U103	77-78-1	Sulfuric acid, dimethyl ester
U189	1314-80-3	Sulfur phosphide (R)
See F027	93-76-5	2,4,5-T
U207	95-94-3	1,2,4,5-Tetrachlorobenzene
U208	630-20-6	1,1,1,2-Tetrachloroethane
U209	79-34-5	1,1,2,2-Tetrachloroethane
U210	127-18-4	Tetrachloroethylene
See F027	58-90-2	2,3,4,6-Tetrachlorophenol

§ 261.35   40 CFR Ch. I (7-1-91 Edition)

Hazardous waste No.	Chemical abstracts No.	Substance
U213	109-99-9	Tetrahydrofuran (I)
U214	563-68-8	Thallium(I) acetate
U215	6533-73-9	Thallium(I) carbonate
U216	7791-12-0	Thallium(I) chloride
U216	7791-12-0	Thallium chloride TlCl
U217	10102-45-1	Thallium(I) nitrate
U218	62-55-5	Thioacetamide
U153	74-93-1	Thiomethanol (I,T)
U244	137-26-8	Thioperoxydicarbonic diamide [$(H_2N)C(S)$]$_2S_2$, tetramethyl-
U219	62-56-6	Thiourea
U244	137-26-8	Thiram
U220	108-88-3	Toluene
U221	25376-45-8	Toluenediamine
U223	26471-62-5	Toluene diisocyanate (R,T)
U328	95-53-4	o-Toluidine
U353	106-49-0	p-Toluidine
U222	636-21-5	o-Toluidine hydrochloride
U011	61-82-5	1H-1,2,4-Triazol-3-amine
U227	79-00-5	1,1,2-Trichloroethane
U228	79-01-6	Trichloroethylene
U121	75-69-4	Trichloromonofluoromethane
See F027	95-95-4	2,4,5-Trichlorophenol
See F027	88-06-2	2,4,6-Trichlorophenol
U234	99-35-4	1,3,5-Trinitrobenzene (R,T)
U182	123-63-7	1,3,5-Trioxane, 2,4,6-trimethyl-
U235	126-72-7	Tris(2,3-dibromopropyl) phosphate
U236	72-57-1	Trypan blue
U237	66-75-1	Uracil mustard
U176	759-73-9	Urea, N-ethyl-N-nitroso-
U177	684-93-5	Urea, N-methyl-N-nitroso-
U043	75-01-4	Vinyl chloride
U248	[1] 81-81-2	Warfarin, & salts, when present at concentrations of 0.3% or less
U239	1330-20-7	Xylene (I)
U200	50-55-5	Yohimban-16-carboxylic acid, 11,17-dimethoxy-18-[(3,4,5-trimethoxybenzoyl)oxy]-, methyl ester, (3beta,16beta,17alpha,18beta,20alpha)-
U249	1314-84-7	Zinc phosphide Zn_3P_2, when present at concentrations of 10% or less

[1] CAS Number given for parent compound only.

(Approved by the Office of Management and Budget under control number 2050-0047)

[45 FR 78529, 78541, Nov. 25, 1980]

EDITORIAL NOTE: For FEDERAL REGISTER citations affecting § 261.33, see the List of CFR Sections Affected in the Finding Aids section of this volume.

§ 261.35 Deletion of certain hazardous waste codes following equipment cleaning and replacement.

(a) Wastes from wood preserving processes at plants that do not resume or initiate use of chlorophenolic preservatives will not meet the listing definition of F032 once the generator has met all of the requirements of paragraphs (b) and (c) of this section. These wastes may, however, continue to meet another hazardous waste listing description or may exhibit one or more of the hazardous waste characteristics.

(b) Generators must either clean or replace all process equipment that may have come into contact with chlorophenolic formulations or constituents thereof, including, but not limited to, treatment cylinders, sumps, tanks, piping systems, drip pads, fork lifts, and trams, in a manner that minimizes or eliminates the escape of hazardous waste or constituents, leachate, contaminated drippage, or hazardous waste decomposition products to the ground water, surface water, or atmosphere.

(1) Generators shall do one of the following:

(i) Prepare and follow an equipment cleaning plan and clean equipment in accordance with this section;

Appendix 5

EPA Notification of Hazardous Waste Activity

How to Notify U.S. EPA of Your Waste Activities

I. How to Determine if You Handle a Regulated Hazardous Waste

Persons who generate, transport, treat, store, or dispose of solid wastes are responsible for determining if their solid waste is a hazardous waste regulated under the *Resource Conservation and Recovery Act (RCRA)*. In addition, persons who recycle secondary materials must also determine whether those materials are solid and hazardous wastes under the provisions of RCRA. If you need help making this determination after reading these instructions, contact the addressee listed for your State in Section III. C. of these instructions.

You will need to refer to 40 CFR Part 261 of the *Code of Federal Regulations* (copy enclosed, see Section VII) to help you decide if the waste you handle is regulated under RCRA.*

To determine if you are regulated under RCRA, ask yourself the following questions:

A) Do I Handle A Solid Waste?

Section 261.2 of the *Code of Federal Regulations* (hereafter referred to as CFR) defines "solid waste" as any discarded material that is not excluded under Section 261.4(a) or that is not excluded by variance granted under Sections 260.30 and 260.31. A discarded material is any material which is:

1) abandoned, as explained in 261.2(b); or

2) recycled, as explained in 261.2(c); or

3) considered inherently waste-like as explained in 261.2(c).

B) Has My Solid Waste Been Excluded From The Regulations Under Section 261.4?

The list of general exclusions can be found in Section 261.4 of the CFR. If the solid waste that you handle has been excluded, either by rule or special variance, then you do not need to notify U.S. EPA for that waste.

If your solid waste was not excluded from regulation, you need to determine if it is a hazardous waste that U.S. EPA regulates. The U.S. EPA regulates hazardous waste two ways:

1) by specifically listing the waste and assigning it a unique EPA Waste Code Number; or

2) by regulating it because it possesses any of four hazardous characteristics and assigning it a generic EPA Waste Code Number.

C) Is My Solid Waste Specifically Listed as a Hazardous Waste?

Sections 261.31 — 261.33 of the CFR identify certain solid wastes that U.S. EPA has specifically listed as hazardous. Persons who handle listed hazardous waste are subject to regulation and must notify U.S. EPA of their activities unless they are exempted as discussed below. Refer to this section of the CFR (enclosed as Section VII) to see if your waste is included as a "listed waste."

D) Does My Solid Waste Possess a Hazardous Characteristic?

Even if your waste is not specifically listed as a hazardous waste, it may still be hazardous because it exhibits certain hazardous characteristics. These characteristics are —

1) Ignitability;

2) Corrosivity;

3) Reactivity; and

4) Toxicity Characteristic.

Sections 261.20 through 261.24 of the CFR explain what each of the characteristics is and outlines the testing procedures you should use to determine if your waste meets these characteristics. Persons who handle characteristic waste that is regulated must notify U.S. EPA of their activities unless they are exempted, as discussed below. If you are handling a newly regulated waste (see Toxicity Characteristics Rule in Section VIII) and have already notified EPA prior to that activity **and already have an EPA Identification Number,** no re-notification is required.

E) Has My Hazardous Waste Been Exempted From The Regulations?

The list of exemptions can be found in 261.5 and 261.6(a)(3) of the CFR. If the hazardous waste that you handle has been exempted, then you do not need to notify U.S. EPA for that waste.

---
* *Many States have requirements that vary from the Federal regulations. These State regulations may be more strict than the Federal requirements by identifying additional wastes as hazardous, or may not yet include all wastes currently regulated under RCRA. It is your responsibility to comply with all regulations that apply to you. For more information on state requirements, you are strongly urged to contact the appropriate addressee listed for your State in Section III of these instructions.*

II. How To Determine if You Must Notify U.S. EPA of Your Waste-as-Fuel Activities

Persons who market or burn hazardous waste or used oil (and any material produced from or otherwise containing hazardous waste or used oil) for energy recovery are required to notify U.S. EPA (or their State agency if the State is authorized to operate its own hazardous waste program) and obtain a U.S. EPA Identification Number unless they are exempt as outlined below (see Subparts D and E of 40 CFR Part 266). Hazardous waste and used oil are considered to be burned for energy recovery if they are burned in a boiler or industrial furnace that is not regulated as a hazardous waste incinerator under Subpart O of 40 CFR Parts 264 or 265.

Even if you have previously notified U.S. EPA of hazardous waste activities and have a U.S. EPA Identification Number, you must renotify to identify your waste-as-fuel activities. (You do not have to renotify for those activities you previously notified for, only for any newly regulated activities.) If you have previously notified, be sure to complete Item I "First or Subsequent Notification," by marking an "X" in the box for subsequent notification. Fill in your U.S. EPA Identification Number in the spaces provided. (Your U.S. EPA Identification Number will not change.)

Who is Exempt From Waste-As-Fuel Notification Requirements?

1) *Ordinary Generators (and initial transporters)*: Generators (and initial transporters who pick up used oil or hazardous waste from generators) are not marketers subject to the notification requirement *if they do not market hazardous waste fuel or used oil fuel directly to a burner.* In such situations, it is the recipient of that fuel who makes the decision to market the materials as a fuel, (typically after processing or blending), and it is the recipient who must notify.

In addition, used oil generators or initial transporters who send their oil to a person who processes or blends it to produce used oil fuel and who incidently burns used oil to provide energy for the processing or blending are also exempt from the notification requirement. This is because such persons are generally considered to be primarily fuel processors and marketers, but only incidental burners.

2) *Persons who Market or Burn Specification Used Oil Fuel*: Used oil fuel that meets the specification provided under 40 CFR 266.40(e) is essentially exempt from the regulations. *However, the person who first claims that the used oil meets the specification is subject to the notification and certain other requirements.* The burner (or any subsequent marketer) is not required to notify.

3) *Used Oil Generators Operating Used-Oil-Fired Space Heaters*: Persons who burn their used oil (and used oil received from individuals who are do-it-yourself oil changers) in used-oil-fired space heaters are exempt from the notification requirement provided that the device is vented to the outdoors.

4) *Specific Exemptions Provided by 40 CFR 261.6*: The rules provide conditional exemptions for several specific waste-derived fuels under 261.6(a)(3), including fuels produced by petroleum refineries that recycle refinery hazardous waste, and coke and coal tar derived from coal coking wastes by the iron and steel industry. Marketers and burners of these exempted fuels are not subject to the notification requirement.

III. How to File EPA Form 8700-12, "Notification of Regulated Waste Activity"

If your waste activity is regulated under RCRA, you must notify the U.S. EPA of your activities and obtain a U.S. EPA Identification Number. You can satisfy both of these requirements by completing and signing the enclosed notification form and mailing it to the appropriate address listed in Part C of this section.

Per the Hazardous Waste Import Regulations, 40 CFR 262.60, *foreign generators should not apply for a Federal I.D. number*. These regulations state that when filling out a U.S. manifest, you must include the name and address of the foreign generator, and the name and address and EPA I.D. number of the importer. Please contact the U.S. firms involved with your shipments and determine which firm will serve as importer.

If this is a subsequent notification, you need to complete Items I, II, III, VI, VII, VIII and X and any other sections that are being added to (i.e., newly regulated activities) or altered (i.e., installation contact). All other sections may be left blank.

A) How Many Forms Should I File?

A person who is subject to the hazardous waste regulations and/or the waste-as-fuel regulations under RCRA should submit one notification form per site or location. If you conduct hazardous waste activities at more than one location, you must submit a separate form for each location. (If you previously notified for hazardous waste activities and are now notifying for waste-as-fuel activities at the same location, you must submit a second form, but your U.S. EPA Identification Number will remain the same).

If you only transport hazardous waste and do not generate, market, burn, treat, store, or dispose of these wastes, you may submit one form which covers all transportation activities your company conducts. This form should be sent to the appropriate address (listed in Part C) that serves the State where your company has its headquarters or principal place of business. However, if you are a transporter who also generates, treats, stores, or disposes of hazardous wastes, you must complete and submit separate notification forms to cover each location.

B) Can I Request That This Information Be Kept Confidential?

All information you submit in a notification can be released to the public, according to the Freedom of Information Act, unless it is determined to be confidential by U.S. EPA pursuant to 40 CFR Part 2. Since notification information is very general, the U.S. EPA believes it is unlikely that any information in your notification could qualify to be protected from release. However, you may make a claim of confidentiality by printing the word "CONFIDENTIAL" on both sides of the Notification Form and on any attachments.

EPA will take action on the confidentiality claims in accordance with 40 CFR Part 2.

C) Where Should I Send My Completed Form?

Listed alphabetically, on the following pages, are the addresses and phone numbers of the proper contacts in each State where you can get additional information and more forms, and where you should mail your completed forms. As shown here, the U.S. EPA and many States have arranged for the States to answer your questions and receive completed forms. In a few instances, the workload is shared between U.S. EPA and the State, or handled by U.S. EPA alone. *To avoid delay and confusion, follow the directions for your State very carefully.*

> *Estimated burden: Public reporting burden for this collection of information is estimated to be 3.5 hours, including time for reviewing instructions, searching existing data sources, gathering and maintaining the data needed, and completing and reviewing the collection of information. Send comments regarding the burden estimate or any other aspect of this collection of information, including suggestions for reducing this burden, to Chief, Information Policy Branch, PM-223, U.S. Environmental Protection Agency, 401 M St., S.W., Washington, D.C. 20460; and to the Paperwork Reduction Project (2050-0028), Office of Management and Budget, Washington, D.C. 20503.*

IV. Line-by-Line Instructions for Completing EPA Form 8700-12

Type or print in black ink all items except Item X, "Signature," leaving a blank box between words. The boxes are spaced at 1/4" intervals which accommodate elite type (12 characters per inch). When typing, hit the space bar twice between characters. If you print, place each character in a box. Abbreviate if necessary to stay within the number of boxes allowed for each Item. If you must use additional sheets, indicate clearly the number of the Item on the form to which the information on the separate sheet applies.

Note: When submitting a **subsequent notification** form, notifiers must complete in their entirety Items I, II, III, VI, VII, VIII and X. Other sections that are being added to (i.e., newly regulated activities) or altered (i.e., installation contact) must also be completed. All other sections may be left blank.

Item I — Installation's EPA ID Number:

Place an "X" in the appropriate box to indicate whether this is your first or a subsequent notification *for this site*. If you have filed a previous notification, enter the EPA Identification Number assigned to this site in the boxes provided. Leave EPA ID Number blank if this is your first notification *for this site*.

Note: When the owner of a facility changes, the new owner must notify U.S. EPA of the change, even if the previous owner already received a U.S. EPA Identification Number. Because the U.S. EPA ID Number is "site-specific," the new owner will keep the existing ID number. If the facility moves to another location, the owner/operator must notify EPA of this change. In this instance a new U.S. EPA Identification Number will be assigned, since the facility has changed locations.

Items II and III — Name and Location of Installation:

Complete Items II and III. Please note that the address you give for Item III, "Location of Installation," must be a physical address, *not a post office box or route number*.

County Code and Name: Give the county code, if known. If you do not know the county code, enter the county name, from which EPA can automatically generate the county code. If the county name is unknown contact the local Post Office. To obtain a list of county codes, contact the National Technical Information Service, U.S. Department of Commerce, Springfield, Virginia, 22161 or at (703) 487-4650. The list of codes is contained in the Federal Information Processing Standards Publication (FIPS PUB) number 6-3.

Item IV — Installation Mailing Address:

Please enter the Installation Mailing Address. If the Mailing Address and the Location of Installation (Item III) are the same, you can print "Same" in box for Item IV.

Item V — Installation Contact:

Enter the name, title, and business telephone number of the person who should be contacted regarding information submitted on this form.

Item VI — Installation Contact Address:

A) **Code:** If the contact address is the same as the location of installation address listed in Item III or the installation mailing address listed in Item IV, place an "X" in the appropriate box to indicate where the contact may be reached. If the location of installation address, the installation mailing address, and the installation contact address are all the same, mark the "Location" box. If an "X" is entered in either the location or mailing box, Item VI. B. should be left blank.

B) **Address:** Enter the contact address <u>only</u> if the contact address is different from either the location of installation address (Item III) or the installation mailing address (Item IV), and Item VI. A. was left blank.

Item VII — Ownership:

A) Name: Enter the name of the legal owner(s) of the installation, including the property owner. Also enter the address and phone number where this individual can be reached. Use the comment section in XI or additional sheets if necessary to list more than one owner.

B) Land Type: Using the codes listed below, indicate in VII. B. the code which best describes the current legal status of the land on which the facility is located:

F = Federal
S = State
I = Indian
P = Private
C = County
M = Municipal*
D = District
O = Other

Note: If the Land Type is best described as Indian, County or District, please use those codes. Otherwise, use Municipal.

C) Owner Type: Using the codes listed below, indicate in VII. C. the code which best describes the legal status of the current owner of the facility:

F = Federal
S = State
I = Indian
P = Private
C = County
M = Municipal*
D = District
O = Other

Note: If the Owner Type is best described as Indian, County or District, please use those codes. Otherwise, use Municipal.

D) Change of Owner Indicator: *(If this is your installation's first notification, leave Item VII. D. blank and skip to Item VIII. If this is a subsequent notification, complete Item VII. D. as directed below.)*

If the owner of this facility has changed since the facility's original notification, place an "X" in the box marked "Yes" and enter the date the owner changed.

If the owner of this facility has not changed since the facility's original notification, place an "X" in the box marked "No" and skip to Item VIII.

If an additional owner(s) has been added or replaced since the facility's original notification, place an "X" in the box marked "Yes". Use the comment section in XI to list any additional owners, the dates they became owners, and which owner(s) (if any) they replaced. If necessary attach a separate sheet of paper.

Item VIII — Type of Regulated Waste Activity:

A) Hazardous Waste Activity: Mark an "X" in the appropriate box(es) to show which hazardous waste activities are going on **at this installation**.

1) **Generator:** If you generate a hazardous waste that is identified by characteristic or listed in 40 CFR Part 261, mark an "X" in the appropriate box for the quantity of non-acutely hazardous waste that is generated per calendar month. If you generate acutely hazardous waste please refer to 40 CFR Part 262 for further information.

2) **Transporter:** If you transport hazardous waste, indicate if it is your own waste, for commercial purposes, or mark both boxes if both classifications apply. Mark an "X" in each appropriate box to indicate the method(s) of transportation you use. Transporters do not have to complete Item IX of this form, but must sign the certification in Item X. The Federal regulations for hazardous waste transporters are found in 40 CFR Part 263.

3) **Treater/Storer/Disposer:** If you treat, store or dispose of regulated hazardous waste, then mark an "X" in this box. You are reminded to contact the appropriate addressee listed for your State in Section III. C. of this package to request Part A of the RCRA Permit Application. The Federal regulations for hazardous waste facility owners/operators are found in 40 CFR Parts 264 and 265.

4) **Hazardous Waste Fuel:** If you market hazardous waste fuel, place an "X" in the appropriate box(es). If you burn hazardous waste fuel on-site, place an "X" in the appropriate box and indicate the type(s) of combustion devices in which hazardous waste fuel is burned. (Refer to definition section for complete description of each device).

Note: Generators are required to notify for waste-as-fuel activities only if they market directly to the burner.

"Other Marketer" is defined as any person, other than a generator marketing hazardous waste, who markets hazardous waste fuel.

5) **Underground Injection Control:** If you generate and/or treat, store or dispose of hazardous waste, place an "X" in the box if an injection well is located at your installation. "Underground Injection" means the subsurface emplacement of fluids through a bored, drilled or driven well; or through a dug well, where the depth of the dug well is greater than the largest surface dimension.

B) **Used Oil Fuel Activities**

Mark an "X" in the appropriate box(es) to indicate which used oil fuel activities are taking place at this installation.

1) **Off-Specification Used Oil Fuel:** If you market off-specification used oil, place an "X" in the appropriate box(es). If you burn used oil fuel place an "X" in the box(es) below to indicate type(s) of combustion devices in which off-specification used oil fuel is burned. (Refer to definition section for complete description of each device).

Note: Used oil generators are required to notify only if marketing directly to the burner.

"Other Marketer" is defined as any person, other than a generator marketing his or her used oil, who markets used oil fuel.

2) **Specification Used Oil Fuel:** If you are the first to claim that the used oil meets the specification established in 40 CFR 266.40(e) and is exempt from further regulation, you must mark an "X" in this box.

Item IX — Description of Regulated Wastes:

Note: Only persons involved in hazardous waste activity (Item VIII A.) need to complete this item. Transporters requesting a U.S. EPA Identification Number do not need to complete this item, but must sign the "Certification" in Item X.

You will need to refer to 40 CFR Part 261 (enclosed as Section VII) in order to complete this section. Part 261 identifies those wastes that EPA defines as hazardous. If you need help completing this section, please contact the appropriate addressee for your State as listed in Section III. C. of this package.

A) **Characteristics of Nonlisted Hazardous Wastes:** If you handle hazardous wastes which are not listed in 40 CFR Part 261, Subpart D but do exhibit a characteristic of hazardous waste as defined in 40 CFR Part 261, Subpart C, you should describe these wastes by the EPA hazardous waste number for the characteristic. Place an "X" in the box next to the characteristic of the wastes that you handle. If you mark "4. Toxicity Characteristic," please list the specific EPA hazardous waste number for the specific contaminant(s) in the box(es) provided. Refer to Section VIII to determine the appropriate hazardous waste number.

B) **Listed Hazardous Wastes:** If you handle hazardous wastes that are listed in 40 CFR Part 261, Subpart D, enter the appropriate 4-digit numbers in the boxes provided.

Note: If you handle more than 12 listed hazardous wastes, please continue listing the waste codes on the extra sheet provided at the end of this booklet. If it is used, attach the additional page to the rest of the form before mailing it to the appropriate EPA Regional or State Office.

C) **Other Wastes:** If you handle other wastes or State regulated wastes that have a waste code, enter the appropriate code number in the boxes provided.

Item X — Certification:

This certification must be signed by the owner, operator, or an authorized representative of your installation. An "authorized representative" is a person responsible for the overall operation of the facility (i.e., a plant manger or superintendent, or a person of equal responsibility). *All notifications must include this certification to be complete.*

Item XI — Comments:

Use this space for any additional comments.

V. Definitions

The following definitions are included to help you to understand and complete the Notification Form:

ACT or RCRA means the Solid Waste Disposal Act, as amended by the Resource Conservation and Recovery Act of 1976, as amended by the Hazardous and Solid Waste Amendments of 1984, 42 U.S.C. Section 6901 *et seq.*

Authorized Representative means the person responsible for the overall operation of the facility or an operational unit (i.e., part of a facility), e.g., superintendent or plant manager, or person of equivalent responsibility.

Boiler means an enclosed device using controlled flame combustion and having the following characteristics:

(1) the unit has physical provisions for recovering and exporting energy in the form of steam, heated fluids, or heated gases;

(2) the unit's combustion chamber and primary energy recovery section(s) are of integral design (i.e., they are physically formed into one manufactured or assembled unit);

(3) the unit continuously maintains an energy recovery efficiency of at least 60 percent, calculated in terms of the recovered energy compared with the thermal value of the fuel;

(4) the unit exports and utilizes at least 75 percent of the recovered energy, calculated on an annual basis (excluding recovered heat used internally in the same unit, for example, to preheat fuel or combustion air or drive fans or feedwater pumps); and

(5) the unit is one which the Regional Administrator has determined on a case-by-case basis, to be a boiler after considering the standards in 40 *CFR* 260.32.

Burner means the owner or operator of any boiler or industrial furnace that burns hazardous waste fuel for energy recovery and that is not regulated as a RCRA hazardous waste incinerator.

Disposal means the discharge, deposit, injection, dumping, spilling, leaking, or placing of any solid waste or hazardous waste into or on any land or water so that such solid waste or hazardous waste or any constituent thereof may enter the environment or be emitted into the air or discharged into any waters, including ground waters.

Disposal Facility means a facility or part of a facility at which hazardous waste is intentionally placed into or on any land or water, and at which waste will remain after closure.

EPA Identification (I.D.) Number means the number assigned by EPA to each generator, transporter, and treatment, storage, or disposal facility.

Facility means all contiguous land, structures, other appurtenances, and improvements on the land, used for treating, storing, or disposing of hazardous waste. A facility may consist of several treatment, storage, or disposal operational units (e.g., one or more landfills, surface impoundments, or combinations of them).

Generator means any person, by site, whose act or process produces hazardous waste identified or listed in 40 *CFR* Part 261.

Hazardous Waste means a hazardous waste as defined in 40 *CFR* 261.3.

Hazardous Waste Fuel means hazardous waste and any fuel that contains hazardous waste that is burned for energy recovery in a boiler or industrial furnace that is not subject to regulation as a RCRA hazardous waste incinerator. However, the following hazardous waste fuels are subject to regulation as used oil fuels:

(1) Used oil fuel burned for energy recovery that is also a hazardous waste solely because it exhibits a characteristic of hazardous waste identified in Subpart C of 40 *CFR* Part 261; and

(2) Used oil fuel mixed with hazardous wastes generated by a small quantity generator subject to 40 *CFR* 261.5.

Industrial Boiler means a boiler located on the site of a facility engaged in a manufacturing process where substances are transformed into new products, including the component parts of products, by mechanical or chemical processes.

Industrial Furnace means any of the following enclosed devices that are integral components of manufacturing processes and that use controlled flame combustion to accomplish recovery of materials or energy: cement kilns, lime kilns, aggregate kilns (including asphalt kilns), phosphate kilns, coke ovens, blast furnaces, smelting furnaces, refining furnaces, titanium dioxide chloride process oxidation reactors, methane reforming furnaces, pulping liquor recovery furnaces, combustion devices used in the recovery of sulfer values from spent sulfuric acid, and other devices as the Administrator may add to this list.

Marketer means a person who markets hazardous waste fuel or used oil fuel. However, the following marketers are not subject to waste-as-fuel requirements (including notification) under Subparts D and E of 40 CFR Part 266:

(1) Generators and initial transporters (i.e., transporters who receive hazardous waste or used oil directly from generators including initial transporters who operate transfer stations) who do not market directly to persons who burn the fuels; and

(2) Persons who market used oil fuel that meets the specification provided under 40 CFR 266.40(e) and who are not the first to claim the oil meets the specification.

Municipality means a city, village, town, borough, county, parish, district, association, Indian tribe or authorized Indian tribal organization, designated and approved management agency under Section 208 of the Clean Water Act, or any other public body created by or under State law and having jurisdiction over disposal of sewage, industrial wastes, or other wastes.

Off-Specification Used Oil Fuel means used oil fuel that does not meet the specification provided under 40 CFR 266.40(e).

Operator means the person responsible for the overall operation of a facility.

Owner means a person who owns a facility or part of a facility, including landowner.

Small Quantity Exemption means small quantities of hazardous waste that are exempt from the requirements of 40 CFR 266.108.

Smelter Deferral means that the mandate in section 3000(g) to regulate facilities burning hazardous waste for energy recovery as may be necessary to protect human health and the environment does not apply to devices burning for the purpose of material recovery.

Specification Used Oil Fuel means used oil fuel that meets the specification provided under 40 CFR 266.40(e).

Storage means the holding of hazardous waste for a temporary period, at the end of which the hazardous waste is treated, disposed of, or stored elsewhere.

Transportation means the movement of hazardous waste by air, rail, highway, or water.

Transporter means a person engaged in the off-site transportation of hazardous waste by air, rail, highway, or water.

Treatment means any method, technique, or process, including neutralization, designed to change the physical, chemical, or biological character or composition of any hazardous waste so as to neutralize such waste, or so as to recover energy or material resources from the waste, or so as to render such waste nonhazardous, or less hazardous; safer to transport, store or dispose of; or amenable for recovery, amenable for storage, or reduced in volume. Such term includes any activity or processing designed to change the physical form or composition of hazardous waste so as to render it nonhazardous.

Underground Injection Control means the subsurface emplacement of fluids through a bored, drilled or driven well; or through a dug well, where the depth of the dug well is greater than the largest surface dimension.

Used Oil means any oil that has been refined from crude oil, used, and as a result of such use, is contaminated by physical or chemical impurities. Wastes that contain oils that have not been used (e.g., fuel oil storage tank bottom clean-out wastes) are not used oil unless they are mixed with used oil.

Used Oil Fuel means any used oil burned (or destined to be burned) for energy recovery including any fuel produced from used oil by processing, blending or other treatment, and that does not contain hazardous waste (other than that generated by a small quantity generator and exempt from regulation as hazardous waste under provisions of 40 CFR 261.5). Used oil fuel may itself exhibit a characteristic of hazardous waste and remain subject to regulation as used oil fuel provided it is not mixed with hazardous waste.

Utility Boiler means a boiler that is used to produce electricity, steam or heated or cooled air or other gases or fluids for sale.

Waste Fuel means hazardous waste fuel or off-specification used oil fuel.

VI. EPA Hazardous Waste Numbers for Waste Streams Commonly Generated by Small Quantity Generators

The Environmental Protection Agency recognizes that generators of small quantities of hazardous waste, many of which are small businesses, may not be familiar with the manner in which hazardous waste materials are identified in the Code of Federal Regulations. This insert has been assembled in order to aid small quantity generators in determining for their wastes the EPA Hazardous Waste Numbers that are needed to complete the "Notification of Regulated Waste Activity," Form 8700–12.

This insert is composed of two tables. Table 1 lists eighteen general industry categories that contain small quantity generators. For each of these categories, commonly generated hazardous waste streams are identified. Table 2 lists EPA Hazardous Waste Numbers for each waste stream identified in Table 1.

To use this insert:

1. Locate your industry in Table 1 to identify the waste streams common to your activities.

2. Find each of your waste streams in Table 2, and review the more detailed descriptions of typical wastes to determine which waste streams actually result from your activities.

3. If you determine that a waste stream does apply to you, report the 4–digit EPA Hazardous Waste Number in Item IX. B. of Form 8700–12, "Notification of Regulated Waste Activity."

The industries and waste streams described here do not provide a comprehensive list but rather serve as a guide to potential small quantity generators in determining which of their wastes, if any, are hazardous. Except for the pesticide category, this insert does not include EPA Hazardous Waste Numbers for commercial chemical products that are hazardous when discarded unused. These chemicals and their EPA Hazardous Waste Number are listed in 40 CFR 261.33.

If the specific Hazardous Waste Number that should be applied to your waste stream is unclear, please refer to 40 CFR Part 261, reprinted in Section VII of this notification package. In those cases where more than one Hazardous Waste Number is applicable, all should be used. If you have any questions, or if you are unable to determine the proper EPA Hazardous Waste Numbers for your wastes, contact your state hazardous waste management agency as listed in Section III of this notification package, or the RCRA/Superfund Hotline at 1–800–424–9346.

Table 1

Typical Waste Streams Produced by Small Quantity Generators

Industry	Waste Streams
Laboratories	Acids/Bases Heavy Metals/Inorganics Ignitable Wastes Reactives Solvents
Printing and Allied Industries	Acids/Bases Heavy Metals/Inorganics Ink Sludges Spent Plating Wastes Solvents
Pesticide End Users and Application	Heavy Metals/Inorganics Services Pesticides Solvents
Construction	Acids/Bases Ignitable Wastes Solvents
Equipment Repair	Acids/Bases Ignitable Wastes Lead Acid Batteries Solvents
Furniture/Wood Manufacturing and Refinishing	Ignitable Wastes Solvents

Other Manufacturing:
    1) Textiles
    2) Plastics
    3) Leather
	Heavy Metals/Inorganics
	Solvents

Laundries and Dry Cleaners	Dry Cleaning Filtration Residues Solvents
Educational and Vocational Shops	Acids/Bases Ignitable Wastes Pesticides Reactives Solvents
Building Cleaning and Maintenance	Acid/Bases Solvents
Vehicle Maintenance	Acids/Bases Heavy Metals/Inorganics Ignitable Wastes Lead Acid Batteries Solvents

Table 1 (continued)

Typical Waste Streams Produced by Small Quantity Generators

Industry	Waste Streams
Wood Preserving	Preserving Agents
Motor Freight Terminals and Railroad Transportation	Acids/Bases Heavy Metals/Inorganics Ignitable Wastes Lead Acid Batteries Solvents
Funeral Services	Solvents (formaldehyde)
Metal Manufacturing	Acids/Bases Cyanide Wastes Heavy Metals/Inorganics Ignitable Wastes Reactives Solvents Spent Plating Wastes
Chemical Manufacturers	Acids/Bases Cyanide Wastes Heavy Metals/Inorganics Ignitable Wastes Reactives Solvents
Cleaning Agents and Cosmetics	Acids/Bases Heavy Metals/Inorganics Ignitable Wastes Pesticides Solvents
Formulators	Acids/Bases Cyanide Wastes Heavy Meals/Inorganics Ignitable Wastes Pesticides Reactives Solvents

Table 2

Typical Waste Streams and EPA Hazardous Waste Numbers

ACIDS/BASES:

Acids, bases or mixtures having a pH less than or equal to 2 or greater than or equal to 12.5, or liquids that corrode steel at a rate greater than 0.25 inches per year, are considered to be corrosive (for a complete description of corrosive wastes, see 40 CFR 261.22, Characteristic of Corrosivity). All corrosive materials and solutions have the EPA Hazardous Waste Number of D002. The following are some examples of the more commonly used corrosives:

Examples of Corrosive Waste Streams

Acetic Acid	Oleum
Ammonium Hydroxide	Perchloric Acid
Chromic Acid	Phosphoric Acid
Hydrobromic Acid	Potassium Hydroxide
Hydrochloric Acid	Sodium Hydroxide
Hydrofluoric Acid	Sulfuric Acid
Nitric Acid	

DRY CLEANING FILTRATION RESIDUES:

Cooked powder residue (perchloroethylene plants only), still residues and spent cartridge filters containing perchloroethylene or valclene are hazardous and have an EPA Hazardous Waste Number of F002.

Still residues containing petroleum solvents with a flash point less than 140°F are also considered hazardous, and have an EPA Hazardous Waste Number of D001.

HEAVY METALS/INORGANICS:

Heavy Metals and other inorganic waste materials exhibit the characteristic of TCLP Toxicity and are considered hazardous if the extract from a representative sample of the waste has any of the specific constituent concentrations as shown in 40 CFR 261.24, Table 1. This may include dusts, solutions, wastewater treatment sludges, paint wastes, waste inks and other such materials which contain heavy metals/inorganics (note that wastewater treatment sludges from electroplating operations containing nickel and cyanide, are identified as F006). The following are TCLP Toxic:

Waste Stream	EPA Hazardous Waste Number
Arsenic	D004
Barium	D005
Cadmium	D006
Chromium	D007
Lead	D008
Mercury	D009
Selenium	D010
Silver	D011

IGNITABLE WASTES:

Ignitable wastes include any flammable liquids, nonliquids, and contained gases that have a flashpoint less than 140F (for a complete description of ignitable wastes, see 40 CFR 261.21, Characteristic of ignitability). Examples are spent solvents (see also solvents), solvent still bottoms, ignitable paint wastes (paint removers, brush cleaners

Table 2 (continued)

and stripping agents), epoxy resins and adhesives (epoxies, rubber cements and marine glues), and waste inks containing flammable solvents. Unless otherwise specified, all ignitable wastes have an EPA Hazardous Waste Number of D001.

Some commonly used ignitable compounds are:

Waste Stream	EPA Hazardous Waste Number
Acetone	F003
Benzene	D001
n-Butyl Alcohol	F003
Chlorobenzene	F002
Cychlohexanone	F003
Ethyl Acetate	F003
Ethylbenzene	F003
Ethyl Ether	F003
Ethylene Dichloride	D001
Methanol	F003
Methyl Isobutyl Ketone	F003
Petroleum Distillates	D001
Xylene	F003

INK SLUDGES CONTAINING CHROMIUM AND LEAD:

This includes solvent washes and sludges, caustic washes and sludges, or waster washes and sludges from cleaning tubs and equipment used in the formulation of ink from pigments, driers, soaps, and stabilizers containing chromium and lead. All ink sludges have an EPA Hazardous Waste Number of K086.

LEAD ACID BATTERIES:

Used lead acid batteries should be reported on the notification form only if they are not recycled. Used lead acid batteries that are recycled do not need to be counted in determining the quantity of waste that you generate per month, nor do they require a hazardous waste manifest when shipped off your premises. (Note: Special requirements do apply if you recycle your batteries on your own premises — see 40 CFR Part 266.)

Waste Stream	EPA Hazardous Waste Number
Lead Dross	D008
Spent Acids	D002
Lead Acid Batteries	D008, D002

ORGANIC WASTES:

See Section VIII, Table 1-Maximum Concentration of Contaminants for the Toxicity Characteristic for a list of constituents and regulatory levels.

PESTICIDES:

Pesticides, pesticide residues, washing and rinsing solutions and dips which contain constituent concentrations at or above Toxicity Characteristic regulatory levels (see Section VIII) are hazardous waste. Pesticides that have an oral LD50 toxicity (rat) < 50 mg/kg, inhalation LC50 toxicity (rat) < 2 mg/L or a dermal LD 50 toxicity (rabbit)

< 200 mg/kg, are hazardous materials. The following pesticides would be hazardous waste if they are technical grade, unused and disposed. For a more complete listing, see 40 CFR 261.32-33 for specific listed pesticides, discarded commercial chemical products, and other wastes, wastewaters, sludges, and by-products from pesticide production. (Note that while many of these pesticides are not longer in common use, they are included here for those cases where they may be found in storage.)

Table 2 (continued)

Waste Stream	EPA Hazardous Waste Number
Aldicarb	P070
Aldrin	P004
Amitrole	U011
Arsenic Pentoxide	P011
Arsenic Trioxide	P012
Cacodylic Acid	U136
Carbamic Acid, Methylnitroso- Ethyl Ester	U178
Chlordane	U036
Copper Cyanides	P029
1,2-Dibromo-3-Chloropropane	U066
1,2-Dichloropropane	U083
1,3-Dichloropropene	U084
2,4-Dichlorophenoxy Acetic Acid	U240
DDT	U061
Dieldrin	P037
Dimethoate	P044
Dimethylcarbamoyl Chloride	U097
Dinitrocresol	P047
Dinoseb	P020
Disodium Monmomethane arsonate	D004
Disulfoton	P039
Endosulfan	P050
Endrin	P051
Ethylmercuric Chloride	D009
Famphur	P097
Nepthachlor	P059
Hexachlorobenzene	U127
Kepone	U142
Lindane	U129
2-Methoxy Mercuric Chloride	D009
Methoxychlor	D014
Methyl Parahtion	P071
Monosodium Methanearsonate	D004
Nicotine	P075
Parathion	P089
Pentachloronitrobenzene	U185
Pentachlorophenol	U242
Phenylmercuir Acetate	D009
Phorate	P094

Table 2 (continued)

Strychnine	P108
2,4,5-Trichlorophenoxy Acetic Acid	U232
2-(2,4,5-Trichlorophenoxy)-Propionic Acid	U233
Thallium Sulfate	P115
Thiram	U244
Toxaphene	P123
Warfarin	U248

SOLVENTS:

Spent solvents, solvent still bottoms or mixtures containing solvents are often hazardous. This includes solvents used in degreasing and paint brush cleaning, and distillation residues from reclamation. The following are some commonly used hazardous solvents (see also ignitable wastes for other hazardous solvents, and 40 CFR 261.31 for most listed hazardous waste solvents):

Waste Stream	EPA Hazardous Waste Number	
Benzene	D001	
Carbon Disulfide	F005	
Carbon Tetrachloride	F001	
Chlorobenzene	F002	
Cresols	F004	
Cresylic Acid	F004	
O-Dichlorobenzene	F002	
Ethanol	D001	
Ethylene Dichloride	D001	
Isobutanol	F005	
Isopropanol	D001	
Kerosene	D001	
Methyl Ethyl Ketone	F005	
Methylene Chloride	F001	(Sludges)
	F002	(Still Bottoms)
Naphtha	D001	
Nitrobenzene	F004	
Petroleum Solvents (Flash-point less than 140F)	D001	
Pyridine	F005	
1,1,1-Trichloroethane	F001	(Sludges)
	F002	(Still Bottoms)
Tetrachloroethylene	F001	(Sludges)
	F002	(Still Bottoms)
Toluene	F005	
Trichloroethylene	F001	(Sludges)
	F002	(Still Bottoms)
Trichlorofluoromethane	F002	
Trichlorotrifluoroethane	F002	
White Spirits	D001	

Table 2 (continued)

REACTIVES:

Reactive wastes include reactive materials or mixtures which are unstable, react violently with or form explosive mixtures with water, generate toxic gases or vapors when mixed with water (or when exposed to pH conditions between 2 and 12.5 in the case of cyanide or sulfide bearing wastes), or are capable of detonation or explosive reaction when irritated or heated (for a complete description of reactive wastes, see 40 CFR 261.23, Characteristic of reactivity). Unless otherwise specified, all reactive wastes have an EPA Hazardous Waste Number of D003. The following materials are commonly considered to be reactive:

Waste Stream	EPA Hazardous Waste Number
Acetyl Chloride	D003
Chromic Acid	D003
Cyanides	D003
Organic Peroxides	D003
Perchlorates	D003
Permanganates	D003
Hypochlorites	D003
Sulfides	D003

SPENT PLATING AND CYANIDE WASTES:

Spent plating wastes contain cleaning solutions and plating solutions with caustics, solvents, heavy metals and cyanides. Cyanide wastes may also be generated from heat treatment operations, pigment production and manufacturing of anti-caking agents. Plating wastes are generally Hazardous Waste Numbers F006–F009. Heat treatment wastes are generally Hazardous Waste Numbers F010–F012. See 40 CFR 261.31 for a more complete description of plating wastes.

WOOD PRESERVING AGENTS:

Compounds or mixtures used in wood preserving, including the wastewater treatment sludge from wastewater treatment operations, are considered hazardous. Bottom sediment sludges from the treatment of wastewater processes that use creosote or pentachlorophenol are hazardous, and have an EPA Hazardous Waste Number of K001. Unless otherwise indicated, specific wood preserving components are:

Waste Stream	EPA Hazardous Waste Number
Chromated Copper Arsenate	D004
Creosote	K001
Pentachlorophenol	K001

188 RCRA REGULATORY COMPLIANCE GUIDE

Please print or type with ELITE type (12 characters per inch) in the unshaded areas only

Form Approved. OMB No. 2050-0028. Expires 9-30-92
GSA No. 0246-EPA-OT

Please refer to the *Instructions for Filing Notification* before completing this form. The information requested here is required by law *(Section 3010 of the Resource Conservation and Recovery Act).*

⊕EPA **Notification of Regulated Waste Activity**
United States Environmental Protection Agency

Date Received (For Official Use Only)

I. Installation's EPA ID Number *(Mark 'X' in the appropriate box)*
- A. First Notification
- B. Subsequent Notification *(complete item C)*
- C. Installation's EPA ID Number

II. Name of Installation *(Include company and specific site name)*

III. Location of Installation *(Physical address not P.O. Box or Route Number)*

Street

Street (continued)

City or Town | State | ZIP Code

County Code | County Name

IV. Installation Mailing Address *(See instructions)*

Street or P.O. Box

City or Town | State | ZIP Code

V. Installation Contact *(Person to be contacted regarding waste activities at site)*

Name (last) | (first)

Job Title | Phone Number *(area code and number)*

VI. Installation Contact Address *(See Instructions)*

A. Contact Address: Location / Mailing | B. Street or P.O. Box

City or Town | State | ZIP Code

VII. Ownership *(See Instructions)*

A. Name of Installation's Legal Owner

Street, P.O. Box, or Route Number

City or Town | State | ZIP Code

Phone Number *(area code and number)* | B. Land Type | C. Owner Type | D. Change of Owner Indicator: Yes / No | (Date Changed) Month / Day / Year

EPA Form 8700-12 (Rev. 9-92) Previous edition is obsolete.   -1-   Continue on reverse

APPENDIX 5—EPA NOTIFICATION

Please print or type with ELITE type (12 characters per inch) in the unshaded areas only

Form Approved. OMB No. 2050-0028. Expires 9-30-92
GSA No. 0246-EPA-OT

ID – For Official Use Only

VIII. Type of Regulated Waste Activity (Mark 'X' in the appropriate boxes. Refer to Instructions.)

A. Hazardous Waste Activity

1. Generator (See Instructions)
   - a. Greater than 1000kg/mo (2,200 lbs.)
   - b. 100 to 1000 kg/mo (220 – 2,200 lbs.)
   - c. Less than 100 kg/mo (220 lbs.)
2. Transporter (indicate Mode in boxes 1-5 below)
   - a. For own waste only
   - b. For commercial purposes

Mode of Transportation
   1. Air
   2. Rail
   3. Highway
   4. Water
   5. Other – specify

3. Treater, Storer, Disposer (at installation) Note: A permit is required for this activity; see instructions.
4. Hazardous Waste Fuel
   - a. Generator Marketing to Burner
   - b. Other Marketers
   - c. Boiler and/or Industrial Furnace
     1. Smelter Deferral
     2. Small Quantity Exemption

   Indicate Type of Combustion Device(s)
     1. Utility Boiler
     2. Industrial Boiler
     3. Industrial Furnace
5. Underground Injection Control

B. Used Oil Fuel Activities

1. Off-Specification Used Oil Fuel
   - a. Generator Marketing to Burner
   - b. Other Marketer
   - c. Burner – indicate device(s) – Type of Combustion Device
     1. Utility Boiler
     2. Industrial Boiler
     3. Industrial Furnace

2. Specification Used Oil Fuel Marketer (or On-site Burner) Who First Claims the Oil Meets the Specification

IX. Description of Regulated Wastes (Use additional sheets if necessary)

A. Characteristics of Nonlisted Hazardous Wastes. Mark 'X' in the boxes corresponding to the characteristics of nonlisted hazardous wastes your installation handles. (See 40 CFR Parts 261.20 - 261.24)

1. Ignitable (D001)
2. Corrosive (D002)
3. Reactive (D003)
4. Toxicity Characteristic (D000)

(List specific EPA hazardous waste number(s) for the Toxicity characteristic contaminant(s))

B. Listed Hazardous Wastes. (See 40 CFR 261.31 - 33. See instructions if you need to list more than 12 waste codes.)

1. | 2. | 3. | 4. | 5. | 6.
7. | 8. | 9. | 10. | 11. | 12.

C. Other Wastes. (State or other wastes requiring a handler to have an I.D. number. See instructions.)

1. | 2. | 3. | 4. | 5. | 6.

X. Certification

I certify under penalty of law that this document and all attachments were prepared under my direction or supervision in accordance with a system designed to assure that qualified personnel properly gather and evaluate the information submitted. Based on my inquiry of the person or persons who manage the system, or those persons directly responsible for gathering the information, the information submitted is, to the best of my knowledge and belief, true, accurate, and complete. I am aware that there are significant penalties for submitting false information, including the possibility of fine and imprisonment for knowing violations.

Signature	Name and Official Title (type or print)	Date Signed

XI. Comments

Note: Mail completed form to the appropriate EPA Regional or State Office. (See Section III of the booklet for addresses.)

EPA Form 8700-12 (Rev. 9-92) Previous edition is obsolete.

RCRA REGULATORY COMPLIANCE GUIDE

Please print or type with ELITE type (12 characters per inch) in the unshaded areas only

Form Approved. OMB No. 2050-0028. Expires 9-30-92
GSA No. 0246-EPA-OT

ID – For Official Use Only

IX. Description of Regulated Wastes Continued *(Additional sheet)*

B. Listed Hazardous Wastes. (See 40 CFR 261.31 – 33. Use this page only if you need to list more than 12 waste codes.)

13	14	15	16	17	18
19	20	21	22	23	24
25	26	27	28	29	30
31	32	33	34	35	36
37	38	39	40	41	42
43	44	45	46	47	48
49	50	51	52	53	54
55	56	57	58	59	60
61	62	63	64	65	66
67	68	69	70	71	72
73	74	75	76	77	78
79	80	81	82	83	84
85	86	87	88	89	90
91	92	93	94	95	96
97	98	99	100	101	102
103	104	105	106	107	108
109	110	111	112	113	114
115	116	117	118	119	120

EPA Form 8700-12 (Rev. 9-92) Previous edition is obsolete.

Appendix 6

Tier One Form

TIER ONE INSTRUCTIONS

GENERAL INFORMATION

Submission of this form is required by Title III of the Superfund Amendments and Reauthorization Act of 1986, Title III Section 312, Public Law 99-499, codified at 42 U.S.C. Section 11022.

CERTIFICATION
The owner or operator or the officially designated representative of the owner or operator must certify that all information included in the Tier I submission is true, accurate, and complete. On the Tier I form, enter your full name and official title. Sign your name and enter the current date. Also, enter the total number of pages in the submission, including all attachments.

The purpose of this form is to provide State and local officials and the public with information on the general types and locations of hazardous chemicals present at your facility during the past year.

YOU MUST PROVIDE ALL INFORMATION REQUESTED ON THIS FORM.

You may substitute the Tier Two form for this Tier One form. (The Tier Two form provides detailed information and must be submitted in response to a specific request from State or local officials.)

WHO MUST SUBMIT THIS FORM
Section 312 of Title III requires that the owner or operator of a facility submit this form if, under regulations implementing the Occupational Safety and Health Act of 1970, the owner or operator is required to prepare or have available Material Safety Data Sheets (MSDS) for hazardous chemicals present at the facility. MSDS requirements are specified in the Occupational Safety and Health Administration (OSHA) Hazard Communication Standard, found in Title 29 of the Code of Federal Regulations at §1910.1200.

This form does not have to be submitted if all of the chemicals located at your facility are excluded under Section 311(e) of Title III or if the weight of each covered hazardous chemical never equals or exceeds the minimum threshold listed in Title III Section 312 during the reporting year.

WHAT CHEMICALS ARE INCLUDED
You must report the information required on this form for every hazardous chemical for which you are required to prepare or have available an MSDS under the Hazard Communication Standard, unless the chemicals are excluded under Section 311(e) of Title III or they are below the minimum reporting thresholds.

WHAT CHEMICALS ARE EXCLUDED
Section 311(e) of Title III excludes the following substances:

(i) Any food, food additive, color additive, drug, or cosmetic regulated by the Food and Drug Administration;

(ii) Any substance present as a solid in any manufactured item to the extent exposure to the substance does not occur under normal conditions of use;

(iii) Any substance to the extent it is used for personal, family, or household purposes, or is present in the same form and concentration as a product packaged for distribution and use by the general public;

(iv) Any substance to the extent it is used in a research laboratory or a hospital or other medical facility under the direct supervision of a technically qualified individual;

(v) Any substance to the extent it is used in routine agricultural operations or is a fertilizer held for sale by a retailer to the ultimate customer.

OSHA regulations, Section 1910.1200(b), stipulate exemptions from the requirement to prepare or have available an MSDS.

REPORTING THRESHOLDS
Minimum thresholds have been established for Tier One/Tier Two reporting under Title III, Section 312. These thresholds are as follows:

For Extremely Hazardous Substances (EHSs) designated under section 302 of Title III, the reporting threshold is 500 pounds (or 227 kg.) or the threshold planning quantity (TPQ), whichever is lower;

For all other hazardous chemicals for which facilities are required to have or prepare an MSDS, the minimum reporting threshold is 10,000 pounds (or 4,540 kg.).

You need to report hazardous chemicals that were present at your facility at any time during the previous calendar year at levels that equal or exceed these thresholds. For instructions on threshold determinations for components of mixtures, see "What About Mixtures?" on page 3 of these instructions.

WHEN TO SUBMIT THIS FORM
Owners or operators of facilities that have hazardous chemicals on hand in quantities equal to or greater than set threshold levels must submit either Tier One or Tier Two Forms by March 1.

WHERE TO SUBMIT THIS FORM
Send one completed Inventory form to each of the following organizations:

1. Your State emergency response commission
2. Your local emergency planning committee
3. The fire department with jurisdiction over your facility.

PENALTIES
Any owner or operator of a facility who fails to submit or supplies false Tier One information shall be liable to the United States for a civil penalty of up to $25,000 for each such violation. Each day a violation continues shall constitute a separate violation. In addition, any citizen may commence a civil action on his or her own behalf against any owner or operator who fails to submit Tier One information.

APPENDIX 6—TIER ONE FORM

INSTRUCTIONS

Please read these instructions carefully. Print or type all responses.

You may use the Tier Two form as a worksheet for completing Tier One. Filling in the Tier Two chemical information section should help you assemble your Tier One responses.

If your responses require more than one page, fill in the page number at the top of the form.

REPORTING PERIOD
Enter the appropriate calendar year, beginning January 1 and ending December 31.

FACILITY IDENTIFICATION
Enter the complete name of your facility (and company identifier where appropriate).

Enter the full street address or state road. If a street address is not available, enter other appropriate identifiers that describe the physical location of your facility (e.g., longitude and latitude). Include city, county, state, and zip code.

Enter the primary Standard Industrial Classification (SIC) code and the Dun & Bradstreet number for your facility. The financial officer of your facility should be able to provide the Dun & Bradstreet number. If your firm does not have this information, contact the State or regional office of Dun & Bradstreet to obtain your facility number or have one assigned.

OWNER/OPERATOR
Enter the owner's or operator's full name, mailing address, and phone number.

EMERGENCY CONTACT
Enter the name, title, and work phone number of at least one local person or office that can act as a referral if emergency responders need assistance in responding to a chemical accident at the facility.

Provide an emergency phone number where such emergency information will be available 24 hours a day, every day. This requirement is mandatory. The facility must make some arrangement to ensure that a 24 hour contact is available.

IDENTICAL INFORMATION
Check the box indicating identical information, located below the emergency contacts on the Tier One form, if the current information being reported is identical to that submitted last year. Chemical descriptions, amounts, and locations must be provided in this year's form, even if the information is identical to that submitted last year.

PHYSICAL AND HEALTH HAZARDS
Descriptions, Amounts, and Locations

This section requires aggregate information on chemicals by hazard categories as defined in 40 CFR 370.2. The two health hazard categories and three physical hazard categories are a consolidation of the 23 hazard categories defined in the OSHA Hazard Communication Standard, 29 CFR 1910.1200. For each hazard type, indicate the total amounts and general locations of all applicable chemicals present at your facility during the past year.

Hazard Category Comparison
For Reporting Under Sections 311 and 312

EPA's Hazard Categories	OSHA's Hazard Categories
Fire Hazard	Flammable Combustion Liquid Pyrophoric Oxidizer
Sudden Release of Pressure	Explosive Compressed Gas
Reactive	Unstable Reactive Organic Peroxide Water Reactive
Immediate (Acute) Health Hazards	Highly Toxic Toxic Irritant Sensitizer Corrosive Other hazardous chemicals with an adverse effect with short term exposure
Delayed (Chronic) Health Hazard	Carcinogens Other hazardous chemicals with an adverse effect with long term exposure

- What units should I use?

  Calculate all amounts as *weight in pounds*. To convert gas or liquid volume to weight in pounds, multiply by an appropriate density factor.

INSTRUCTIONS

Please read these instructions carefully. Print or type all responses.

- **What about mixtures?**

  If a chemical is part of a mixture, you have the option of reporting either the weight of the entire mixture or only the portion of the mixture that is a particular hazardous chemical (e.g., if a hazardous solution weighs 100 lbs. but is composed of only 5% of a particular hazardous chemical, you can indicate either 100 lbs. of the mixture or 5 lbs. of the hazardous chemical).

  The option used for each mixture must be consistent with the option used in your Section 311 reporting.

  Because EHSs are important to Section 303 planning, EHSs have lower thresholds. The amount of an EHS at a facility (both pure EHS substances and EHSs in mixtures) must be aggregated for purposes of threshold determination. It is suggested that the aggregation calculation be done as a first step in making the threshold determination. Once you determine whether a threshold has been reached for an EHS, you should report either the total weight of the EHS at your facility, or the weight of each mixture containing the EHS.

- **Where do I count a chemical that is a fire and reactive physical hazard and an immediate (acute) health hazard?**

  Add the chemical's weight to your totals for all three hazard categories and include its location in all three categories. Many chemicals fall into more than one hazard category.

MAXIMUM AMOUNT

The amounts of chemicals you have on hand may vary throughout the year. The peak weights -- greatest single-day weights during the year -- are added together in this column to determine the maximum weight for each hazard type. Since the peaks for different chemicals often occur on different days, this maximum amount will seem artificially high.

To complete this and the following sections, you may choose to use the Tier Two form as a worksheet.

To determine the Maximum Amount:

1. List all of your reportable hazardous chemicals individually.
2. For each chemical...
   a. Indicate all physical and health hazards that the chemical presents. Include all chemicals, even if they are present for only a short period of time during the year.
   b. Estimate the maximum weight in pounds that was present at your facility on any single day of the reporting period.

3. For each hazard type -- beginning with Fire and repeating for all physical and health hazard types...
   a. Add the maximum weights of all chemicals you indicated as the particular hazard type.
   b. Look at the Reporting Ranges at the bottom of the Tier One form. Find the appropriate range value code.
   c. Enter this range value as the Maximum Amount.

EXAMPLE:

You are using the Tier Two form as a worksheet and have listed raw weights in pounds for each of your hazardous chemicals. You have marked an X in the Immediate (acute) hazard column for phenol and sulfuric acid. The maximum amount raw weight you listed were 10,000 lbs. and 500 lbs. respectively. You add these together to reach a total of 10,500 lbs. Then you look at the Reporting Range at the bottom of your Tier One form and find that the value of 04 corresponds to 10,500 lbs. Enter 04 as your Maximum Amount for Immediate (acute) hazards materials.

You also marked an X in the Fire hazard box for phenol. When you calculate your Maximum Amount totals for fire hazards, add the 10,000 lb. weight again.

AVERAGE DAILY AMOUNT

This column should represent the average daily amount of chemicals *of each hazard type* that were present at or above applicable thresholds at your facility at any point during the year.

To determine this amount:

1. List all of your reportable hazardous chemicals individually (same as for Maximum Amount).
2. For each chemical...
   a. Indicate all physical and health hazards that the chemical presents (same as for Maximum Amount).
   b. Estimate the average weight in pounds that was present at your facility throughout the year. To do this, total all daily weights and divide by the number of days the chemical was present on the site.
3. For each hazard type -- beginning with Fire and repeating for all physical and health hazards...
   a. Add the average weights of all chemicals you indicated for the particular hazard type.
   b. Look at the Reporting Ranges at the bottom of the Tier One form. Find the appropriate range value code.
   c. Enter this range value as the Average Daily Amount.

INSTRUCTIONS

Please read these instructions carefully. Print or type all responses.

EXAMPLE:

You are using the Tier Two form, and have marked an X in the Immediate (acute) hazard column for nicotine and phenol. Nicotine is present at your facility 100 days during the year, and the sum of the daily weights is 100,000 lbs. By dividing 100,000 lbs. by 100 days on-site, you calculate an Average Daily Amount of 1,000 lbs. for nicotine. Phenol is present at your facility 50 days during the year, and the sum of the daily weights is 10,000 lbs. By dividing 10,000 lbs. by 50 days on-site, you calculate an Average Daily Amount of 200 lbs. for phenol. You then add the two average daily amounts together to reach a total of 1,200 lbs. Then you look at the Reporting Range on your Tier One form and find that the value 03 corresponds to 1,200 lbs. Enter 03 as your Average Daily Amount for Immediate (acute) Hazard.

You also marked an X in the Fire hazard column for phenol. When you calculate your Average Daily Amount for fire hazards, use the 200 lb. weight again.

NUMBER OF DAYS ON-SITE

Enter the greatest number of days that a single chemical within that hazard category was present on-site.

EXAMPLE:

At your facility, nicotine is present for 100 days and phosgene is present for 150 days. Enter 150 in the space provided.

GENERAL LOCATION

Enter the general location within your facility where each hazard may be found. General locations should include the names or identifications of buildings, tank fields, lots, sheds, or other such areas.

For each hazard type, list the locations of all applicable chemicals. As an alternative you may also attach a site plan and list the site coordinates related to the appropriate locations. If you do so, check the Site Plan box.

EXAMPLE:

On your worksheet you have marked an X in the Fire hazard column for acetone and butane. You noted that these are kept in steel drums in Room C of the Main Building, and in pressurized cylinders in Storage Shed 13, respectively. You could enter Main Building and Storage Shed 13 as the General Locations of your fire hazards. However, you choose to attach a site plan and list coordinates. Check the Site Plan box at the top of the column and enter site coordinates for the Main Building and Storage Shed 13 under General Locations.

If you need more space to list locations, attach an additional Tier One form and continue your list on the proper line. Number all pages.

CERTIFICATION

Instructions for this section are included on page one of these instructions.

RCRA REGULATORY COMPLIANCE GUIDE

Revised June 1990

Page ____ of ____ pages
Form Approved OMB No. 2050-0072

Tier One — EMERGENCY AND HAZARDOUS CHEMICAL INVENTORY
Aggregate Information by Hazard Type

FOR OFFICIAL USE ONLY

ID #: ____
Date Received: ____

Important: Read instructions before completing form

Reporting Period From January 1 to December 31, 19____

Facility Identification

Name: ____
Street: ____
City: ____ County: ____ State: ____ Zip: ____

SIC Code: ☐☐☐☐
Dun & Brad Number: ☐☐-☐☐☐-☐☐☐☐

Emergency Contacts

Name: ____
Title: ____
Phone: ( ) ____
24 Hour Phone: ( ) ____

Name: ____
Title: ____
Phone: ( ) ____
24 Hour Phone: ( ) ____

Owner/Operator

Name: ____
Mail Address: ____
Phone: ( ) ____

☐ Check if information below is identical to the information submitted last year.

Physical Hazards

Hazard Type	Max Amount*	Average Daily Amount*	Number of Days On-Site	General Location
Fire	☐☐	☐☐	☐☐☐	____
Sudden Release of Pressure	☐☐	☐☐	☐☐☐	____
Reactivity	☐☐	☐☐	☐☐☐	____

☐ Check if site plan is attached

Health Hazards

Hazard Type	Max Amount*	Average Daily Amount*	Number of Days On-Site	General Location
Immediate (acute)	☐☐	☐☐	☐☐☐	____
Delayed (Chronic)	☐☐	☐☐	☐☐☐	____

Certification *(Read and sign after completing all sections)*

I certify under penalty of law that I have personally examined and am familiar with the information submitted in pages one through ____, and that based on my inquiry of those individuals responsible for obtaining the information, I believe that the submitted information is true, accurate and complete.

Name and official title of owner/operator OR owner/operator's authorized representative: ____

Signature: ____   Date signed: ____

* Reporting Ranges

Range Code	Weight Range in Pounds From...	To...
01	0	99
02	100	999
03	1000	9,999
04	10,000	99,999
05	100,000	999,999
06	1,000,000	9,999,999
07	10,000,000	49,999,999
08	50,000,000	99,999,999
09	100,000,000	499,999,999
10	500,000,000	999,999,999
11	1 billion	higher than 1 billion

Appendix 7

Tier Two Form

TIER TWO INSTRUCTIONS

GENERAL INFORMATION

Submission of this Tier Two form (when requested) is required by Title III of the Superfund Amendments and Reauthorization Act of 1986, Section 312, Public Law 99-499, codified at 42 U.S.C. Section 11022. The purpose of this Tier Two form is to provide State and local officials and the public with specific information on hazardous chemicals present at your facility during the past year.

CERTIFICATION

The owner or operator or the officially designated representative of the owner or operator must certify that all information included in the Tier Two submission is true, accurate, and complete. On the first page of the Tier Two report, enter your full name and official title. Sign your name and enter the current date. Also, enter the total number of pages included in the Confidential and Non-Confidential Information Sheets as well as all attachments. An original signature is required on at least the first page of the submission. Submissions to the SERC, LEPC, and fire department must each contain an original signature on at least the first page. Subsequent pages must contain either an original signature, a photocopy of the original signature, or a signature stamp. Each page must contain the date on which the original signature was affixed to the first page of the submission and the total number of pages in the submission.

YOU MUST PROVIDE ALL INFORMATION REQUESTED ON THIS FORM TO FULFILL TIER TWO REPORTING REQUIREMENTS.

This form may also be used as a worksheet for completing the Tier One form or may be submitted in place of the Tier One form.

WHO MUST SUBMIT THIS FORM

Section 312 of Title III requires that the owner or operator of a facility submit this Tier Two form if so requested by a State emergency response commission, a local emergency planning committee, or a fire department with jurisdiction over the facility.

This request may apply to the owner or operator of any facility that is required, under regulations implementing the Occupational Safety and Health Act of 1970, to prepare or have available a Material Safety Data Sheet (MSDS) for a hazardous chemical present at the facility. MSDS requirements are specified in the Occupational Safety and Health Administration (OSHA) Hazard Communication Standard, found in Title 29 of the Code of Federal Regulations at §1910.1200.

This form does not have to be submitted if all of the chemicals located at your facility are excluded under Section 311(e) of Title III.

WHAT CHEMICALS ARE INCLUDED

If you are submitting Tier Two forms in lieu of Tier One, you must report the required information on this Tier Two form for each hazardous chemical present at your facility in quantities equal to or greater than established threshold amounts (discussed below), unless the chemicals are excluded under Section 311(e) of Title III. Hazardous chemicals are any substance for which your facility must maintain an MSDS under OSHA's Hazard Communication Standard.

If you elect to submit Tier One rather than Tier Two, you may still be required to submit Tier Two information upon request.

WHAT CHEMICALS ARE EXCLUDED

Section 311(e) of Title III excludes the following substances:

(i) Any food, food additive, color additive, drug, or cosmetic regulated by the Food and Drug Administration;

(ii) Any substance present as a solid in any manufactured item to the extent exposure to the substance does not occur under normal conditions of use;

(iii) Any substance to the extent it is used for personal, family, or household purposes, or is present in the same form and concentration as a product packaged for distribution and use by the general public;

(iv) Any substance to the extent it is used in a research laboratory or a hospital or other medical facility under the direct supervision of a technically qualified individual;

(v) Any substance to the extent it is used in routine agricultural operations or is a fertilizer held for sale by a retailer to the ultimate customer.

OSHA regulations, Section 1910.1200(b), stipulate exemptions from the requirement to prepare or have available an MSDS.

REPORTING THRESHOLDS

Minimum thresholds have been established for Tier One/Tier Two reporting under Title III, Section 312. These thresholds are as follows:

For Extremely Hazardous Substances (EHSs) designated under section 302 of Title III, the reporting threshold is 500 pounds (or 227 kg.) or the threshold planning quantity (TPQ), whichever is lower;

For all other hazardous chemicals for which facilities are required to have or prepare an MSDS, the minimum reporting threshold is 10,000 pounds (or 4,540 kg.).

You need to report hazardous chemicals that were present at your facility at any time during the previous calendar year at levels that equal or exceed these thresholds. For instructions on threshold determinations for components of mixtures, see "What About Mixtures?" on page 2 of these instructions.

A requesting official may limit the responses required under Tier Two by specifying particular chemicals or groups of chemicals. Such requests apply to hazardous chemicals regardless of established thresholds.

APPENDIX 7—TIER TWO FORM

INSTRUCTIONS

Please read these instructions carefully. Print or type all responses.

WHEN TO SUBMIT THIS FORM

Owners or operators of facilities that have hazardous chemicals on hand in quantities equal to or greater than set threshold levels must submit either Tier One or Tier Two forms by March 1.

If you choose to submit Tier One, rather than Tier Two, be aware that you may have to submit Tier Two information later, upon request of an authorized official. You must submit the Tier Two form within 30 days of receipt of a written request.

WHERE TO SUBMIT THIS FORM

Send either a completed Tier One form or Tier Two form(s) to each of the following organizations:

1. Your State Emergency Response Commission.
2. Your Local Emergency Planning Committee.
3. The fire department with jurisdiction over your facility.

If a Tier Two form is submitted in response to a request, send the completed form to the requesting agency.

PENALTIES

Any owner or operator who violates any Tier Two reporting requirements shall be liable to the United States for a civil penalty of up to $25,000 for each such violation. Each day a violation continues shall constitute a separate violation.

If your Tier Two responses require more than one page use additional forms and fill in the page number at the top of the form.

REPORTING PERIOD

Enter the appropriate calendar year, beginning January 1 and ending December 31.

FACILITY IDENTIFICATION

Enter the full name of your facility (and company identifier where appropriate).

Enter the full street address or state road. If a street address is not available, enter other appropriate identifiers that describe the physical location of your facility (e.g., longitude and latitude). Include city, county, state, and zip code.

Enter the primary Standard Industrial Classification (SIC) code and the Dun & Bradstreet number for your facility. The financial officer of your facility should be able to provide the Dun & Bradstreet number. If your firm does not have this information, contact the State or regional office of Dun & Bradstreet to obtain your facility number or have one assigned.

OWNER/OPERATOR

Enter the owner's or operator's full name, mailing address, and phone number.

EMERGENCY CONTACT

Enter the name, title, and work phone number of at least one local person or office who can act as a referral if emergency responders need assistance in responding to a chemical accident at the facility.

Provide an emergency phone number where such emergency information will be available 24 hours a day, every day. This requirement is mandatory. The facility must make some arrangement to ensure that a 24 hour contact is available.

IDENTICAL INFORMATION

Check the box indicating identical information, located below the emergency contacts on the Tier Two form, if the current chemical information being reported is identical to that submitted last year. Chemical descriptions, hazards, amounts, and locations must be provided in this year's form, even if the information is identical to that submitted last year.

CHEMICAL INFORMATION: Description, Hazards, Amounts, and Locations

The main section of the Tier Two form requires specific information on amounts and locations of hazardous chemicals, as defined in the OSHA Hazard Communication Standard.

If you choose to indicate that all of the information on a specific hazardous chemical is identical to that submitted last year, check the appropriate optional box provided at the right side of the storage codes and locations on the Tier Two form. Chemical descriptions, hazards, amounts, and locations must be provided even if the information is identical to that submitted last year.

- What units should I use?

    Calculate all amounts as *weight in pounds*. To convert gas or liquid volume to weight in pounds, multiply by an appropriate density factor.

- What about mixtures?

    If a chemical is part of a mixture, *you have the option* of reporting either the weight of the entire mixture or only the portion of the mixture that is a particular hazardous chemical (e.g., if a hazardous solution weighs 100 lbs. but is composed of only 5% of a particular hazardous chemical, you can indicate either 100 lbs. of the mixture *or* 5 lbs. of the chemical).

    The option used for each mixture must be consistent with the option used in your Section 311 reporting.

    Because EHSs are important to Section 303 planning, EHSs have lower thresholds. The amount of an EHS at a facility (both pure EHS substances and EHSs in mixtures) must be aggregated for purposes of threshold determination. It is suggested that the aggregation calculation be done as a first step in making the threshold determination. Once you determine whether a threshold for an EHS has been reached, you should report either the total weight of the EHS at your facility, or the weight of each mixture containing the EHS.

CHEMICAL DESCRIPTION

1. Enter the Chemical Abstract Service registry number (CAS). For mixtures, enter the CAS number of the mixture as a whole if it has been assigned a number distinct from its constituents. For a mixture that has no CAS number, leave this item blank or report the CAS numbers of as many constituent chemicals as possible.

   If you are withholding the name of a chemical in accordance with criteria specified in Title III, Section 322, enter the generic class or category that is structurally descriptive of the chemical (e.g., list toulene diisocyanate as organic isocyanate) and check the box marked Trade Secret. Trade secret information should be submitted to EPA and must include a substantiation. Please refer to EPA's final regulation on trade secrecy (53 FR 28772, July 29, 1988) for detailed information on how to submit trade secrecy claims.

2. Enter the chemical name or common name of each hazardous chemical.

3. Check box for *ALL* applicable descriptors: pure or mixture; *and* solid, liquid, or gas; and whether the chemical is or contains an EHS.

4. If the chemical is a mixture containing an EHS, enter the chemical name of each EHS in the mixture.

EXAMPLE:
You have pure chlorine gas on hand, as well as two mixtures that contain liquid chlorine. You write "chlorine" and enter the CAS number. Then you check "pure" *and* "mix" -- as well as "liquid" *and* "gas".

PHYSICAL AND HEALTH HAZARDS

For each chemical you have listed, check all the physical and health hazard boxes that apply. These hazard categories are defined in 40 CFR 370.2. The two health hazard categories and three physical hazard categories are a consolidation of the 23 hazard categories defined in the OSHA Hazard Communication Standard, 29 CFR 1910.1200.

Hazard Category Comparison For Reporting Under Sections 311 and 312

EPA's Hazard Categories	OSHA's Hazard Categories
Fire Hazard	Flammable Combustion Liquid Pyrophoric Oxidizer
Sudden Release of Pressure	Explosive Compressed Gas
Reactive	Unstable Reactive Organic Peroxide Water Reactive
Immediate (Acute) Health Hazards	Highly Toxic Toxic Irritant Sensitizer Corrosive
	Other hazardous chemicals with an adverse effect with short term exposure
Delayed (Chronic) Health Hazard	Carcinogens
	Other hazardous chemicals with an adverse effect with long term exposure

MAXIMUM AMOUNT

1. For each hazardous chemical, estimate the greatest amount present at your facility on any single day during the reporting period.

2. Find the appropriate range value code in Table I.

3. Enter this range value as the Maximum Amount.

Table I REPORTING RANGES

Range Value	Weight Range in Pounds From...	To...
01	0	99
02	100	999
03	1,000	9,999
04	10,000	99,999
05	100,000	999,999
06	1,000,000	9,999,999
07	10,000,000	49,999,999
08	50,000,000	99,999,999
09	100,000,000	499,999,999
10	500,000,000	999,999,999
11	1 billion	higher than 1 billion

If you are using this form as a worksheet for completing Tier One, enter the actual weight in pounds in the shaded space below the response blocks. Do this for both Maximum Amount and Average Daily Amount.

EXAMPLE:

You received one large shipment of a solvent mixture last year. The shipment filled five 5,000-gallon storage tanks. You know that the solvent contains 10% benzene, which is a hazardous chemical.

You figure that 10% of 25,000 gallons is 2,500 gallons. You also know that the density of benzene is 7.29 pounds per gallon, so you multiply 2,500 gallons by 7.29 pounds per gallon to get a weight of 18,225 pounds.

Then you look at Table I and find that the range value 04 corresponds to 18,225. You enter 04 as the Maximum Amount.

(If you are using the form as a worksheet for completing a Tier One form, you should write 18,255 in the shaded area.)

AVERAGE DAILY AMOUNT

1. For each hazardous chemical, estimate the average weight in pounds that was present at your facility during the year.

   To do this, total all daily weights and divide by the number of days the chemical was present on the site.

2. Find the appropriate range value in Table I.

3. Enter this range value as the Average Daily Amount.

EXAMPLE:

The 25,000-gallon shipment of solvent you received last year was gradually used up and completely gone in 315 days. The sum of the daily volume levels in the tank is 4,536,000 gallons. By dividing 4,536,000 gallons by 315 days on-site, you calculate an average daily amount of 14,400 gallons.

You already know that the solvent contains 10% benzene, which is a hazardous chemical. Since 10% of 14,400 is 1,440, you figure that you had an average of 1,440 gallons of benzene. You also know that the density of benzene is 7.29 pounds per gallon, so you multiply 1,440 by 7.29 to get a weight of 10,500 pounds.

Then you look at Table I and find that the range value 04 corresponds to 10,500. You enter 04 as the Average Daily Amount.

(If you are using the form as a worksheet for completing a Tier One form, you should write 10,500 in the shaded area.)

NUMBER OF DAYS ON-SITE

Enter the number of days that the hazardous chemical was found on-site.

EXAMPLE:

The solvent composed of 10% benzene was present for 315 days at your facility. Enter 315 in the space provided.

STORAGE CODES AND STORAGE LOCATIONS

List all non-confidential chemical locations in this column, along with storage types/conditions associated with each location. Please note that a particular chemical may be located in several places around the facility. Each row of boxes followed by a line represents a unique location for the same chemical.

Storage Codes: Indicate the types and conditions of storage present.

a. Look at Table II. For each location, find the appropriate storage type and enter the corresponding code in the first box.

b. Look at Table III. For each location, find the appropriate storage types for pressure and temperature conditions. Enter the applicable pressure code in the second box. Enter the applicable temperature code in the third box.

Table II – STORAGE TYPES

CODES	Types of Storage
A	Above ground tank
B	Below ground tank
C	Tank inside building
D	Steel drum
E	Plastic or non-metallic drum
F	Can
G	Carboy
H	Silo
I	Fiber drum
J	Bag
K	Box
L	Cylinder
M	Glass bottles or jugs
N	Plastic bottles or jugs
O	Tote bin
P	Tank wagon
Q	Rail car
R	Other

Table III – PRESSURE AND TEMPERATURE CONDITIONS

CODES	Storage Conditions
	(PRESSURE)
1	Ambient pressure
2	Greater than ambient pressure
3	Less than ambient pressure
	(TEMPERATURE)
4	Ambient temperature
5	Greater than ambient temperature
6	Less than ambient temperature but not cryogenic
7	Cryogenic conditions

EXAMPLE:

The benzene in the main building is kept in a tank inside the building, at ambient pressure and less than ambient temperature.

Table II shows you that the code for a tank inside a building is C. Table III shows you that the code for ambient pressure is 1, and the code for less than ambient temperature is 6.

You enter: [C][1][6]

Storage Locations:

Provide a brief description of the precise location of the chemical, so that emergency responders can locate the area easily. You may find it advantageous to provide the optional site plan or site coordinates as explained below.

For each chemical, indicate at a minimum the building or lot. Additionally, where practical, the room or area may be indicated. You may respond in narrative form with appropriate site coordinates or abbreviations.

If the chemical is present in more than one building, lot, or area location, continue your responses down the page as needed. If the chemical exists everywhere at the plant site simultaneously, you may report that the chemical is ubiquitous at the site.

Optional attachments: If you choose to attach one of the following, check the appropriate Attachments box at the bottom of the Tier Two form.

   a. *A site plan* with site coordinates indicated for buildings, lots, areas, etc. throughout your facility.
   b. *A list of site coordinate abbreviations* that correspond to buildings, lots, areas, etc. throughout your facility.
   c. *A description of dikes and other safeguard measures* for storage locations throughout your facility.

EXAMPLE:

You have benzene in the main room of the main building, and in tank 2 in tank field 10. You attach a site plan with coordinates as follows: main building = G-2, tank field 10 = B-6. Fill in the Storage Location as follows:

B-6 [Tank 2]   G-2 [Main Room]

CONFIDENTIAL INFORMATION

Under Title III, Section 324, you may elect to withhold location information on a specific chemical from disclosure to the public. If you choose to do so:

- Enter the word "confidential" in the Non-Confidential Location section of the Tier Two form on the first line of the storage locations.
- On a separate Tier Two Confidential Location Information Sheet, enter the name and CAS number of each chemical for which you are keeping the location confidential.
- Enter the appropriate location and storage information, as described above for non-confidential locations.
- Attach the Tier Two Confidential Location Information Sheet to the Tier Two form. This separates confidential locations from other information that will be disclosed to the public.

CERTIFICATION

Instructions for this section are included on page one of these instructions.

APPENDIX 7—TIER TWO FORM 203

Tier Two

EMERGENCY AND HAZARDOUS CHEMICAL INVENTORY

Specific Information by Chemical

Page ___ of ___ pages
Form Approved OMB No. 2050-0072

Facility Identification

Name _____
Street _____
City _____ County _____ State ___ Zip ___
SIC Code ____ Dun & Brad Number _____

FOR OFFICIAL USE ONLY
ID # _____
Date Received _____

Owner/Operator Name

Name _____
Mail Address _____
Phone ( ) _____

Emergency Contact

Name _____ Title _____
Phone ( ) _____ 24 Hr. Phone ( ) _____
Name _____ Title _____
Phone ( ) _____ 24 Hr. Phone ( ) _____

Reporting Period From January 1 to December 31, 19___

Important: Read all instructions before completing form

Confidential Location Information Sheet

☐ Check if information below is identical to the information submitted last year.

Storage Codes and Locations
(Confidential)

| Container Type | Pressure | Temperature | Storage Locations (Optional) |

CAS # ☐☐☐☐☐☐☐☐☐ Chem. Name _____ ☐

CAS # ☐☐☐☐☐☐☐☐☐ Chem. Name _____ ☐

CAS # ☐☐☐☐☐☐☐☐☐ Chem. Name _____ ☐

Certification (Read and sign after completing all sections)

I certify under penalty of law that I have personally examined and am familiar with the information submitted in pages one through ___ and that based on my inquiry of those individuals responsible for obtaining the information, I believe that the submitted information is true, accurate, and complete.

Name and official title of owner/operator OR owner/operator's authorized representative _____
Signature _____ Date signed _____

Optional Attachments

☐ I have attached a site plan
☐ I have attached a list of site coordinate abbreviations
☐ I have attached a description of dikes and other safeguard measures

Tier Two EMERGENCY AND HAZARDOUS CHEMICAL INVENTORY

Specific Information by Chemical

Form blank — Tier Two Emergency and Hazardous Chemical Inventory reporting form (Revised November 1990, Form Approved OMB No. 2050-0072), containing sections for Facility Identification, Owner/Operator Name, Emergency Contact, Reporting Period, Chemical Description, Physical and Health Hazards, Inventory, Storage Codes and Locations (Non-Confidential), Optional Attachments, and Certification.

Appendix 8
Appendix E to Hazard Communication Standard (Advisory) - Guidelines for Employer Compliance

The Hazard Communication Standard (HCS) is based on a simple concept - - that employees have both a need and right to know the hazards and identities of the chemicals they are exposed to when working. They also need to know what protective measures are available to prevent adverse effects from occurring.

The HCS is designed to provide employees with the information they need. Knowledge acquired under the HCS will help employers provide safer workplaces for their employees. When employers have information about the chemicals being used, they can take steps to reduce exposures, substitute less hazardous materials, and establish proper work practices. These efforts will help prevent the occurrence of work-related illnesses and injuries caused by chemicals.

The HCS addresses the issues of evaluating and communicating hazards to workers. Evaluation of chemical hazards involves a number of technical concepts, and is a process that requires the professional judgment of experienced experts. That's why the HCS is designed so that employers who simply use chemicals, rather than produce or import them, are not required to evaluate the hazards of those chemicals. Hazard determination is the responsibility of the producers and importers of the materials. Producers and importers of chemicals are then required to provide the hazard information to employers that purchase their products.

Employers that don't produce or import chemicals need only focus on those parts of the rule that deal with establishing a workplace program and communicating information to their workers. This appendix is a general guide for such employers to help them determine what's required under the rule. It does not supplant or substitute for the regulatory provisions, but rather provides a simplified outline of the steps an average employer would follow to meet those requirements.

1. <u>Becoming Familiar With the Rule</u>

OSHA has provided a simple summary of the HCS in a pamphlet entitled "Chemical Hazard Communication," OSHA Publication Number 3084. Some employers prefer to begin to become familiar with the rule's requirements by reading this pamphlet. A copy may be obtained from your local OSHA Area Office, or by contacting the OSHA Publications Office at (202) 523-9667. The standard is long, and some parts of it are technical, but the basic concepts are simple. In fact, the requirements reflect what many employers have been doing for years. You may find that you are already largely in compliance with many of the provisions, and will simply have to modify your existing programs somewhat. It you are operating in an OSHA-approved State Plan State, you must comply with the State's requirements, which may be different than those of the Federal rule. Many of the State Plan States had hazard communication or "right-to-know" laws prior to promulgation of the Federal rule. Employers in State Plan States should contact their State OSHA offices for more information regarding applicable requirements.

The HCS requires information to be prepared and transmitted regarding all hazardous chemicals. The HCS covers both physical hazards (such as flammability), and health hazards (such as irritation, lung damage, and cancer). Most chemicals used in the workplace have some hazard potential, and this will be covered by the rule.

One difference between this rule and many others adopted by OSHA is that this one is performance-oriented. That means that you have the flexibility to adapt the rule to the needs of your workplace, rather than having to follow specific, rigid requirements. It also means that you have to exercise more judgment to implement an appropriate and effective program.

The standard's design is simple. Chemical manufacturers and importers must evaluate the hazards of the chemicals they produce or import. Using that information, they must then prepare labels for containers, and more detailed technical bulletins called material safety data sheets (MSDS).

Chemical manufacturers, importers, and distributors of hazardous chemicals are all required to provide the appropriate labels and material safety data sheets to the employers to which they ship the chemicals. The information is to be provided automatically. Every container of hazardous chemicals you receive must be labeled, tagged, or marked with the required information. Your supplier must also send you a properly completed material safety data sheet (MSDS) at the time of the first shipment of the chemical, and with the next shipment after the MSDS is updated with new and significant information about the hazards.

You can rely on the information received from your supplier. You have no independent duty to analyze the chemical or evaluate the hazards of it. Employers that "use" hazardous chemicals must have a program to ensure the information is provided to exposed employees. "Use" means to package, handle, react, or transfer. This is an intentionally broad scope, and includes any situation where a chemical is present in such a way that employees may be exposed under normal conditions of use or in a foreseeable emergency.

The requirements of the rule that deal specifically with the hazard communication program are found in the standard in paragraphs (e), written hazard communication program; (f), labels and other forms of warning; (g), material safety data sheets; and (h), employee information and training. The requirements of these paragraphs should be the focus on your attention. Concentrate on becoming familiar with them, using paragraphs (b), scope and application, and (c), definitions, as references when needed to help explain the provisions.

There are two types of work operations where the coverage of the rule is limited. These are laboratories and operations where chemicals are only handled in sealed containers (e.g., a warehouse). The limited provisions for these workplaces can be found in paragraph (b), scope and application. Basically, employers having these types of work operations need only keep labels on containers as they are received; maintain material safety data sheets that are received, and give employees access to them; and provide information and training for employees. Employers do not have to have written hazard communication programs and lists of chemicals for these types of operations.

The limited coverage of laboratories and sealed container operations addresses the obligation of an employer to the workers in the operations involved, and does not affect the employer's duties as a distributor of chemicals. For example, a distributor may have warehouse operations where employees would be protected under the limited sealed container provisions. In this situation, requirements for obtaining and maintaining MSDSs are limited to providing access to those received with containers while the substance is in the workplace, and requesting MSDSs when employees request access for those not received with the containers. However, as a distributor of hazardous chemicals, that employer will still have responsibilities for providing MSDSs to downstream customers at the time of the first shipment and when the MSDS is updated. Therefore, although they may not be required for the employees in the work operation, the distributor may, nevertheless, have to have MSDSs to satisfy other requirements of the rule.

2. Identify Responsible Staff

Hazard communication is going to be a continuing program in your facility. Compliance with the HCS is not a "one shot deal." In order to have a successful program, it will be necessary to assign responsibility for both the initial and ongoing activities that have to be undertaken to comply with the rule. In some cases, these activities may already be part of current job assignments. For example, site supervisors are frequently responsible for on-the-job training sessions. Early identification of the responsible employees, and involvement of them in the development of your plan of action, will result in a more effective program design. Evaluation of the effectiveness of your program will also be enhanced by involvement of affected employees.

For any safety and health program, success depends on commitment at every level of the organization. This is particularly true for hazard communication, where success requires a change in behavior. This will only occur if employers understand the program, and are committed to its success, and if employees are motivated by the people presenting the information to them.

3. Identify Hazardous Chemicals in the Workplace

The standard requires a list of hazardous chemicals in the workplace as part of the written hazard communication program. The list will eventually serve as an inventory of everything for which an MSDS must be maintained. At this point, however, preparing the list will help you complete the rest of the program

APPENDIX 8—HAZARD COMMUNICATION STANDARD 209

since it will give you some idea of the scope of the program required for compliance in your facility.

The best way to prepare a comprehensive list is to survey the workplace. Purchasing records may also help, and certainly employers should establish procedures to ensure that in the future purchasing procedures result in MSDSs being received before a material is used in the workplace. The broadest possible perspective should be taken when doing the survey. Sometimes people think of "chemicals" as being only liquids in containers. The HCS covers chemicals in all physical forms - - liquids, solids, gases, vapors, fumes, and mists - - whether they are "contained" or not. The hazardous nature of the chemical and the potential for exposure are the factors which determine whether a chemical is covered. If it's not hazardous, it's not covered. If there is no potential for exposure (e.g., the chemical is inextricably bound and cannot be released), the rule does not cover the chemical.

Look around. Identify chemicals in containers, including pipes, but also think about chemicals generated in the work operations. For example, welding fumes, dusts, and exhaust fumes are all sources of chemical exposures. Read labels provided by suppliers for hazard information. Make a list of all chemicals in the workplace that are potentially hazardous. For your own information and planning, you may also want to note on the list the location(s) of the products within the workplace, and an indication of the hazards as found on the label. This will help you as you prepare the rest of your program.

Paragraph (b), scope and application, includes exemptions for various chemicals or workplace situations. After compiling the complete list of chemicals, you should review paragraph (b) to determine if any of the items can be eliminated from the list because they are exempted materials. For example, food, drugs, and cosmetics brought into the workplace for employee consumption are exempt. So rubbing alcohol in the first aid kit would not be covered.

Once you have compiled as complete a list as possible of the potentially hazardous chemicals in the workplace, the next step is to determine if you have received material safety data sheets for all of them. Check your files against the inventory you have just compiled. If any are missing, contact your supplier and request one. It is a good idea to document these requests, either by copy of a letter or a note regarding telephone conversations. If you have MSDSs for chemicals that are not on your list, figure out why. Maybe you don't use the chemical anymore. Or maybe you missed it in your survey. Some suppliers do provide MSDSs for products that are not hazardous. These do not have to be maintained by you.

You should not allow employees to use any chemicals for which you have not received an MSDS. The MSDS provides information you need to ensure proper protective measures are implemented prior to exposure.

4. Preparing and Implementing a Hazard Communication Program

All workplaces where employees are exposed to hazardous chemicals must have a written plan which describes how the standard will be implemented in that facility. Preparation of a plan is not just a paper exercise - - all of the elements must be implemented in the workplace in order to be in compliance with the rule. See paragraph (e) of the standard for the specific requirements regarding written hazard communication programs. The only work operations which do not have to comply with the written plan requirements are laboratories and work operations where employees only handle chemicals in sealed containers. See paragraph (b), scope and application, for the specific requirements for these two types of workplaces.

The plan does not have to be lengthy or complicated. It is intended to be a blueprint for implementation of your program - - an assurance that all aspects of the requirements have been addressed. Many trade associations and other professional groups have provided sample programs and other assistance materials to affected employers. These have been very helpful to many employers since they tend to be tailored to the particular industry involved. You may wish to investigate whether your industry trade groups have developed such materials.

Although such general guidance may be helpful, you must remember that the written program has to reflect what you are doing in your workplace. Therefore, if you use a generic program it must be adapted to address the facility it covers. For example, the written plan must list the chemicals present at the site, indicate who is to be responsible for the various aspects of the program in your facility, and indicate where written materials will be made available to employees.

If OSHA inspects your workplace for compliance with the HCS, the OSHA compliance officer will ask to see your written plan at the outset of the inspection. In general, the following items will be considered in evaluating your program. The written program must describe how the requirements for labels and other forms of warning, material safety data sheets, and employee information and training, are going to be met in your facility. The following discussion provides the type of information compliance officers will be looking for to decide whether these elements of the hazard communication program have been properly addressed:

A. *Labels and Other Forms of Warning*

In-plant containers of hazardous chemicals must be labeled, tagged, or marked with the identity of the material and appropriate hazard warnings. Chemical manufacturers, importers, and distributors are required to ensure that every container of hazardous chemicals they ship is appropriately labeled with such information and with the name and address of the producer or other responsible party. Employers purchasing chemicals can rely on the labels provided by their suppliers. If the material is subsequently transferred by the employer from a labeled container to another container, the employer will have to label that container unless it is subject to the portable container exemption. See paragraph (f) for specific labeling requirements.

The primary information to be obtained from an OSHA-required label is an identity for the material, and appropriate hazard warnings. The identity is any term which appears on the label, the MSDS, and the list of chemicals, and thus links these three sources of information. The identity used by the supplier may be a common or trade name ("Black Magic Formula"), or a chemical name (1,1,1-trichloroethane). The hazard warning is a brief statement of the hazardous effects of the chemical ("flammable," "causes lung damage"). Labels frequently contain other information, such as precautionary measures ("do not use near open flame"), but this information is provided voluntarily and is not required by the rule. Labels must be legible, and prominently displayed. There are no specific requirements for size or color, or any specified text.

With these requirements in mind, the compliance officer will be looking for the following types of information to ensure that labeling will be properly implemented in your facility:
1. Designation of person(s) responsible for ensuring labeling of in-plant containers;
2. Designation of person(s) responsible for ensuring labeling of any shipped containers;
3. Description of labeling system(s) used;
4. Description of written alternatives to labeling of in-plant containers (if used); and,
5. Procedures to review and update label information when necessary.

Employers that are purchasing and using hazardous chemicals - - rather than producing or distributing them - - will primarily be concerned with ensuring that every purchased container is labeled. If materials are transferred into other containers, the employer must ensure that these are labeled as well, unless they fall under the portable container exemption (paragraph (f)(7)). In terms of labeling systems, you can simply choose to use the labels provided by your suppliers on the containers. These will generally be verbal text labels, and do not usually include numerical rating systems or symbols that require special training. The most important thing to remember is that this is a continuing duty - - all in-plant containers of hazardous chemicals must always be labeled. Therefore, it is important to designate someone to be responsible for ensuring that the labels are maintained as required on the containers in your facility, and that newly purchased materials are checked for labels prior to use.

B. *Material Safety Data Sheets*

Chemical manufacturers and importers are required to obtain or develop a material safety data sheet for each hazardous chemical they produce or import. Distributors are responsible for ensuring that their customers are provided a copy of these MSDSs. Employers must have an MSDS for each hazardous chemical which they use. Employers may rely on the information received from their suppliers. The specific requirements for material safety data sheets are in paragraph (g) of the standard.

APPENDIX 8—HAZARD COMMUNICATION STANDARD

There is no specified format for the MSDS under the rule, although there are specific information requirements. OSHA has developed a non-mandatory format, OSHA Form 174, which may be used by chemical manufacturers and importers to comply with the rule. The MSDS must be in English. You are entitled to receive from your supplier a data sheet which includes all of the information required under the rule. If you do not receive one automatically, you should request one. If you receive one that is obviously inadequate, with, for example, blank spaces that are not completed, you should request an appropriately completed one. If your request for a data sheet or for a corrected data sheet does not produce the information needed, you should contact your local OSHA Area Office for assistance in obtaining the MSDS.

The role of MSDSs under the rule is to provide detailed information on each hazardous chemical, including its potential hazardous effects, its physical and chemical characteristics, and recommendations for appropriate protective measures. This information should be useful to you as the employer responsible for designing protective programs, as well as to the workers. If you are not familiar with material safety data sheets and with chemical terminology, you may need to learn to use them yourself. A glossary of MSDS terms may be helpful in this regard. Generally speaking, most employers using hazardous chemicals will primarily be concerned with MSDS information regarding hazardous effects and recommended protective measures. Focus on the sections of the MSDS that are applicable to your situation.

MSDSs must be readily accessible to employees when they are in their work areas during their workshifts. This may be accomplished in many different ways. You must decide what is appropriate for your particular workplace. Some employers keep the MSDSs in a binder in a central location (e.g., in the pick-up truck on a construction site). Others, particularly in workplaces with large numbers of chemicals, computerize the information and provide access through terminals. As long as employees can get the information when they need it, any approach may be used. The employees must have access to the MSDSs themselves - - simply having a system where the information can be read to them over the phone is only permitted under the mobile worksite provision, paragraph (g)(9), when employees must travel between workplaces during the shift. In this situation, they have access to the MSDSs prior to leaving the primary worksite, and when they return, so the telephone system is simply an emergency arrangement.

In order to ensure that you have a current MSDS for each chemical in the plant as required, and that employee access is provided, the compliance officers will be looking for the following types of information in your written program:

1. Designation of person(s) responsible for obtaining and maintaining the MSDSs;
2. How such sheets are to be maintained in the workplace (e.g., in notebooks in the work area(s) or in a computer with terminal access), and how employees can obtain access to them when they are in their work area during the work shift;
3. Procedures to follow when the MSDS is not received at the time of the first shipment;
4. For producers, procedures to update the MSDS when new and significant health information is found; and,
5. Description of alternatives to actual data sheets in the workplace, if used.

For employers using hazardous chemicals, the most important aspect of the written program in terms of MSDSs is to ensure that someone is responsible for obtaining and maintaining the MSDSs for every hazardous chemical in the workplace. The list of hazardous chemicals required to be maintained as part of the written program will serve as an inventory. As new chemicals are purchased, the list should be updated. Many companies have found it convenient to include on their purchase orders the name and address of the person designated in their company to receive MSDSs.

C. *Employee Information and Training*

Each employee who may be "exposed" to hazardous chemicals when working must be provided information and trained prior to initial assignment to work with a hazardous chemical, and whenever the hazard changes. "Exposure" or "exposed" under the rule means that "an employee is subjected to a hazardous chemical in the course of employment through any route of entry (inhalation, ingestion, skin contact or absorption, etc.) and includes potential (e.g., accidental or possible) exposure." See paragraph (h) of the standard for specific requirements. Information and training may be done either by individual chemical, or by categories of hazards (such as flammability or carcinogenicity). If there are only a few chemicals in the workplace, then you may want to discuss each one individually. Where there are large numbers of chemicals, or the chemicals change frequently, you will probably want to train generally, based on the hazard categories (e.g., flammable liquids, corrosive materials, carcinogens). Employees will have access to the substance-specific information on the labels and MSDSs.

Information and training is a critical part of the hazard communication program. Information regarding hazards and protective measures is provided to workers through written labels and material safety data sheets. However, through effective information and training, workers will learn to read and understand such information, determine how it can be obtained and used in their own workplaces, and understand the risks of exposure to the chemicals in their workplaces as well as the ways to protect themselves. A properly conducted training program will ensure comprehension and understanding. It is not sufficient to either just read material to the workers, or simply hand them material to read. You want to create a climate where workers feel free to ask questions. This will help you to ensure that the information is understood. You must always remember that the underlying purpose of the HCS is to reduce the incidence of chemical source illnesses and injuries. This will be accomplished by modifying behavior through the provision of hazard information and information about protective measures. If your program works, you and your workers will better understand the chemical hazards within the workplace. The procedures you establish regarding, for example, purchasing, storage, and handling of these chemicals will improve, and thereby reduce the risks posed to employees exposed to the chemical hazards involved. Furthermore, your workers' comprehension will also be increased, and proper work practices will be followed in your workplace.

If you are going to do the training yourself, you will have to understand the material and be prepared to motivate the workers to learn. This is not always an easy task, but the benefits are worth the effort. More information regarding appropriate training can be found in OSHA Publication No. 2254 which contains voluntary training guidelines prepared by OSHA's Training Institute. A copy of this document is available from OSHA's Publications Office at (202) 523-9667.

In reviewing your written program with regard to information and training, the following items need to be considered:
1. Designation of person(s) responsible for con-ducting training;
2. Format of the program to be used (audiovisuals, classroom instruction, etc.);
3. Elements of the training program (should be consistent with the elements in paragraph (h) of the HCS); and
4. Procedure to train new employees at the time of their initial assignment to work with a hazardous chemical, and to train employees when a new hazard is introduced into the workplace.

The written program should provide enough details about the employer's plans in this area to assess whether or not a good faith effort is being made to train employees. OSHA does not expect that every worker will be able to recite all of the information about each chemical in the workplace. In general, the most important aspects of training under the HCS are to ensure that employees are aware that they are exposed to hazardous chemicals, that they know how to read and use labels and material safety data sheets, and that, as a consequence of learning this information, they are following the appropriate protective measures established by the employer. OSHA compliance officers will be talking to employees to determine if they have received training, if they know they are exposed to hazardous chemicals, and if they know where to obtain substance-specific information on labels and MSDSs.

The rule does not require employers to maintain records of employee training, but many employers choose to do so. This may help you monitor your own program to ensure that all employees are appropriately trained. If you already have a training program, you may simply have to supplement it with whatever additional information is required under the HCS. For example, construction employers that are already in compliance with the construction training standard (29 CFR 1926.21) will have little extra training to do.

An employer can provide employees information and training through whatever means found appropriate and protective. Although there would always have to be some training on-site (such as informing employees of the location and availability of the written program and MSDSs), employee training may be satisfied in part by general training about the requirements of the HCS and about chemical hazards on the job which is provided by, for example, trade associations, unions, colleges, and professional schools. In addition, previous training, education and experience of a worker may relieve the employer of some of the burdens of informing and training that worker. Regardless of the method relied upon, however, the employer is always ultimately responsible for ensuring that employees are adequately trained. If the compliance officer finds that the training is deficient, the employer will be cited for the deficiency regardless of who actually provided the training on behalf of the employer.

D. *Other Requirements*

In addition to these specific items, compliance officers will also be asking the following questions in assessing the adequacy of the program: Does a list of the hazardous chemicals exist in each work area or at a central location? Are methods the employer will use to inform employees of the hazards of non-routine tasks outlined? Are employees informed of the hazards associated with chemicals contained in unlabeled pipes in their work areas? On multi-employer worksites, has the employer provided other employers with information about labeling systems and precautionary measures where the other employers have employees exposed to the initial employer's chemicals? Is the written program made available to employees and their designated representatives?

If your program adequately addresses the means of communicating information to employees in your workplace, and provides answers to the basic questions outlined above, it will be found to be in compliance with the rule.

5. <u>Checklist for Compliance</u>

The following checklist will help to ensure you are in compliance with the rule:

___ Obtained a copy of the rule
___ Read and understood the requirements
___ Assigned responsibility for tasks
___ Prepared an inventory of chemicals
___ Ensured containers are labeled
___ Obtained MSDS for each chemical

___ Prepared written program
___ Made MSDSs available to workers
___ Conducted training of workers
___ Established procedures to maintain current program
___ Established procedures to evaluate effectiveness

6. Further Assistance

If you have a question regarding compliance with the HCS, you should contact your local OSHA Area Office for assistance. In addition, each OSHA Regional Office has a Hazard Communication Coordinator who can answer your questions. Free consultation services are also available to assist employers, and information regarding these services can be obtained through the Area and Regional offices as well. The telephone number for the OSHA office closest to you should be listed in your local telephone directory. If you are not able to obtain this information, you may contact OSHA's Office of Information and Consumer Affairs at (202) 523-8151 for further assistance in identifying the appropriate contacts.

Appendix 9

Sample Material Safety Data Sheet

Material Safety Data Sheet

May be used to comply with OSHA's Hazard Communication Standard. 29 CFR 1910.1200. Standard must be consulted for specific requirements.

U.S. Department of Labor
Occupational Safety and Health Administration
(Non-Mandatory Form)
Form Approved
OMB No. 1218-0072

IDENTITY (As Used on Label and List)	Note: Blank spaces are not permitted. If any item is not applicable, or no information is available, the space must be marked to indicate that.

Section I

Manufacturer's Name	Emergency Telephone Number
Address (Number, Street, City, State, and ZIP Code)	Telephone Number for Information
	Date Prepared
	Signature of Preparer (optional)

Section II — Hazardous Ingredients/Identity Information

Hazardous Components (Specific Chemical Identity; Common Name(s))	OSHA PEL	ACGIH TLV	Other Limits Recommended	% (optional)

Section III — Physical/Chemical Characteristics

Boiling Point		Specific Gravity (H_2O = 1)	
Vapor Pressure (mm Hg.)		Melting Point	
Vapor Density (AIR = 1)		Evaporation Rate (Butyl Acetate = 1)	
Solubility in Water			
Appearance and Odor			

Section IV — Fire and Explosion Hazard Data

Flash Point (Method Used)	Flammable Limits	LEL	UEL
Extinguishing Media			
Special Fire Fighting Procedures			
Unusual Fire and Explosion Hazards			

(Reproduce locally)

OSHA 174, Sept. 1985

APPENDIX 9—MATERIAL SAFETY DATA SHEET

Section V — Reactivity Data

Stability	Unstable		Conditions to Avoid
	Stable		

Incompatibility (Materials to Avoid)

Hazardous Decomposition or Byproducts

Hazardous Polymerization	May Occur		Conditions to Avoid
	Will Not Occur		

Section VI — Health Hazard Data

Route(s) of Entry: Inhalation?   Skin?   Ingestion?

Health Hazards (Acute and Chronic)

Carcinogenicity: NTP?   IARC Monographs?   OSHA Regulated?

Signs and Symptoms of Exposure

Medical Conditions Generally Aggravated by Exposure

Emergency and First Aid Procedures

Section VII — Precautions for Safe Handling and Use

Steps to Be Taken in Case Material Is Released or Spilled

Waste Disposal Method

Precautions to Be Taken in Handling and Storing

Other Precautions

Section VIII — Control Measures

Respiratory Protection (Specify Type)

Ventilation	Local Exhaust	Special
	Mechanical (General)	Other

Protective Gloves	Eye Protection

Other Protective Clothing or Equipment

Work/Hygienic Practices

Appendix 10

Toxic Chemical Release Inventory Reporting Form R*

* *This form was for use in meeting reporting requirements for calendar year January through December 1991. Because this Form R was not made available by EPA until June 1992, the usual July 1st deadline for submission of this Form R was extended to September 1, 1992. Significant new requirements were added with this version of the Form R, which are fully explained in the instructions that follow. Form R is subject to revision each year. The Form R for reporting activity during calendar year 1992 should be available from EPA in early 1993.*

Important Changes in the Section 313 Requirements for Reporting Year 1991

Reporting requirements for calendar year 1991 (reports due July 1, 1992) differ from previous years:

(1) The following chemicals have been specifically delisted and are not covered for the 1991 reporting year:

Chemical Name	CAS Number
Terephthalic acid	100-21-0
Melamine	108-78-1
*C.I. Pigment Blue 15	147-14-8
Sodium hydroxide (solution)	1310-73-2
*C.I. Pigment Green 7	1328-53-6
Aluminum oxide (non-fibrous forms)	1344-28-1
C.I. Acid Blue 9 diammonium salt	2650-18-2
C.I. Acid Blue 9 disodium salt	3844-45-9
Sodium sulfate (solution)	7757-82-6
Titanium dioxide	13463-67-7
*C.I. Pigment Green 36	14302-13-7

*These substances were delisted from the "Copper Compounds" category.

(2) The following chemicals have been added to the toxic chemical list and are covered for the 1991 reporting year:

Chemical Name	CAS Number
Bromotrifluoromethane (Halon 1301)	75-63-8
Trichlorofluoromethane (CFC-11)	75-69-4
Dichlorodifluoromethane (CFC-12)	75-71-8
Dichlorotetrafluoroethane (CFC-114)	76-14-2
Monochloropentafluoroethane (CFC-115)	76-15-3
Dibromotetrafluoroethane (Halon 2402)	124-73-2
Bromochlorodifluoromethane (Halon 1211)	353-59-3

(3) Reporting in Part II, Section 8, "Source Reduction and Recycling Activities," is now mandatory under the Pollution Prevention Act of 1990. All facilities required to file Form R are now required to report any source reduction and recycling activity engaged in during the reporting year. See the instructions for Part II, Section 8 for information about the new requirements.

(4) Toxic chemicals that are used for energy recovery purposes now have a separate reporting data element. If the reported toxic chemical is actually used for energy recovery and has a significant heat of combustion value, that activity will be reported as energy recovery. If the toxic chemical is incinerated with no recovery of energy, or if the heat of combustion value of the toxic chemical is too low to contribute significantly to energy recovery, the activity will be considered waste treatment.

(5) The de minimis exemption has been revised; beneficiation activities are no longer excluded from this exemption. Under any circumstances, toxic chemicals received in mixtures or trade name products under the de minimis value of one percent, or 0.1 percent if carcinogenic, are exempted from threshold determinations and release calculations.

Once a listed toxic chemical exceeds its de minimis level, however, all releases occurring after that point are subject to reporting.

(6) A TRI facility identification number has been assigned to each facility that previously submitted Form R reports. This identification number is designed to simplify locating facility reports. All facilities which submitted a Form R previously will receive a section 313 compliance package that includes a self-adhesive mailing label with the TRI facility identification number. If this package does not contain a mailing label or you have misplaced it, contact the Emergency Planning and Community Right-to-Know Information Hotline for help in determining your TRI facility identification number.

(7) The toll-free telephone number for the Emergency Planning and Community Right-to-Know Information Hotline, 1-800-535-0202, is now accessible throughout the U.S., including Washington, D.C., and Alaska. The toll telephone number has been changed to 703-920-9877.

Important Changes to Form R for 1991

The Form R for reporting year 1991 contains many changes. The changes were made to consolidate related data elements and clarify reporting requirements. The following changes have been made for the 1991 reporting year (reports due on or before July 1, 1992):

- The format of Form R has been changed to make the data readable by the Optical Character Recognition (OCR) Scanner.

- Part II, Section 8 of Form R, "Source Reduction and Recycling Activities," contains data elements mandated by the Pollution Prevention Act of 1990 (PPA).

- Form R now consists of two parts:
  - Part I. Facility Identification Information (pages 1-2); and
  - Part II. Chemical-Specific Information (pages 3-9).

  Part II of previous Form Rs, "Off-Site Locations to which Toxic Chemicals are Transferred in Wastes," has been incorporated into Part II, Section 6 of this year's form, "Transfers of the Toxic Chemical in Wastes to Off-Site Locations." This change allows location information and transfer amounts to be reported together. Part III of previous Form Rs, "Chemical-Specific Information," is now Part II. Part IV of previous From Rs, "Supplemental Information," has been eliminated.

- A space was added to page 1 for indicating if the form being submitted is a revision.

- Space has been made available to enter the toxic chemical name and TRI facility identification number on every page of Form R (minimum of 9 pages per Form R). These spaces are designed to help ensure correct reporting by facilities and correct data entry by EPA. They are not required data elements.

- On page 1, Part I, Section 4.1, space has been added for including the reporting facility's mailing address if it differs from the street address.

- The data elements for entering the names of receiving streams and water bodies have been incorporated into Part II, Section 5, "Releases of the Toxic Chemical to the Environment On-Site." As a result, the amount released will appear next to the name of the receiving stream or water body.

- In Part II, Section 5, "Releases of the Toxic Chemical to the Environment On-Site," and Part II, Section 6, "Transfers of the Toxic Chemical in Wastes to Off-Site Locations," the range reporting columns have been removed. Space has been added to enter either an estimate or a code representing one of the three reporting ranges.

- Information on off-site transfers for recycling and energy recovery is included in Part II, Section 6, "Transfers of the Toxic Chemical in Wastes to Off-Site Locations." Section 6 has been modified to allow for more than one operation code (i.e., waste treatment, disposal, recycling, or energy recovery) and more than one amount to be entered per location.

- Section 6.1, "Discharges to Publicly Owned Treatment Works" now contains two parts: 6.1.A, "Total Quantity Transferred to POTWs and Basis of Estimate," and 6.1.B, "POTW Name and Location Information." If you transfer a toxic chemical in wastes to more than one POTW, enter the total transfers to all POTWs in section 6.1.A.1, and in section 6.1.A.2 enter the basis of estimate for the total amount transferred. In section 6.1.B, list the name and location of all POTWs that received the toxic chemical in wastes.

- If additional space is needed for completing Sections 5.3, "Discharges to Receiving Stream or Water Body" and Section 6, "Transfers of the Toxic Chemical in Wastes to Off-Site Locations," pages 5 and 6 should be photocopied, and the extra pages submitted.

- Part II, Section 7A, "On-Site Waste Treatment Methods and Efficiency," has been expanded and now is the only data element on page 7 of Form R. If additional space is needed for Section 7A (On-Site Waste Treatment Methods and Efficiency), this page may be photocopied, and the extra pages submitted. Page 8 contains two new required data elements: Section 7B, "On-Site Energy Recovery Processes," and Section 7C, "On-Site Recycling Processes."

- Page 9 consists of the required PPA data elements. Section 8, "Source Reduction and Recycling Activities," is now a required section of Form R. See Part II, Section 8 for the data elements.

A. General Information

Submission of EPA Form R, the Toxic Chemical Release Inventory (TRI) Reporting Form, is required by section 313 of the Emergency Planning and Community Right-to-Know Act (EPCRA, or Title III of the Superfund Amendments and Reauthorization Act of 1986), Public Law 99-499. The information contained in Form R constitutes a "report," and the submission of a report to the appropriate authorities constitutes "reporting."

Reporting is required to provide the public with information on the releases of listed toxic chemicals in their communities and to provide EPA with release information to assist the Agency in determining the need for future regulations. Facilities must report the quantities of both routine and accidental releases of listed toxic chemicals, as well as the maximum amount of the listed toxic chemical on-site during the calendar year and the amount contained in wastes transferred off-site.

The Pollution Prevention Act, passed into law in October, 1990 (Pub. L. 101-508), added reporting requirements to Form R. These new requirements will affect all facilities required to submit Form R under section 313 of EPCRA. The new data, which is described in the preceding section, "Important Changes to Form R for 1991," will be required beginning with reports for calendar year 1991 (first reports due to EPA and States by July 1, 1992).

A completed Form R must be submitted for each toxic chemical manufactured, processed, or otherwise used at each covered facility as described in the reporting rule in 40 CFR Part 372 (originally published February 16, 1988, in the Federal Register). These instructions supplement and elaborate on the requirements in the reporting rule. Together with the reporting rule, they constitute the reporting requirements. All references in these instructions are to sections in the reporting rule unless otherwise indicated.

A.1 How to Assemble a Complete Report

The Toxic Chemical Release Reporting Form, EPA Form R, consists of two parts:

- Part I, Facility Identification Information (pages 1 and 2); and
- Part II, Chemical-Specific Information (pages 3-9).

Most of the information required in Part I of Form R can be completed, photocopied, and attached to each chemical-specific report. However, Part I of each Form R submitted must have an original signature on the certification statement and the trade secret designation must be entered as appropriate. Part II must be completed separately for each toxic chemical or chemical category. Because a complete Form R consists of at least 9 unique pages, any submissions containing less than 9 unique pages is not a valid submission.

A complete report for any listed toxic chemical that is not claimed as a trade secret consists of the following completed parts:

- Part I with an original signature on the certification statement (Section 2); and
- Part II (Section 8 is now mandatory).

Staple all 9 pages of each report together. If you check yes on Part II, Section 8.12, you may attach additional information on pollution prevention activities at your facility.

A.2 Trade Secret Claims

For any toxic chemical whose identity is claimed as a trade secret, you must submit to EPA two versions of the substantiation form as prescribed in 40 CFR Part 350, published July 29, 1988, in the Federal Register (53 FR 28772) as well as two versions of Form R. One set of forms, the "unsanitized" version, should provide the actual identity of the toxic chemical. The other set of forms, the "sanitized" version, should provide only a generic identity of the toxic chemical. If EPA deems the trade secret substantiation form valid, only the sanitized set of forms will be made available to the public.

Use the order form in this document to obtain copies of the rule and substantiation form. Further explanation of the trade secret provisions is provided in Part I, Sections 2.1 and 2.2, and Part II, Section 1.3, of the instructions.

In summary, a complete report to EPA for a toxic chemical claimed as a trade secret must include all of the following:

- A completed "unsanitized" version of a Form R report including the toxic chemical identity (staple the pages together);

- A "sanitized" version of a completed Form R report in which the toxic chemical identity items (Part II, Sections 1.1 and 1.2) have been left blank but in which a generic chemical name has been supplied (Part II, Section 1.3) (staple the pages together);

- A completed "unsanitized" version of a trade secret substantiation form (staple the pages together); and

- A "sanitized" version of a completed trade secret substantiation form (staple the pages together).

Securely fasten all four reports together.

Some states also require submission of both sanitized and unsanitized reports for toxic chemicals whose identity is claimed as a trade secret. Others require only a sanitized version. Facilities may jeopardize the trade secret status of a toxic chemical by submitting an unsanitized version of Form R to a state agency or Indian tribe that does not require unsanitized forms. You may identify an individual State's submission requirements by contacting the appropriate state-designated Section 313 contact (see Appendix F).

A.3 Recordkeeping

Sound recordkeeping practices are essential for accurate and efficient TRI reporting. It is in the facility's interest, as well as EPA's, to maintain records properly.

Facilities must keep a copy of each Form R report filed for at least three years from the date of submission. These reports will be of use in subsequent years when completing future Form R reports.

Facilities must also maintain those documents, calculations, worksheets, and other forms upon which they relied to gather information for prior Form R reports. In the event of a problem with data elements on a facility's Form R, EPA may request documentation from the facility that supports the information reported. In the future, EPA may conduct data quality reviews of past Form R submissions. An essential component of this process would be to review a facility's records for accuracy and reliability.

A partial list of records, organized by year, that a facility should maintain include:

- Previous years' Form Rs;
- Section 313 Reporting Threshold Worksheets;
- Engineering calculations and other notes;
- Purchase records from suppliers;
- Inventory data;
- EPA (NPDES) permits;
- EPCRA Section 312, Tier II Reports;
- Monitoring records;
- Flowmeter data;
- RCRA Hazardous Waste Generator's Report;
- Pretreatment reports filed by the facility with the local government;
- Invoices from waste management companies;
- Manufacturer's estimates of treatment efficiencies;
- RCRA Manifests; and
- Process diagrams that indicate emissions and releases.

A.4 When the Report Must be Submitted

The report for any calendar year must be submitted on or before July 1 of the following year (e.g., the report for calendar year 1991, January-December, must be submitted on or before July 1, 1992).

Voluntary Revision of a Previous Submission

Voluntary revisions must be submitted on a Form R identical to the version originally submitted to EPA for that reporting year. The Emergency Planning and Community Right-to-Know Information Hotline can help you identify the version of Form R used for each reporting year.

For the 1991 reporting year only, enter "X" in the space marked "Enter 'X' here if this is a revision" on page 1 of the form if you are making a voluntary revision to a previous Form R submission. If you have obtained the Document Control Number (DCN) of the original submission from EPA, enter that number in red ink in any available space on page 1 of the form. Enter the revised data to the Form R and circle all changes from the original submission in red ink. Sign the certification statement and provide a current date.

For reporting years prior to 1991, there are two options for making voluntary revisions. The first is to submit a photocopy of the original Form R submission (from your file), with corrections made in red ink. Write the words "VOLUNTARY REVISION", and the Document Control Number (DCN), if available, on page 1 of the Form R, and re-sign and re-date the certification statement on page 1.

The second is to obtain a blank Form R for the reporting year affected by the correction(s). Complete all data elements on this Form, but circle with red ink those data elements that you have changed. A cover letter should be included to clarify exactly which voluntary revisions you have made.

Send the entire completed or revised Form R report to EPA and the appropriate state agency (or the designated official of an Indian tribe). Submissions for the next calendar year are not considered revisions of a previous year's data.

A.5 Where to Send the Form R

Form R submissions must be sent to both EPA and the State (or the designated official of an Indian tribe). If a Form R is not received by both EPA and the State (or the designated official of an Indian tribe), the submitter is considered out of compliance and subject to enforcement action.

Send reports to EPA by mail to:

> EPCRA Reporting Center
> P.O. Box 23779
> Washington, D.C. 20026-3779
> Attn: Toxic Chemical Release Inventory

To submit a Form R via hand delivery or certified mail, please call the Emergency Planning and Community Right-to-Know Information Hotline to obtain the street address of the EPCRA Reporting Center.

In addition, you must also send a copy of the report to the State in which the facility is located. ("State" also includes: the District of Columbia, the Commonwealth of Puerto Rico, Guam, American Samoa, the U.S. Virgin Islands, the Northern Mariana Islands, and any other territory or possession over which the U.S. has jurisdiction.) Refer to Appendix F for the appropriate State submission addresses.

Facilities located on Indian land should send a copy to the Chief Executive Officer of the applicable Indian tribe. Some tribes have entered into a cooperative agreement with States; in this case, Form R submissions should be sent to the entity designated in the cooperative agreement.

Submission of section 313 reports in magnetic media and computer-generated facsimile formats has been approved by EPA. EPA has developed a package called the "Toxic Chemical Release Inventory Reporting System." The easy-to-use diskette comes with complete instructions for its use. It also provides prompts and messages to help you report according to EPA instructions. For copies of the diskette you may call the EPCRA Hotline.

Many firms are offering computer software to assist facilities in producing magnetic media submissions or computer-generated facsimiles of Form R reports. To ensure accuracy, EPA will only accept magnetic media submissions and computer generated facsimiles that meet basic specifications established by EPA. To determine if software offered by a firm meets these specifications, EPA reviews and approves all software upon request. Call the Emergency Planning and Community Right-to-Know Information Hotline to identify the software that has been approved by EPA for the current reporting year.

It should be noted, however, that some States may accept only hard copies of Form R. If this is the case, a magnetic media or computer-generated facsimile may be unacceptable.

A.6 How to Obtain Forms and Other Information

A copy of Form R is included in this booklet. Remove this form and produce as many photocopies as needed. Related guidance documents may be obtained from:

> Section 313 Document Distribution Center
> P.O. Box 12505
> Cincinnati, OH 45212

See Appendix I for the document request form and more information on available documents.

Questions about completing Form R may be directed to the Emergency Planning and Community Right-to-Know Information Hotline at the following address or telephone numbers.

Emergency Planning and Community
Right-to-Know Information Hotline
U.S. Environmental Protection Agency
401 M Street, S.W. (OS-120)
Washington, DC 20460

(800) 535-0202 or (703) 920-9877
from 8:30 am - 7:30 pm Eastern Time
(Mon-Fri, except Federal Holidays.)

EPA Regional Staff may also be of assistance. Refer to Appendix G for a list of EPA Regional Offices.

A.7 Who Must Submit this Form

Section 313 of EPCRA requires that reports be filed by owners and operators of facilities that meet all three of the following criteria:

- The facility has 10 or more full-time employees; and

- The facility is included in Standard Industrial Classification (SIC) Codes 20 through 39; and

- The facility manufactures (defined to include importing), processes, or otherwise uses any listed toxic chemical in quantities equal to or greater than the established threshold in the course of a calendar year.

B. How to Determine if Your Facility Must Submit EPA Form R

(See Figure 1 for more information.)

B.1 Full-Time Employee Determination

A "full-time employee," for purposes of section 313 reporting, is defined as 2,000 work hours per year. This definition is dependent only upon the number of hours worked by all employees for the facility during the calendar year and not the number of persons working. To determine the number of full-time employees working for your facility, add up the hours worked by all employees during the calendar year, including contract employees and sales and support staff working for the facility, and divide the total by 2,000 hours. In other words, if the total number of hours worked by all employees is 20,000 hours or more, your facility meets the ten employee threshold.

Examples include:

- A facility consists of 11 employees who each worked 1500 hours for the facility in a calendar year. Consequently, the total number of hours worked by all employees for the facility during the calendar year is 16,500 hours. The number of full-time employees for this facility is equal to 16,500 hours divided by 2,000 hours per full-time employee, or 8.3 full-time employees. Therefore, even though 11 persons worked for this facility during the calendar year, the number of hours worked is equivalent to 8.3 full-time employees. This facility does not meet the employee criteria and is not subject to section 313 reporting.

- Another facility consists of 6 workers and 3 sales staff. The 6 workers each worked 2,000 hours for the facility in the calendar year. The sales staff also each worked 2,000 hours in the calendar year although they may have been on the road half of the year. In addition, 5 contract employees were hired for a period during which each worked 400 hours for the facility. The total number of hours is equal to the time worked by the workers at the facility (12,000 hours), plus the time worked by the sales staff for the facility (6,000 hours), plus the time worked by the contract employees at the facility (2,000 hours), or 20,000 hours. Dividing the 20,000 hours by 2,000 yields 10 full-time employees. This facility has met the full-time employee criteria and may be subject to reporting if the other criteria are met.

B.2 Primary SIC Code Determination

Standard Industrial Classification (SIC) codes 20-39 are covered by the rule and are listed in Table I. The first two digits of a 4-digit SIC code define a major business sector, while the last two digits denote a facility's specialty within the major sector. If you are not familiar with the SIC codes that apply to your facility, contact your trade association, Chamber of Commerce, or legal counsel. For a detailed description of 4-digit SIC codes, refer to the "Standard Industrial Classification Manual 1987." Clothbound editions are available in most major libraries or may be ordered through the National Technical Information Service, 5285 Port Royal Road, Springfield, VA, 22161, (703) 487-4650. The access number for the clothbound manual is PB87-100012, and the price is $30.00.

Section 313 requires that reports be filed by "facilities," which are defined as "all buildings, equipment, structures, and other stationary items which are located on a single site or on contiguous or adjacent sites and which are owned or operated by the same person." The SIC code system, however, classifies businesses not as "facilities," but as "establishments," which are defined as "distinct and separate economic activities [that] are performed at a single physical location."

Guidelines for using these definitions to determine primary SIC codes for facilities are presented in the following subsections.

B.2.a Multi-Establishment Facilities

Your facility may include multiple establishments that have different SIC codes. If so, calculate the value of the products produced or shipped from each establishment within the facility and then use the following rule to determine if your facility meets the SIC code criterion:

- If the total value of the products shipped from or produced at establishments with primary SIC codes between 20 and 39 is greater than 50 percent of the value of the entire facility's products and services, the entire facility meets the SIC code criterion.

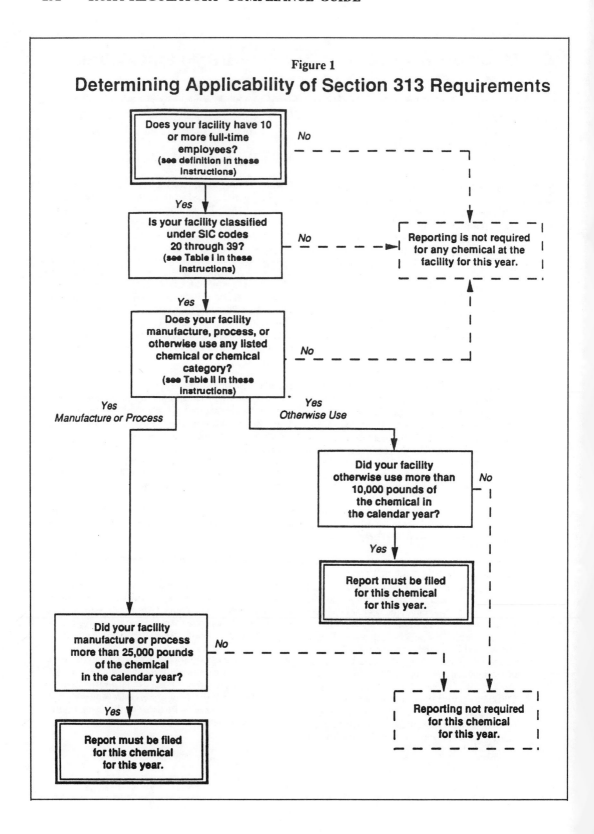

- If any one establishment with a primary SIC code between 20 and 39 produces or ships products whose value exceeds the value of products and services produced or shipped by any other establishment within the facility, the facility also meets the SIC code criterion.

The value of production attributable to a particular establishment may be isolated by subtracting the value of products obtained from other establishments within the same facility that are incorporated into its final products. This procedure eliminates the potential for "double counting" production in situations where establishments are engaged in sequential production activities at a single facility.

Examples include:

- One establishment in a gold mining facility is engaged primarily in the exploration of gold deposits, developing mines, and mining gold. This establishment deploys several means to mine the gold, including crushing, grinding, gravity concentration, froth flotation, amalgamation, cyanidation, and the production of bullion at the mine and mill sites (these processes are classified under SIC code 1041). All of the ore discovered through this establishment is delivered to a second establishment which is primarily engaged in rolling, drawing, and extruding the gold for sale and distribution. The smeltering establishment in the facility is classified under SIC code 3339. The facility could calculate the value of production for each establishment separately (both SIC code 1041 and 3339 having separate values). Alternatively, the facility could determine the value of the smelter operation by subtracting the value of the ore produced from the value of entire facility's production (Gross value of facility - SIC code 1041 value = Value for SIC code 3999).

- A food processing establishment in a facility processes crops grown at the facility in a separate establishment. The facility could base the value of the products of each establishment on the total production value of each establishment. Alternatively, the facility could first determine the value of the crops grown at the agricultural establishment, and then calculate the contribution of the food processing establishment by subtracting the crop value from the total value of the product shipped from the processing establishment. (Value of product shipped from processing - crop value = value of processing establishment)

A covered multi-establishment facility must make toxic chemical threshold determinations and, if required, must report all relevant information about releases, source reduction, recycling, and waste treatment associated with a listed toxic chemical for the entire facility, even from establishments that are not in SIC codes 20-39. EPA realizes, however, that certain establishments in a multi-establishment facility can be, for all practical purposes, separate business units. Therefore, individual establishments may report releases separately, provided that the total releases for the whole facility is represented by the sum of releases reported by the separate establishments.

B.2.b Auxiliary Facilities

An auxiliary facility is one that supports another facility's activities (e.g., research and development laboratories, warehouses, storage facilities, and waste-treatment facilities). An auxiliary facility can assume the SIC code of another covered facility if its primary function is to service that other covered facility's operations. Thus, a separate warehouse facility (i.e., one not located within the physical boundaries of a covered facility) may become a covered facility because it services a facility in SIC codes 20-39. Auxiliary facilities that are in SIC codes 20-39 are required to report if they meet the employee criterion and reporting thresholds for manufacture, process, or otherwise use. Auxiliary establishments that are part of a multi-establishment facility must be factored into threshold determinations for the facility as a whole.

B.2.c Facility-Related Exemptions

<u>Laboratories:</u> Listed toxic chemicals that are manufactured, processed, or otherwise used in laboratory activities at a covered facility under the direct supervision of a technically qualified individual do not have to be considered for threshold and release calculations. However, pilot plant scale and specialty chemical production do not qualify for this laboratory activities exemption.

<u>Property Owners:</u> You are not required to report if you merely own real estate on which a facility covered by this rule is located; that is, you have no other business interest in the operation of that facility (e.g., your company owns an industrial park). The operator of that facility, however, is subject to reporting requirements.

B.3 Activity Determination

B.3.a Definitions of "Manufacture," "Process," and "Otherwise Use"

Manufacture: The term "manufacture" means to produce, prepare, compound, or import a listed toxic chemical. (See Part II, Section 3.1 of these instructions for further clarification.)

Import is defined as causing the toxic chemical to be imported into the customs territory of the United States. If you order a listed toxic chemical (or a mixture containing the chemical) from a foreign supplier, then you have imported the chemical when that shipment arrives at your facility directly from a source outside of the United States. By ordering the chemical, you have "caused it to be imported," even though you may have used an import brokerage firm as an agent to obtain the toxic chemical.

The term manufacture also includes coincidental production of a toxic chemical (e.g., as a byproduct or impurity) as a result of the manufacture, processing, otherwise use, or treatment of other chemical substances. In the case of coincidental production of an impurity (i.e., a toxic chemical that remains in the product that is distributed in commerce), the de minimis limitation, discussed in Section B.4.b of these instructions, applies. The de minimis limitation does not apply to byproducts (e.g., a toxic chemical that is separated from a process stream and further processed or disposed). Certain listed toxic chemicals may be manufactured as a result of wastewater treatment or other treatment processes. For example, neutralization of acid wastewater can result in the coincidental manufacture of ammonium nitrate (solution).

Example 1: Coincidental Manufacture

Your company, a nitric acid manufacturer, uses ammonia in a waste treatment system to neutralize an acidic wastewater stream containing nitric acid. The reaction of the ammonia and nitric acid produces an ammonium nitrate solution. Ammonium nitrate solution is a listed toxic chemical, as are nitric acid and ammonia. Your facility thus otherwise uses ammonia as a reactant and manufactures ammonium nitrate solution as a byproduct. If the ammonium nitrate solution is produced in a quantity that exceeds the threshold (e.g., 25,000 pounds for the reporting year), the facility must report for the ammonium nitrate solution. If more than 10,000 pounds of ammonia is added to the wastewater treatment system, then the facility must report for ammonia.

Process: The term "process" means the preparation of a listed toxic chemical, after its manufacture, for distribution in commerce. Processing is usually the intentional incorporation of a toxic chemical into a product (see Part II, Section 3.2 of these instructions for further clarification). Processing includes preparation of the toxic chemical in the same physical state or chemical form as that received by your facility, or preparation that produces a change in physical state or chemical form. The term also applies to the processing of a mixture or other trade name product (see Section B.4.b of these instructions) that contains a listed toxic chemical as one component.

Otherwise Use: The term "otherwise use" encompasses any activity involving a listed toxic chemical at a facility that does not fall under the definitions of "manufacture" or "process." A chemical that is otherwise used by a

---

Example 2: Typical Process and Manufacture Activities

- Your company receives toluene, a listed toxic chemical, from another facility, and reacts the toluene with air to form benzoic acid. Your company processes toluene and manufactures benzoic acid. Benzoic acid, however, is not a listed toxic chemical and thus does not trigger reporting requirements.

- Your facility combines toluene purchased from a supplier with various materials to form paint. Your facility processes toluene.

- Your company receives a nickel compound (nickel compound is a listed toxic chemical category) as a bulk solid and performs various size-reduction operations (e.g., grinding) before packaging the compound in 50 pound bags. Your company processes the nickel compound.

- Your company receives a prepared mixture of resin and chopped fiber to be used in the injection molding of plastic products. The resin contains a listed toxic chemical that becomes incorporated into the plastic. Your facility processes the toxic chemical.

> **Example 3: Otherwise Use**
>
> When your facility cleans equipment with toluene, you are otherwise using toluene. Your facility also separates two components of a mixture by dissolving one component in toluene, and subsequently recovers the toluene from the process for reuse or disposal. Your facility otherwise uses toluene.

facility is not intentionally incorporated into a product distributed in commerce (see Part II, Section 3.3 of these Instructions for further clarification).

B.3.b Activity Exemptions

Use Exemptions. Certain uses of listed toxic chemicals are specifically exempted:

- use as a structural component of the facility;
- use in routine janitorial or facility grounds maintenance;
- personal uses by employees or other persons;
- use of products containing toxic chemicals for the purpose of maintaining motor vehicles operated by the facility; or
- use of toxic chemicals contained in intake water (used for processing or non-contact cooling) or in intake air (used either as compressed air or for combustion).

Article Exemptions. Quantities of a listed toxic chemical contained in an article do not have to be factored into threshold or release determinations when that article is processed or otherwise used at your facility. An article is defined as a manufactured item that is formed to a specific shape or design during manufacture, that has end-use functions dependent in whole or in part upon its shape or design during end-use, and that does not release a toxic chemical under normal conditions of the processing or otherwise use of that item at the facility.

If the processing or otherwise use of similar articles results in a total release of less than 0.5 pounds of a toxic chemical in a calendar year to any environmental media, EPA will allow this release quantity to be rounded to zero, and the manufactured items remain exempt as articles. EPA requires facilities to round off and report all estimates to the nearest whole number. The 0.5-pound limit does not apply to each individual article, but applies to the sum of all releases from processing or otherwise use of like articles.

The article exemption applies to the normal processing or otherwise use of an article. **It does not apply to the manufacture of an article.** Toxic chemicals processed into articles produced at a facility must be factored into threshold and release determinations.

A closed item containing toxic chemicals (e.g., a transformer containing PCBs) that does not release the toxic chemicals during normal use is considered an article if a facility uses the item as intended and the toxic chemicals are not released. If a facility services the closed item (e.g., a transformer) by replacing the toxic chemicals, the toxic chemicals added during the reporting year must be counted in threshold and release calculations.

> **Example 4: Article Exemption**
>
> - Lead that is incorporated into a lead acid battery is processed to manufacture the battery, and therefore must be counted toward threshold and release determinations. However, the use of the lead acid battery elsewhere in the facility does not have to be counted. Disposal of the battery after its use does not constitute a "release"; thus, the battery remains an article.
>
> - Metal rods that are extruded into wire are not articles because their form changes during processing.
>
> - If an item used in the facility is fragmented, the item is still an article if those fragments being discarded remain identifiable as the article (e.g., recognizable pieces of a cylinder, pieces of wire). For instance, an 8-foot piece of wire is broken into two 4-foot pieces of wire, without releasing any toxic chemicals. Each 4-foot piece is identifiable as a piece of wire; therefore, the article status for these pieces of wire remains intact.
>
> - Toxic chemicals received in the form of pellets are not articles because the pellet form is simply a convenient form for further processing of the material.

When the processing or otherwise use of an item generates fumes, dust, filings, or grindings, the article exemption is not applicable. The toxic chemical(s) in the item must be counted toward the appropriate threshold determination, and the fumes, dust, filings, and grindings must be reported as releases or wastes. Scrap pieces that are recognizable as an article do not constitute a release.

B.3.c Activity Qualifiers

Table II contains the list of individual toxic chemicals and categories of chemicals subject to 1991 calendar year reporting. Some of the toxic chemicals listed in Table II have parenthetic qualifiers listed next to them. A toxic chemical that is listed without a qualifier is subject to reporting in all forms in which it is manufactured, processed, and otherwise used.

Fume or dust. Three of the metals on the list (aluminum, vanadium, and zinc) contain the qualifier "fume or dust." Fume or dust refers to dry forms of these metals but does not refer to "wet" forms such as solutions or slurries. As explained in Section B.3a of these instructions, the term manufacture includes the generation of a toxic chemical as a byproduct or impurity. In such cases, a facility should determine if, for example, it generated more than 25,000 pounds of aluminum fume or dust in 1991 as a result of its activities. If so, the facility must report that it manufactures "aluminum (fume or dust)." Similarly, there may be certain technologies in which one of these metals is processed in the form of a fume or dust to make other toxic chemicals or other products for distribution in commerce. In reporting releases, the facility would only report releases of the fume or dust.

EPA considers dusts to consist of solid particles generated by any mechanical processing of materials including crushing, grinding, rapid impact, handling, detonation, and decrepitation of organic and inorganic materials such as rock, ore, and metal. Dusts do not tend to flocculate, except under electrostatic forces. A fume is an airborne dispersion consisting of small solid particles created by condensation from a gaseous state, in distinction to a gas or vapor. Fumes arise from the heating of solids such as lead. The condensation is often accompanied by a chemical reaction, such as oxidation. Fumes flocculate and sometimes coalesce.

Manufacturing qualifiers. Two of the entries to the section 313 toxic chemical list contain a qualifier relating to manufacture. For isopropyl alcohol, the qualifier is "manufacturing -- strong acid process." For saccharin, the qualifier simply is "manufacturing." For isopropyl alcohol, the qualifier means that only facilities manufacturing isopropyl alcohol by the strong acid process are required to report. In the case of saccharin, only manufacturers of the toxic chemical are subject to the reporting requirements. A facility that processes or otherwise uses either toxic chemical would not be required to report for those toxic chemicals. In both cases, supplier notification does not apply because only manufacturers, not users, of the toxic chemical must report.

Solutions. Two substances on the list, ammonium nitrate and ammonium sulfate, are qualified by the term "solution," which refers to the physical state of these toxic chemicals. Solid, molten, and pelletized forms of these toxic chemicals are exempt from threshold and release determinations. Only facilities that manufacture, process, or otherwise use these toxic chemicals in the form of a solution are required to report. Supplier notification applies only if the toxic chemical is distributed as a solution.

Phosphorus (yellow or white). The listing for phosphorus is qualified by the term "yellow or white." This means that only manufacturing, processing, or otherwise use of phosphorus in the yellow or white chemical form triggers reporting. Conversely, manufacturing, processing, or otherwise use of "black" or "red" phosphorus does not trigger reporting. Supplier notification also applies only to distribution of yellow or white phosphorus.

Asbestos (friable). The listing for asbestos is qualified by the term "friable," referring to the physical characteristic of being able to be crumbled, pulverized, or reducible to a powder with hand pressure. Only manufacturing, processing, or otherwise use of asbestos in the friable form triggers reporting. Supplier notification applies only to distribution of mixtures or trade name products containing friable asbestos.

Aluminum Oxide (fibrous forms). The listing for aluminum oxide is qualified by the term "fibrous forms." Fibrous refers to a man-made form of aluminum oxide that is processed to produce strands or filaments which can be cut to various lengths depending on the application. Only manufacturing, processing, or otherwise use of aluminum oxide in the fibrous form triggers reporting. Supplier notification applies only to distribution of mixtures or trade name products containing fibrous forms of aluminum oxide.

B.4 Threshold Determination

Section 313 reporting is required if threshold quantities are exceeded. Separate thresholds apply to the amount of the toxic chemical that is manufactured, processed, or otherwise used.

You must submit a report for any listed toxic chemical that is manufactured or processed at your facility in excess of the following threshold:

- 25,000 pounds during the course of a calendar year.

You must submit a report if the quantity of a listed toxic chemical that is otherwise used at your facility exceeds:

- 10,000 pounds during the course of a calendar year.

B.4.a How to Determine If Your Facility Has Exceeded Thresholds

To determine whether your facility has exceeded a section 313 reporting threshold, compare quantities of listed toxic chemicals that you manufacture, process, or otherwise use to the respective thresholds for those activities. A worksheet is provided in Figure 2 to assist facilities in determining whether they exceed any of the reporting thresholds. This worksheet also provides a format for maintaining reporting facility records. Use of this worksheet is not required and the completed worksheet(s) should not accompany Form R reports submitted to EPA and the State.

Complete a separate worksheet for each section 313 toxic chemical or chemical category. Base your threshold determination for listed toxic chemicals with qualifiers only on the quantity of the toxic chemical satisfying the qualifier.

Use of the worksheet is divided into three steps:

Step 1 allows you to record the gross amount of the toxic chemical or chemical category involved in activities throughout the facility. Pure forms as well as the amounts of the toxic chemical or chemical category present in mixtures or trade name products must be considered. The types of activity (i.e., manufacturing, processing, or otherwise using) for which the toxic chemical is used must be identified because separate thresholds apply to each of these activities. A record of the information source(s) used should be kept. Possible information sources include purchase records, inventory data, and calculations by a process engineer. The data collected in Step 1 will be totalled for each activity to identify the overall amount of the toxic chemical or chemical category manufactured (including imported), processed, or otherwise used.

Step 2 allows you to identify uses of the toxic chemical or chemical category that were included in Step 1 but are exempt under section 313. Do not include in Step 2 exempt forms of the toxic chemical not included in the calculations in Step 1. For example, if freon contained in the building's air conditioners was not reported in Step 1, you would not include the amount as exempt in Step 2. Step 2 is intended for use when one form or use of the toxic chemical is exempt while other forms require reporting. Note the type of exemption for future reference. Also identify, if applicable, the fraction or percentage of the toxic chemical present that is exempt. Add the amounts in each activity to obtain a subtotal for exempted amounts of the toxic chemical or chemical categories at the facility.

Step 3 involves subtracting the result of Step 2 from the results of Step 1 for each activity. Compare this net sum to the applicable activity threshold. If the threshold is met or exceeded for any of the three activities, a facility must submit a Form R for that toxic chemical or chemical category. This worksheet should be retained in either case to document your determination for reporting or not reporting, but should not be submitted with the report. Do not sum quantities of the toxic chemical that are manufactured, processed, and otherwise used at your facility, because each of these activities requires a separate threshold determination. For example, if in a calendar year you processed 20,000 pounds of a chemical and you otherwise used 6,000 pounds of that same toxic chemical, your facility has not met or exceeded any applicable threshold and thus is not required to report for that chemical.

You must submit a report if you exceed any threshold for any listed toxic chemical or chemical category. For example, if your facility processes 22,000 pounds of a listed toxic chemical and also otherwise uses 16,000 pounds of that same toxic chemical, it has exceeded the otherwise used threshold (10,000 pounds) and your facility must report even though it did not exceed the process threshold. However, in preparing your reports, you must consider all non-exempted activities and all releases of the toxic chemical from your facility, not just releases from the otherwise use activity.

Figure 2
OPTIONAL SECTION 313 REPORTING THRESHOLD WORKSHEET

Facility Name: _____
Toxic Chemical or Chemical Category: _____   Date Worksheet Prepared: _____
Reporting Year: _____   Prepared By: _____

Step 1. Identify amounts of the toxic chemical manufactured, processed, or otherwise used.

Mixture Name or Other Identifier	Information Source	Percent by Weight	Total Weight (in lbs)	Amount of the Listed Toxic Chemical by Activity (in lbs.):		
				Manufactured	Processed	Otherwise Used
1.						
2.						
3.						
4.						
5.						
6.						
7.						
Subtotal:				(A) _____ lbs.	(B) _____ lbs.	(C) _____ lbs.

Step 2. Identify exempt forms of the toxic chemical that have been included in Step 1.

Mixture Name as Listed Above	Applicable Exemption	Note Fraction or Percent Exempt (if Applicable)	Exempt Amount of the Toxic Chemical from Above (in lbs.):		
			Manufactured	Processed	Otherwise Used
1.					
2.					
3.					
4.					
5.					
6.					
7.					
Subtotal:			(A_1) _____ lbs.	(B_1) _____ lbs.	(C_1) _____ lbs.

Step 3. Calculate the amount subject to threshold: $(A - A_1)$ _____ lbs. $(B - B_1)$ _____ lbs. $(C - C_1)$ _____ lbs.
Compare to thresholds for section 313 reporting.   25,000 lbs.   25,000 lbs.   10,000 lbs.

If any threshold is met, reporting is required for all activities. Do not submit this worksheet with Form R. Retain for your records.

Also note that threshold determinations are based upon the actual amounts of a toxic chemical manufactured, processed, or otherwise used over the course of the calendar year. The threshold determination may not relate to the amount of a toxic chemical brought on-site during the calendar year. For example, if a stockpile of 100,000 pounds of a toxic chemical is present on-site but only 20,000 pounds is applied to a process, only the 20,000 pounds processed is counted toward a threshold determination, not the entire 100,000 pounds of the stockpile.

Threshold Determinations for On-Site Reuse/Recycle Operations.

Threshold determinations of listed toxic chemicals that are recycled or reused at the facility are based only on the amount of the toxic chemical that is added during the year, not the total volume in the system. For example, a facility operates a refrigeration unit that contains 15,000 pounds of ammonia at the beginning of the year. The system is charged with 2,000 pounds of ammonia during the year. The facility has therefore "otherwise used" only 2,000 pounds of the covered toxic chemical and is not required to report (unless there are other "otherwise use" activities of ammonia which, when taken together, exceed the reporting threshold). If, however, the whole refrigeration unit was recharged with 15,000 pounds of ammonia during the year, the facility would exceed the otherwise use threshold, and be required to report.

This exemption does not apply to toxic chemicals "recycled" off-site and returned to a facility. Such toxic chemicals returned to a facility are treated as the equivalent of newly purchased material for purposes of section 313 threshold determinations.

Threshold Determinations for Chemical Categories.

A number of chemical compound categories are subject to reporting. See Table II for a listing of these toxic chemical categories. When reporting for one of these toxic chemical categories, all individual members of a category that are manufactured, processed, or otherwise used must be counted. However, threshold determinations must be made separately for each of the three activities. Do not include in these threshold determinations for a category any chemicals that are also specifically listed section 313 toxic chemicals (see Table II) or specific toxic chemicals that have been deleted from the category (e.g., three compounds deleted from copper compound category -- see the introduction to these instructions). Specifically listed toxic chemicals are subject to their own, individual threshold determination.

Threshold determinations for metal-containing compounds present a special case. If, for example, your facility processes several different lead compounds, base your threshold determination on the total weight of all lead compounds processed. However, if your facility processes both the "parent" metal (lead) as well as one or more lead compounds, you must make threshold determinations for both because they are separately listed toxic chemicals. If your facility exceeds thresholds for both the parent metal and compounds of that same metal, EPA allows you to file one combined report (e.g., one report for lead compounds, including lead) because the release information you will report in connection with metal compounds will be the total pounds of the parent metal released.

One other case involving metal compounds should be noted. Some metal compounds may contain more than one listed metal. For example, lead chromate is both a lead compound and a chromium compound. In such cases, if applicable thresholds are exceeded, you are required to file two separate reports, one for lead compounds and one for chromium compounds. Apply the total weight of the lead chromate to the threshold determinations for both lead compounds and chromium compounds. However, only the amount of each parent metal released (not the amount of the compound) would be reported on the appropriate sections of both Form Rs.

B.4.b Mixtures and Trade Name Products

Toxic chemicals contained in mixtures and trade name products must be factored into threshold and release determinations.

If your facility processed or otherwise used mixtures or trade name products during the calendar year, you are required to use the best information available to determine whether the components of a mixture are above the de minimis concentration and, therefore, must be included in threshold and release determinations. If you know that a mixture or trade name product contains a specific toxic chemical, combine the amount of the toxic chemical in the mixture or trade name product with other amounts of the same toxic chemical processed or otherwise used at your facility for threshold and release determinations. If you know that a mixture contains a toxic chemical but no concentration information is provided by the supplier, you do not have to consider the amount of the toxic chemical present in that mixture for purposes of threshold and release determinations.

Example 5: Mixture and Trade Name Products

Scenario #1: Your facility uses 12,000 pounds of an industrial solvent (Solvent X) for equipment cleaning. The Material Safety Data Sheet (MSDS) for the solvent indicates that it contains at least 50 percent methyl ethyl ketone (MEK), a listed toxic chemical; however, it also states that the solvent contains 20 percent non-hazardous surfactants. This is the only MEK-containing chemical used at the facility.

Follow these steps to determine if the quantity of the toxic chemical in solvent x exceeds the threshold for otherwise use.

1) Determine a reasonable maximum concentration for the toxic chemical by subtracting out the non-hazardous surfactants (i.e., 100%-20% = 80%).

2) Determine the midpoint between the known minimum (50%) and the reasonable maximum calculated above (i.e., (80%-50%)/2+50 = 65%).

3) Multiply total weight of Solvent X otherwise used by 65 percent.

   12,000 pounds x 0.65 = 7,800 pounds

4) Because the total amount of MEK otherwise used at the facility was less than the 10,000 pound otherwise use threshold, the facility is not required to file a Form R for MEK.

Scenario #2: Your facility otherwise used 15,000 pounds of Solvent Y to clean printed circuit boards. The MSDS for the solvent lists only that Solvent Y contains at least 80% of a listed toxic chemical which is only identified as chlorinated hydrocarbons.

Follow these steps to determine if the quantity of the toxic chemical in solvent exceeds the threshold for otherwise use.

1) Because the specific chemical is unknown, the Form R will be filed for "chlorinated hydrocarbons." This name will be entered into Part II, Section 2.1, "Mixture Component Identity." (Note: Because your supplier is claiming the toxic chemical identity a trade secret, you do not have to file substantiation forms.)

2) The upper bound limit is assumed to be 100 percent and the lower bound limit is known to be 80 percent. Using this information, the specific concentration is estimated to be 90 percent (i.e., the mid-point between upper and lower limits).

   $(1.0 + 0.80) / 2 = 0.90$

3) The total weight of Solvent Y is multiplied by 90 percent when calculating for thresholds.

   15,000 x 0.90 = 13,500

4) Because the total amount of chlorinated hydrocarbons exceeds the 10,000 pound otherwise used threshold, you must file a Form R for this chemical.

Observe the following guidelines in estimating concentrations of toxic chemicals in mixtures when only limited information is available:

- If you know the lower and upper bound concentrations of a toxic chemical in a mixture, use the midpoint of these two concentrations for threshold determinations.

- If you know only the lower bound concentration, you should subtract out the percentages of any other known components to determine a reasonable upper bound concentration, and then determine a midpoint.

- If you have no information other than the lower bound concentration, calculate a midpoint assuming an upper bound concentration of 100%.

- If you only know the upper bound concentration, you must use it for threshold determinations.

- In cases where you only have a concentration range available, you should use the midpoint of the range extremes.

De Minimis Exemption. A listed toxic chemical does not have to be considered if it is present in a mixture at a concentration below a specified de minimis level. The de minimis level is 1.0%, or 0.1% if the toxic chemical meets the OSHA carcinogen standard. See Table II for the de minimis value associated with each listed toxic chemical. For mixtures that contain more than one member of a listed toxic chemical category, the de minimis level applies to the aggregate concentration of all such members and not to each individually. EPA included the de minimis exemption in the rule as a burden-reducing step, primarily because facilities are not likely to have information on the presence of a toxic chemical in a mixture or trade name product beyond that available in the product's MSDS. The de minimis levels are consistent with OSHA requirements for development of MSDS information concerning composition.

For threshold determinations, the de minimis exemption applies to:

- A listed toxic chemical in a mixture or trade name product received by the facility.

- A listed toxic chemical manufactured during a process where the toxic chemical remains in a mixture or trade name product distributed by the facility.

The de minimis exemption does not apply to:

- A toxic chemical manufactured at the facility that does not remain in a product distributed by the facility. A threshold determination must be made on the annual quantity of the toxic chemical manufactured regardless of the concentration. For example, quantities of formaldehyde created as a result of waste treatment must be applied toward the threshold for "manufacture" of this toxic chemical, regardless of the concentration of this toxic chemical in the waste.

In general, when the de minimis exemption applies to threshold determinations and the concentration of the toxic chemical in the mixture is below the de minimis limitation, then you are not required to report releases associated with the processing or otherwise use of the toxic chemical in that mixture. Note that it is possible to meet the threshold for a toxic chemical on a facility-wide basis, but not be required to calculate releases from a particular process because that process involves only mixtures containing the toxic chemical below the de minimis level.

Application of the de minimis exemption to process streams must also be reviewed. Mixtures containing toxic chemicals can be added to a process or generated within a process. A facility is required to consider and report releases from the process once the de minimis concentration level has been exceeded. All releases of the toxic chemical from the process which occur after the de minimis exemption has been exceeded are then subject to reporting, regardless of whether or not the toxic chemical concentration later falls to a level below the de minimis exemption.

Supplier Notification. Beginning in 1989, suppliers of facilities in SIC codes 20-39 are required to develop and distribute a notice if the mixtures or trade name products they manufacture or process, and subsequently distribute, contain listed toxic chemicals. These notices are distributed to other companies in SIC codes 20-39 or to companies that sell or otherwise distribute the product to facilities in SIC codes 20-39. If a MSDS is not required for the mixture or trade name product, the notification must be in written form (i.e., letter). Otherwise, the notice must be incorporated into or attached to the MSDS for that product. The supplier notification requirement began with the first shipment of a product in 1989 and must accompany the first shipment each year thereafter. In addition, a new or revised notice must be sent if a change occurs in the product which affects the weight percent of a listed toxic chemical or if it is discovered that a previous notice did not properly identify the toxic chemicals or the percentage by weight. For more information on supplier notification, see Appendix D.

If listed toxic chemical concentrations are equal to or above the de minimis cut-off level, your supplier must identify the specific components as they appear in Table II and provide their percentage composition by weight in the mixture or product. If your supplier maintains that the identity of a toxic chemical is a trade secret, a generic identity that is structurally descriptive must be supplied on the notice. A maximum concentration level must be provided if your supplier contends that chemical composition information is a trade secret. In either case, you do not need to make a trade secret claim on behalf of your supplier (unless you consider your use of the proprietary mixture a trade secret). On Form R, identify the toxic chemical you are reporting according to its generic name provided in the notification. (See the instructions for Part II, Section 2 for more information.) If the listed toxic chemical is present below the de minimis level, no notification is required.

C. Instructions for Completing EPA Form R

The following are specific instructions for completing each part of EPA Form R. The number designations of the parts and sections of these instructions correspond to those in Form R unless otherwise indicated.

For all parts of Form R:

1. Type or print information on the form in the units and format requested. Use black ink. (Using blue ink for the certification signature is suggested as a means of indicating its originality.)

2. All information on Form R is required.

3. Do not leave items in Parts I and II on Form R blank unless specifically directed to do so; if an item does not apply to you, enter not applicable, NA, in the space provided. If your information does not fill all the spaces provided for a type of information, enter NA, in the next blank space in the sequence.

4. Report releases, off-site transfers, and recycling activities to the nearest pound. Do not report fractions of pounds.

5. Do not submit an incomplete form. The certification statement (Part I) specifies that the report is complete as submitted. See page 1 of these instructions for the definition of a complete submission.

6. When completing additional pages for Part II of the form, number the additional information sequentially from the prior sections of the form.

7. Indicate your TRI Facility Identification Number and the toxic chemical, toxic chemical category, or generically named toxic chemical on which you are reporting in the space provided in the top right corner of each page of Form R. Completion of this non-mandatory data element will greatly aid your internal recordkeeping and the quality of EPA's data entry process.

Part I. Facility Identification Information

Section 1. Reporting Year

This is the calendar year to which the reported information applies, not the year in which you are submitting the report. Information for the 1991 reporting year must be submitted on or before July 1, 1992.

Section 2. Trade Secret Information

2.1 Are you claiming the chemical identity on page 3 trade secret?

Answer this question only after you have completed the rest of the report. The specific identity of the toxic chemical being reported in Part II, Section 1, may be designated as a trade secret. If you are making a trade secret claim, mark "yes" and proceed to Section 2.2. Only check "yes" if it is your manufacturing, processing, or otherwise use of the toxic chemical whose identity is a trade secret. (See page 1 of these instructions for specific information on trade secrecy claims.) If you checked "no," proceed to Section 3; do not answer Section 2.2.

2.2 If "yes" in 2.1, is this copy sanitized or unsanitized?

Answer this question only after you have completed the rest of the report. Check "sanitized" if this copy of the report is the public version which does not contain the toxic chemical identity but does contain a generic name in its place, and you have claimed the toxic chemical identity trade secret in Part I, Section 2.1. Otherwise, check "unsanitized."

Section 3. Certification

The certification statement must be signed by the owner or operator or a senior official with management responsibility for the person (or persons) completing the form. The owner, operator, or official must certify the accuracy and completeness of the information reported on the form by signing and dating the certification statement. Each report must contain an original signature. Print or type in the space provided the name and title of the person who signs the statement. This certification statement applies to all the information supplied on the form and should be signed only after the form has been completed.

Section 4. Facility Identification

4.1 Facility Name and Location

Enter the name of your facility (plant site name or appropriate facility designation), street address, mailing address, city, county, state, and zip code in the space provided. Do not use a post office box number as the street address. The street address provided should be the location where the toxic chemicals are manufactured, processed, or otherwise used. If your mailing address and street address are the same, enter NA in the space for the mailing address.

If you have submitted a Form R for previous reporting years, a TRI Facility Identification Number has been assigned to your facility. The TRI Facility Identification Number appears (with other facility-specific information) on the peel-off mailing label on the cover of this Toxic Chemical Release Inventory Instructions for 1991 (EPA 700-K-92-002). Remove the mailing label from the back of this document and apply it to the space marked "place label here" in Part I, Section 4.1 of the blank Form R.

If your mailing label is missing information required on Form R, insert that information in the appropriate box in Part I, Section 4.1. For example, if your label contains your street address and not your mailing address, enter your mailing address in the space provided.

If you do not have a mailing label or cannot locate your TRI Facility Identification Number, please contact the Emergency Planning and Community Right-to-Know Information Hotline.

Enter "NA" in the space for the TRI Facility Identification number if this is your first submission of a Form R.

4.2 Full or Partial Facility Indication

A covered facility must report all releases and source reduction and recycling activities of a listed toxic chemical if it meets a reporting threshold for that toxic chemical. However, if the facility is composed of several distinct establishments, EPA allows these establishments to submit separate reports for the toxic chemical as long as all releases of the toxic chemical from the entire facility are accounted for. Indicate in Section 4.2 whether your report is for the entire covered facility as a whole or for part of a covered facility. Check box (a) if the toxic chemical information applies to the entire covered facility. Check box (b) if the toxic chemical information applies only to part of a covered facility.

Section 313 requires reports by "facilities," which are defined as "all buildings, equipment, structures, and other stationary items which are located on a single site or on contiguous or adjacent sites and which are owned or operated by the same person."

The SIC code system defines business "establishments" as "distinct and separate economic activities [that] are performed at a single physical location." Under section 372.30(c) of the reporting rule, you may submit a separate Form R for each establishment, or for groups of establishments in your facility, provided all releases and source reduction and recycling activities involving the toxic chemical from the entire facility are reported. This allows you the option of reporting separately on the activities involving a toxic chemical at each establishment, or group of establishments (e.g., part of a covered facility), rather than submitting a single Form R for that toxic chemical for the entire facility. However, if an establishment or group of establishments does not manufacture, process, or otherwise use or release a toxic chemical, you do not have to submit a report for that establishment or group of establishments. (See also Section B.2.a of these instructions.)

4.3 Technical Contact

Enter the name and telephone number (including area code) of a technical representative whom EPA or State officials may contact for clarification of the information reported on Form R. This contact person does not have to be the same person who prepares the report or signs the certification statement and does not necessarily need to be someone at the location of the reporting facility; however, this person must be familiar with the details of the report so that he or she can answer questions about the information provided.

4.4 Public Contact

Enter the name and telephone number (including area code) of a person who can respond to questions from the public about the report. If you choose to designate the same person as both the technical and the public contact, you may enter "Same as Section 4.3" in this space. This contact person does not have to be the same person who prepares the report or signs the certification statement and does not necessarily need to be someone at the location of the reporting facility. If this space is left blank, the technical contact will be listed as the public contact in the TRI database.

4.5 Standard Industrial Classification (SIC) Code

Enter the appropriate 4-digit primary Standard Industrial Classification (SIC) code for your facility (Table I lists the SIC codes within the 20-39 range). If the report covers more than one establishment, enter the primary 4-digit SIC code for each establishment starting with the primary SIC code for the entire facility. You are required to enter SIC codes only for those establishments within the facility that fall within SIC codes 20 to 39. If you do not know your SIC code, check with your financial office or contact your local Chamber of Commerce or State Department of Labor.

4.6 Latitude and Longitude

Enter the latitudinal and longitudinal coordinates of your facility. Sources of these data include EPA permits (e.g., NPDES permits), county property records, facility blueprints, and site plans. Instructions on how to determine these coordinates can be found in Appendix E. Enter only numerical data. Do not preface numbers with letters such as N or W to denote the hemisphere.

Latitude and longitude coordinates of your facility are very important for pinpointing the location of reporting facilities and are required elements on the Form R. EPA encourages facilities to make the best possible measurements when determining latitude and longitude. As with any other data field, missing, suspect, or incorrect data may generate a Notice of Technical Error to be issued to the facility. (See Appendix C: Common Errors in Completing Form R Reports).

4.7 Dun and Bradstreet Number

Enter the 9-digit number assigned by Dun and Bradstreet (D & B) for your facility or each establishment within your facility. These numbers code the facility for financial purposes. This number may be available from your facility's treasurer or financial officer. You can also obtain the numbers from your local Dun and Bradstreet office (check the telephone book White Pages). If a facility does not subscribe to the D & B service, a "support number" can be obtained from the Dun & Bradstreet center located in Allentown, Pennsylvania, at (215) 882-7748 (8:30 am to 8:00 pm, Eastern Time). If none of your establishments has been assigned a D & B number, enter not applicable, NA, in box (a). If only some of your establishments have been assigned Dun and Bradstreet numbers, enter those numbers in Part I, Section 4.7.

4.8 EPA Identification Number

The EPA I.D. Number is a 12-character number assigned to facilities covered by hazardous waste regulations under the Resource Conservation and Recovery Act (RCRA). Facilities not covered by RCRA are not likely to have an assigned I.D. Number. If your facility is not required to have an I.D. Number, enter not applicable, NA, in box (a). If your facility has been assigned EPA Identification Numbers, you must enter those numbers in the spaces provided in Section 4.8.

4.9 NPDES Permit Number

Enter the numbers of any permits your facility holds under the National Pollutant Discharge Elimination System (NPDES) even if the permit(s) do not pertain to the toxic chemical being reported. This 9-character permit number is assigned to your facility by EPA or the State under the authority of the Clean Water Act. If your facility does not have a permit, enter not applicable, NA, in Section 4.9a.

4.10 Underground Injection Well Code (UIC) Identification Number

If your facility has a permit to inject a waste containing the toxic chemical into Class 1 deep wells, enter the 12-digit Underground Injection Well Code (UIC) identification number assigned by EPA or by the State under the authority of the Safe Drinking Water Act. If your facility does not hold such a permit(s), enter not applicable, NA, in Section 4.10a. You are only required to provide the UIC number for wells that receive the toxic chemical being reported.

Section 5. Parent Company Information

You must provide information on your parent company. For purposes of Form R, a parent company is defined as the highest level company, located in the United States, that directly owns at least 50 percent of the voting stock of your company. If your facility is owned by a foreign entity, enter not applicable, NA, in this space. Corporate names should be treated as parent company names for companies with multiple facility sites. For example, the

Bestchem Corporation is not owned or controlled by any other corporation but has sites throughout the country whose names begin with Bestchem. In this case, Bestchem Corporation would be listed as the "parent" company.

5.1 Name of Parent Company

Enter the name of the corporation or other business entity that is your ultimate US parent company. If your facility has no parent company, check the NA box.

5.2 Parent Company's Dun & Bradstreet Number

Enter the Dun and Bradstreet Number for your ultimate US parent company, if applicable. The number may be obtained from the treasurer or financial officer of the company. If your parent company does not have a Dun and Bradstreet number, check the NA box.

Part II Chemical Specific Information

In Part II, you are to report on:

- The toxic chemical being reported;
- The general uses and activities involving the toxic chemical at your facility;
- Releases of the toxic chemical from the facility to air, water, and land;
- Quantities of the toxic chemical transferred to off-site locations;
- Information for on-site and off-site waste treatment, energy recovery, disposal, and recycling of the toxic chemical; and
- Source reduction activities.

Section 1. Toxic Chemical Identity

1.1 CAS Number

Enter the Chemical Abstracts Service (CAS) registry number in Section 1.1 **exactly** as it appears in Table II for the chemical being reported. CAS numbers are cross-referenced with an alphabetical list of chemical names in Table II of these instructions. If you are reporting one of the toxic chemical **categories** in Table II (e.g., chromium compounds), enter the applicable category code in the CAS number space. Toxic chemical category codes are listed below and can also be found in Table II.

Toxic Chemical Category Codes

N010	Antimony compounds
N020	Arsenic compounds
N040	Barium compounds
N050	Berylium compounds
N078	Cadmium compounds
N084	Clorophenols
N090	Chromium compounds
N096	Cobalt compounds
N100	Copper compounds
N106	Cyanide compounds
N230	Glycol ethers
N420	Lead compounds
N450	Manganese compounds
N458	Mercury compounds
N495	Nickel compounds
N575	Polybrominated biphenyls (PBBs)
N725	Selenium compounds
N740	Silver compounds
N760	Thallium compounds
N982	Zinc compounds

If you are making a trade secret claim, you must report the CAS number or category code on your unsanitized Form R and unsanitized substantiation form. Do not include the CAS number or category code on your sanitized Form R or sanitized substantiation form.

1.2 Toxic Chemical or Chemical Category Name

Enter the name of the toxic chemical or chemical category exactly as it appears in Table II. If the toxic chemical name is followed by a synonym in parentheses, report the chemical by the name that directly follows the CAS number (i.e., not the synonym). If the listed toxic chemical identity is actually a product trade name (e.g., dicofol), the 9th *Collective Index* name is listed below it in brackets. You may report either name in this case.

Do not list the name of a chemical that does not appear in Table II, such as individual members of a reportable toxic chemical category. For example, if you use silver nitrate, **do not** report silver nitrate with its CAS number. Report this chemical as "silver compounds" with its category code, N740.

If you are making a trade secret claim, you must report the specific toxic chemical identity on your unsanitized Form R and unsanitized substantiation form. Do not report the name of the toxic chemical on your sanitized Form R or sanitized substantiation form. Include a generic name in Part II, Section 1.3 of your sanitized Form R report.

EPA requests that the toxic chemical, chemical category, or generic name also be placed in the box marked "Chemical, Category, or Generic Name" in the upper right-hand corner on all pages of Form R. While this space is not a required data element, providing this information will help you in preparing a complete Form R report.

1.3 Generic Chemical Name

Complete Section 1.3 only if you are claiming the specific toxic chemical identity of the toxic chemical as a trade secret and have marked the trade secret block in Part I, Section 2.1 on page 1 of Form R. Enter a generic chemical name that is descriptive of the chemical structure. You must limit the generic name to seventy characters (e.g., numbers, letters, spaces, punctuation) or less. Do not enter mixture names in Section 1.3; see Section 2 below.

In-house plant codes and other substitute names that are not structurally descriptive of the toxic chemical identity being withheld as a trade secret are not acceptable as a generic name. The generic name must appear on both sanitized and unsanitized Form R's, and the name must be the same as that used on your substantiation forms.

Section 2.   Mixture Component Identity

Do not complete this section if you have completed Section 1 of Part II. Report the generic name provided to you by your supplier in this section if your supplier is claiming the chemical identity proprietary or trade secret. Do not answer "yes" in Part I, Section 2.1 on page 1 of the form if you complete this section. You do not need to supply trade secret substantiation forms for this toxic chemical because it is your supplier who is claiming the chemical identity a trade secret.

2.1   Generic Chemical Name Provided by Supplier

<u>Enter the generic chemical name in this section only if the following three conditions apply:</u>

1. You determine that the mixture contains a listed toxic chemical but the only identity you have for that chemical is a generic name;

2. You know either the specific concentration of that toxic chemical component or a maximum or average concentration level; and

3. You multiply the concentration level by the total annual amount of the whole mixture processed or otherwise used and determine that you meet the process or otherwise use threshold for that single, generically identified mixture component.

Example 6: Mixture Containing Unidentified Toxic Chemical

Your facility uses 20,000 pounds of a solvent that your supplier has told you contains 80 percent "chlorinated aromatic," their generic name for a toxic chemical subject to reporting under section 313. You therefore know that you have used 16,000 pounds of some listed toxic chemical which exceeds the "otherwise use" threshold. You would file a Form R and enter the name "chlorinated aromatic" in the space provided in Part II, Section 2.

Section 3.   Activities and Uses of the Toxic Chemical at the Facility

Indicate whether the toxic chemical is manufactured (including imported), processed, or otherwise used at the facility and the general nature of such activities and uses at the facility during the calendar year. Report activities that take place only at your facility, not activities that take place at other facilities involving your products. You must check all the boxes in this section that apply. If you are a manufacturer of the toxic chemical, you must check (a) and/or (b), and at least one of (c), (d), (e), or (f) in Section 3.1. Refer to the definitions of "manufacture," "process," and "otherwise use" in the general information section of these instructions or Part 40, Section 372.3 of the *Code of Federal Regulations* for additional explanations.

3.1   Manufacture the Toxic Chemical

Persons who manufacture (including import) the toxic chemical must check at least one of the following:

a. *Produce* - the toxic chemical is produced at the facility.

b. *Import* - the toxic chemical is imported by the facility into the Customs Territory of the United States. (See Section B.3.a of these instructions for further clarification of import.)

And check at least one of the following:

c. *For on-site use/processing* - the toxic chemical is produced or imported and then further processed or otherwise used at the same facility. If you check this block, you must also check at least one item in Part II, Section 3.2 or 3.3.

d. *For sale/distribution* - the toxic chemical is produced or imported specifically for sale or distribution outside the manufacturing facility.

e. *As a byproduct* - the toxic chemical is produced coincidentally during the production, processing, otherwise use, or disposal of another chemical substance or mixture and, following its production, is separated from that other chemical substance or mixture. Toxic chemicals produced and released as a result of waste treatment or disposal are also considered byproducts.

f. *As an impurity* - the toxic chemical is produced coincidentally as a result of the manufacture, processing, or otherwise use of another chemical but is not separated and remains primarily in the mixture or product with that other chemical.

3.2 Process the Toxic Chemical (incorporative activities)

a. *As a reactant* - A natural or synthetic toxic chemical used in chemical reactions for the manufacture of another chemical substance or of a product. Includes, but is not limited to, feedstocks, raw materials, intermediates, and initiators.

b. *As a formulation component* - A toxic chemical added to a product (or product mixture) prior to further distribution of the product that acts as a performance enhancer during use of the product. Examples of toxic chemicals used in this capacity include, but are not limited to, additives, dyes, reaction diluents, initiators, solvents, inhibitors, emulsifiers, surfactants, lubricants, flame retardants, and rheological modifiers.

c. *As an article component* - A toxic chemical that becomes an integral component of an article distributed for industrial, trade, or consumer use. One example is the pigment components of paint applied to a chair that is sold.

d. *Repackaging* - Processing or preparation of a toxic chemical (or product mixture) for distribution in commerce in a different form, state, or quantity. This includes, but is not limited to, the transfer of material from a bulk container, such as a tank truck to smaller containers such as cans or bottles.

3.3 Otherwise Use the Toxic Chemical (non-incorporative activities)

a. *As a chemical processing aid* - A toxic chemical that is added to a reaction mixture to aid in the manufacture or synthesis of another chemical substance but is not intended to remain in or become part of the product or product mixture. Examples of such toxic chemicals include, but are not limited to, process solvents, catalysts, inhibitors, initiators, reaction terminators, and solution buffers.

b. *As a manufacturing aid* - A toxic chemical that aids the manufacturing process but does not become part of the resulting product and is not added to the reaction mixture during the manufacture or synthesis of another chemical substance. Examples include, but are not limited to, process lubricants, metalworking fluids, coolants, refrigerants, and hydraulic fluids.

c. *Ancillary or other use* - A toxic chemical that is used at a facility for purposes other than aiding chemical processing or manufacturing as described above. Examples include, but are not limited to, cleaners, degreasers, lubricants, fuels, and toxic chemicals used for treating wastes.

Example 7: Activities and Uses of Toxic Chemicals

In the example below, it is assumed that the threshold quantities for manufacture, process, or otherwise use (25,000 pounds, 25,000 pounds, and 10,000 pounds, respectively, for calendar year 1991) have been exceeded and the reporting of listed toxic chemicals is therefore required.

Your facility manufactures sulfuric acid. Fifty percent is sold as a product. The remaining 50 percent is reacted with naphthalene, forming phthalic acid and also producing sulfur dioxide fumes.

- Your company manufactures sulfuric acid, a listed toxic chemical, both for sale/distribution as a commercial product and for on-site use/processing as a feedstock in the phthalic acid production process. Because the sulfuric acid is a reactant, it is also processed. See Figure 3 for how this information would be reported in Part II, Section 3 of Form R.

- Your facility also processes naphthalene, as a reactant to produce phthalic acid, a chemical not on the section 313 list.

Figure 3

SECTION 1. TOXIC CHEMICAL IDENTITY
(Important: DO NOT complete this section if you complete Section 2 below.)

1.1 CAS Number (Important: Enter only one number exactly as it appears on the Section 313 list. Enter category code if reporting a chemical category.)
7664 - 93 - 9

1.2 Toxic Chemical or Chemical Category Name (Important: Enter only one name exactly as it appears on the Section 313 list.)
Sulfuric Acid

1.3 Generic Chemical Name (Important: Complete only if Part I, Section 2.1 is checked "yes." Generic Name must be structurally descriptive.)

SECTION 2. MIXTURE COMPONENT IDENTITY
(Important: DO NOT complete this section if you complete Section 1 above.)

2.1 Generic Chemical Name Provided by Supplier (Important: Maximum of 70 characters, including numbers, letters, spaces, and punctuation.)

SECTION 3. ACTIVITIES AND USES OF THE TOXIC CHEMICAL AT THE FACILITY
(Important: Check all that apply.)

3.1 Manufacture the toxic chemical:
a. [✓] Produce
b. [ ] Import

If produce or import:
c. [✓] For on-site use/processing
d. [ ] For sale/distribution
e. [ ] As a byproduct
f. [ ] As an impurity

3.2 Process the toxic chemical:
a. [✓] As a reactant
b. [ ] As a formulation component
c. [ ] As an article component
d. [ ] Repackaging

3.3 Otherwise use the toxic chemical:
a. [ ] As a chemical processing aid
b. [ ] As a manufacturing aid
c. [ ] Ancillary or other use

Section 4.   Maximum Amount of the Toxic
              Chemical On-Site at Any Time
              During the Calendar Year

For data element 4.1 of Part II, insert the code (see below) that indicates the maximum quantity of the toxic chemical (e.g., in storage tanks, process vessels, on-site shipping containers) at your facility at any time during the calendar year. If the toxic chemical was present at several locations within your facility, use the maximum **total** amount present at the entire facility at any one time.

Weight Range in Pounds

Range Code	From...	To....
01	0	99
02	100	999
03	1,000	9,999
04	10,000	99,999
05	100,000	999,999
06	1,000,000	9,999,999
07	10,000,000	49,999,999
08	50,000,000	99,999,999
09	100,000,000	499,999,999
10	500,000,000	999,999,999
11	1 billion	more than 1 billion

If the toxic chemical present at your facility was part of a mixture or trade name product, determine the maximum quantity of the toxic chemical present at the facility by calculating the weight percent of the toxic chemical only.

Do not include the weight of the entire mixture or trade name product. This data may be found in the Tier II form your facility may have prepared under Section 312 of EPCRA. See Part 40, Section 372.30(b) of the *Code of Federal Regulations* for further information on how to calculate the weight of the toxic chemical in the mixture or trade name product. For toxic chemical categories (e.g., nickel compounds), include all chemical compounds in the category when calculating the maximum amount, using the entire weight of each compound.

Section 5.   Releases of the Toxic Chemical to the
              Environment On-Site

In Section 5, you must account for the total aggregate releases of the toxic chemical to the environment from your facility for the calendar year.

Do not enter the values in Section 5 in gallons, tons, liters, or any measure other than pounds. You must also enter the values as whole numbers. Numbers following a decimal point are not acceptable.

Releases to the environment include emissions to the air, discharges to surface waters, and on-site releases to land and underground injection wells. If you have no releases to a particular media (e.g., stack air), you must check the "NA" box or enter zero; **do not** leave any part of Section 5 blank. Check the box on the last line of this section if you use the additional space for Section 5.3 on page 5 of the Form.

You are not required to count, as a release, quantities of a toxic chemical that are lost due to natural weathering or corrosion, normal/natural degradation of a product, or normal migration of a toxic chemical from a product. For example, amounts of a listed toxic chemical that migrate from plastic products in storage do not have to be counted in estimates of releases of that toxic chemical from the facility. Also, amounts of listed metal compounds (e.g., copper compounds) that are lost due to normal corrosion of process equipment do not have to be considered as releases of copper compounds from the facility.

All releases of the toxic chemical to the air must be classified as either a point or non-point emission, and included in the total quantity reported for these releases in Sections 5.1 and 5.2. Instructions for columns A, B, and C follow the discussions of Sections 5.1 through 5.5.

5.1   Fugitive or Non-Point Air Emissions

Report the total of all releases of the toxic chemical to the air that **are not** released through stacks, vents, ducts, pipes, or any other confined air stream. You must include (1) fugitive equipment leaks from valves, pump seals, flanges, compressors, sampling connections, open-ended lines, etc.; (2) evaporative losses from surface impoundments and spills; (3) releases from building ventilation systems; and (4) any other fugitive or non-point air emissions. Engineering estimates and mass balance calculations (using purchase records, inventories, engineering knowledge or process specifications of the quantity of the toxic chemical entering product, hazardous waste manifests, or monitoring records) may be useful in estimating fugitive emissions.

5.2 Stack or Point Air Emissions

Report the total of all releases of the toxic chemical to the air that occur through stacks, vents, ducts, pipes, or other confined air streams. You must include storage tank emissions. Air releases from air pollution control equipment would generally fall in this category. Monitoring data, engineering estimates, and mass balance calculations may help you to complete this section.

5.3 Discharges to Receiving Streams or Water Bodies

In Section 5.3 you are to enter the name(s) of the stream(s) or water body(ies) to which your facility directly discharges the toxic chemical on which you are reporting. A total of three spaces are provided; however, other streams or water bodies to which the toxic chemical is discharged can be reported in the additional spaces for Section 5.3 found on page 5 of Form R. Enter the name of each receiving stream or surface water body to which the toxic chemical being reported is directly discharged. Report the name of the receiving stream or water body as it appears on the NPDES permit for the facility. If the stream is not covered by a permit, enter the name of the off-site stream or water body by which it is publicly known. Do not list a series of streams through which the toxic chemical flows. Be sure to include the receiving stream(s) or water body(ies) that receive stormwater runoff from your facility. Do not enter names of streams to which off-site treatment plants discharge. Enter "NA" in Section 5.3.1. if you do not discharge the listed toxic chemical to surface water bodies.

Enter the total annual amount of the toxic chemical released from all discharge points at the facility to each receiving stream or water body. Include process outfalls such as pipes and open trenches, releases from on-site wastewater treatment systems, and the contribution from stormwater runoff, if applicable (see instructions for column C below). Do not include discharges to a POTW or other off-site wastewater treatment facilities in this section. These off-site transfers must be reported in Part II, Section 6 of Form R.

Wastewater analyses and flowmeter data may provide the quantities you will need to complete this section.

Discharges of listed acids (e.g., hydrogen fluoride; hydrogen chloride; nitric acid; phosphoric acid; and sulfuric acid) may be reported as zero if the discharges have been neutralized to pH 6 or above. If wastewater containing a listed mineral acid is discharged below pH 6, then releases of the mineral acid must be reported. In this case, pH measurements may be used to estimate the amount of mineral acid released.

If you must report more than three discharges to receiving streams or water bodies, check the box at the bottom of page 4 and enter the additional information on the following page, in Section 5.3, Additional Information on Releases of the Toxic Chemical to the Environment On-Site. In Section 5.3 on page 5, blanks in the data elements are provided so you may continue the numeration you began on page 4.

5.4 Underground Injection On-Site

Enter the total annual amount of the toxic chemical that was injected into all wells, including Class I wells, at the facility. Chemical analyses, injection rate meters, and RCRA Hazardous Waste Generators Reports are good sources for obtaining data that will be useful in completing this section. Check the Not Applicable "NA" box in Section 5.4 if you do not inject the reported toxic chemical into underground wells.

5.5 Releases to Land On-Site

Four predefined subcategories for reporting quantities released to land within the boundaries of the facility are provided. Do not report land disposal at off-site locations in this section. Accident histories and spill records may be useful (e.g., release notification reports required under Section 304 of EPCRA and accident histories required under Section 112(r)(7)(B)(ii) of the Clean Air Act).

5.5.1 Landfill—Typically, the ultimate disposal method for solid wastes is landfilling. Leaks from landfills need not be reported as a release because the amount of the toxic chemical in the landfill has already been reported as a release.

5.5.2 Land treatment/application farming—Land treatment is a disposal method in which a waste containing a listed toxic chemical is applied onto or incorporated into soil. While this disposal method is considered a release to land, any volatilization of listed toxic chemicals into the air occurring during the disposal operation must be included in the total fugitive air releases reported in Part II, Section 5.1 of Form R.

5.5.3 Surface impoundment—A surface impoundment is a natural topographic depression, man-made excavation, or diked area formed primarily of earthen materials (although some may be lined with man-made materials), which is designed to hold an accumulation of liquid wastes or wastes containing free liquids. Examples of surface impoundments are holding, settling, storage, and elevation pits; ponds; and lagoons. If the pit, pond, or lagoon is intended for storage or holding without discharge, it would be considered to be a surface impoundment used as a final disposal method.

Quantities of the toxic chemical released to surface impoundments that are used merely as part of a wastewater treatment process generally must not be reported in this section. However, if the impoundment accumulates sludges containing the toxic chemical, you must include an estimate in this section unless the sludges are removed and otherwise disposed (in which case they should be reported under the appropriate section of the form). For the purposes of this reporting, storage tanks are not considered to be a type of disposal and are not to be reported in this section of Form R.

5.5.4 Other disposal — Includes any amount of a listed toxic chemical released to land that does not fit the categories of landfills, land treatment, or surface impoundment. This other disposal would include any spills or leaks of listed toxic chemicals to land. For example, 2,000 pounds of benzene leaks from a underground pipeline into the land at a facility. Because the pipe was only a few feet from the surface at the erupt point, 30 percent of the benzene evaporates into the air. The 600 pounds released to the air would be reported as a fugitive air release (Part II, Section 5.1) and the remaining 1,400 pounds would be reported as a release to land, other disposal (Part II, Section 5.5.4).

5. Column A   Total Release

Only on-site releases of the toxic chemical to the environment for the calendar year are to be reported in this section of Form R. The total releases from your facility **do not** include transfers or shipments of the toxic chemical from your facility for sale or distribution in commerce, or of wastes to other facilities for waste treatment, recycling, disposal, or energy recovery (see Part II, Section 6 of these Instructions). Both routine releases, such as fugitive air emissions, and accidental or non-routine releases, such as chemical spills, must be included in your estimate of the quantity released. EPA requires no more than two significant digits when reporting releases (e.g., 7,521 pounds would be reported as 7,500 pounds).

Releases of Less Than 1,000 Pounds. For total annual releases or off-site transfers of a toxic chemical from the facility of less than 1,000 pounds, the amount may be reported either as an estimate or by using the range codes that have been developed. The reporting range codes to be used are:

Code	Range (pounds)
A	1-10
B	11-499
C	500-999

Do not enter a range code and an estimate in the same box in column A. Total annual releases or off-site transfers of a toxic chemical from the facility of less than 1 pound may be reported in one of several ways. You should round the value to the nearest pound. If the estimate is 0.5 pounds or greater, you should either enter the range code "A" for "1-10" or enter "1" in column A. If the release is less than 0.5 pounds, you may round to zero and enter "0" in column A.

Note that total annual releases of less than 0.5 pounds from the processing or otherwise use of an article maintain the article status of that item. Thus, if the only releases you have are from processing an article, and such releases are less than 0.5 pounds per year, you are not required to submit a report for that toxic chemical. The 0.5-pound release determination does not apply to just a single article. It applies to the cumulative releases from the processing or otherwise use of the same type of article (e.g., sheet metal or plastic film) that occurs over the course of the calendar year.

Zero Releases. If you have no releases of a toxic chemical to a particular medium, report either NA, not applicable, or 0, as appropriate. Report NA only when there is no possibility a release could have occurred to a specific media or off-site location. If a release to a specific media or off-site location could have occurred, but either did not occur or the annual aggregate release was less than 0.5 pounds, report zero. However, if you report zero releases, a basis of estimate must be provided in column B.

For example, if hydrochloric acid is involved in the facility's processing activities but the facility neutralizes the wastes to a pH of 6 or above, then the facility reports a 0 release for the toxic chemical. If the facility has no underground injection well, "NA" would be written in Part I, Section 4.10 and checked in Part II, Section 5.4 of Form R. Also, if the facility does not landfill the acidic waste, NA would be checked in Part II, Section 5.5.1 of Form R.

Releases of 1,000 Pounds or More. For releases to any medium that amount to 1,000 pounds or more for the year, you must provide an estimate in pounds per year in column A. Any estimate provided in column A should be reported to no more than two significant figures. This estimate should be in whole numbers. Do not use decimal points.

Calculating Releases. To provide the release information required in column A in this section, you must use all readily available data (including relevant monitoring data and emissions measurements) collected at your facility to meet other regulatory requirements or as part of routine plant operations, to the extent you have such data for the toxic chemical.

When relevant monitoring data or emission measurements are not readily available, reasonable estimates of the amounts released must be made using published emission factors, material balance calculations, or engineering calculations. You may not use emission factors or calculations to estimate releases if more accurate data are available.

No additional monitoring or measurement of the quantities or concentrations of any toxic chemical released into the environment, or of the frequency of such releases, beyond that which is required under other provisions of law or regulation or as part of routine plant operations, is required for the purpose of completing Form R.

You must estimate, as accurately as possible, the quantity (in pounds) of the toxic chemical or chemical category that is released annually to each environmental medium. Include only the quantity of the toxic chemical in this estimate. If the toxic chemical present at your facility was part of a mixture or trade name product, calculate only the releases of the toxic chemical, not the other components of the mixture or trade name product. If you are only able to estimate the releases of the mixture or trade name product as a whole, you must assume that the release of the toxic chemical is proportional to its concentration in the mixture or trade name product. See Part 40, Section 372.30(b) of the *Code of Federal Regulations* for further information on how to calculate the concentration and weight of the toxic chemical in the mixture or trade name product.

If you are reporting a toxic chemical category listed in Table II of these instructions rather than a specific toxic chemical, you must combine the release data for all chemicals in the listed toxic chemical category (e.g., all glycol ethers or all chlorophenols) and report the aggregate amount for that toxic chemical category. Do not report releases of each individual toxic chemical in that category separately. For example, if your facility releases 3,000 pounds per year of 2-chlorophenol, 4,000 pounds per year of 3-chlorophenol, and 4,000 pounds per year of 4-chlorophenol to air as fugitive emissions, you should report that your facility releases 11,000 pounds per year of chlorophenols to air as fugitive emissions in Part II, Section 5.1.

For listed toxic chemicals with the qualifier "solution," such as ammonium nitrate, at concentrations of 1 percent (or 0.1 percent in the case of a carcinogen) or greater, the chemical concentrations must be factored into threshold and release calculations because threshold and release amounts relate to the amount of toxic chemical in solution, not the amount of solution.

For metal compound categories (e.g., chromium compounds), report releases of only the parent metal. For example, a user of various inorganic chromium salts would report the total chromium released regardless of the chemical form (e.g., as the original salts, chromium ion, oxide) and exclude any contribution to mass made by other species in the molecule.

5. Column B   Basis of Estimate

For each release estimate, you are required to indicate the principal method used to determine the amount of release reported. You will enter a letter code that identifies the method that applies to the largest portion of the total estimated release quantity.

The codes are as follows:

M- Estimate is based on monitoring data or measurements for the toxic chemical as transferred to an off-site facility.

C- Estimate is based on mass balance calculations, such as calculation of the amount of the toxic chemical in wastes entering and leaving process equipment.

E- Estimate is based on published emission factors, such as those relating release quantity to throughput or equipment type (e.g., air emission factors).

Example 8: Calculating Releases and Transfers

Your facility disposes of 14,000 pounds of lead chromate (PbCrO4.PbO) in an on-site landfill and transfers 16,000 pounds of lead selenite (PbSeO4) to an off-site land disposal facility. You would therefore be submitting three separate reports on the following: lead compounds, selenium compounds, and chromium compounds. However, the quantities you would be reporting would be the pounds of "parent" metal being released or transferred off-site. All quantities are based on mass balance calculations (See Section 5.B for information on Basis of Estimate and Section 6.C for waste treatment or disposal codes and information on transfers of toxic chemicals in wastes). You would calculate releases of lead, chromium, and selenium by first determining the percentage by weight of these metals in the materials you use as follows:

Lead Chromate (PbCrO4.PbO) - Molecular weight = 546.37

    Lead 2 Pb - Molecular weight = 207.2 x 2 = 414.4

    Chromium 1 Cr - Molecular weight = 51.996

    Lead chromate is therefore (% by weight)

$$(414.4/546.37) = 75.85\% \text{ lead and}$$
$$(51.996/546.37) = 9.52\% \text{ chromium}$$

Lead Selenite (PbSeO4) Molecular weight = 350.17

    Lead 1 Pb Molecular weight = 207.2

    Selenium 1 Se Molecular weight = 78.96

    Lead selenite is therefore (% by weight)

$$(207.2/350.17) = 59.17\% \text{ lead and}$$
$$(78.96/350.17) = 22.55\% \text{ selenium.}$$

The total pounds of lead, chromium, and selenium released or transferred from your facility are as follows:

Lead

Release: 0.7585 x 14,000 = 10,619 pounds from lead chromate (round to 11,000 pounds)

Transfer: 0.5917 x 16,000 = 9,467 pounds from lead selenite (round to 9,500 pounds)

Chromium

Release: 0.0952 x 14,000 = 1,333 pounds from lead chromate (round to 1,300 pounds)

Selenium

Transfer: 0.2255 x 16,000 = 3,608 pounds of selenium from lead round to 3,600 pounds)

O- Estimate is based on other approaches such as engineering calculations (e.g., estimating volatilization using published mathematical formulas) or best engineering judgment. This would include applying an estimated removal efficiency to a treatment, even if the composition of the waste before treatment was fully identified through monitoring data.

For example, if 40 percent of stack emissions of the reported toxic chemical were derived using monitoring data, 30 percent by mass balance, and 30 percent by emission factors, you would enter the code letter "M" for monitoring.

If the monitoring data, mass balance, or emission factor used to estimate the release is not specific to the toxic chemical being reported, the form should identify the estimate as based on engineering calculations or best engineering judgment.

If a mass balance calculation yields the flow rate of a waste, but the quantity of reported toxic chemical in the waste is based on solubility data, report "O" because "engineering calculations" were used as the basis of estimate of the quantity of the toxic chemical in the waste.

If the concentration of the toxic chemical in the waste was measured by monitoring equipment and the flow rate of the waste was determined by mass balance, then the primary basis of the estimate is "monitoring" (M). Even though a mass balance calculation also contributed to the estimate, "monitoring" should be indicated because monitoring data was used to estimate the concentration of the waste.

Mass balance (C) should only be indicated if it is **directly** used to calculate the mass (weight) of toxic chemical released. Monitoring data should be indicated as the basis of estimate **only** if the toxic chemical concentration is measured in the waste being released into the environment. Monitoring data should **not** be indicated, for example, if the monitoring data relates to a concentration of the toxic chemical in other process streams within the facility.

It is important to realize that the accuracy and proficiency of release estimation will improve over time. However, submitters are not required to use new emission factors or estimation techniques to revise previous Form R submissions.

5. Column C   Percent From Stormwater

This column relates only to Section 5.3 — discharges to receiving streams or water bodies. If your facility has monitoring data on the amount of the toxic chemical in stormwater runoff (including unchanneled runoff), you must include that quantity of the toxic chemical in your water release in column A **and** indicate the percentage of the total quantity (by weight) of the toxic chemical contributed by stormwater in column C (Section 5.3C).

If your facility has monitoring data on the toxic chemical and an estimate of flow rate, you must use this data to determine the percent stormwater.

If you have monitored stormwater but did not detect the toxic chemical, enter zero (0) in column C. If your facility has no stormwater monitoring data for the chemical, enter not applicable, "NA," in this space on the form.

If your facility does not have periodic measurements of stormwater releases of the toxic chemical, but has submitted chemical-specific monitoring data in permit applications, then these data must be used to calculate the percent contribution from stormwater. Rates of flow can be estimated by multiplying the annual amount of rainfall by the land area of the facility and then multiplying that figure by the runoff coefficient. The runoff coefficient represents the fraction of rainfall that does not seep into the ground but runs off as stormwater. The runoff coefficient is directly related to how the land in the drainage area is used. (See table below.)

Description of Land Area	Runoff Coefficient
Business	
Downtown areas	0.70-0.95
Neighborhood areas	0.50-0.70
Industrial	
Light areas	0.50-0.80
Heavy areas	0.60-0.90
Railroad yard areas	0.20-0.40
Unimproved areas	0.10-0.30
Streets	
Asphaltic	0.70-0.95
Concrete	0.80-0.95
Brick	0.70-0.85
Drives and walks	0.70-0.85
Roofs	0.75-0.95
Lawns: Sandy Soil	
Flat, 2%	0.05-0.10
Average, 2-7%	0.10-0.15
Steep, 7%	0.15-0.20

Example 9: Releases from Stormwater

Your stormwater monitoring data shows that the average concentration of zinc in the stormwater runoff from your facility from a biocide containing a zinc compound is 1.4 milligrams per liter, and the total annual stormwater discharge from the facility is 7.527 million gallons. The total amount of zinc discharged to surface water through the plant wastewater discharge (non-stormwater) is 250 pounds per year. The total amount of zinc discharged with stormwater is:

(7,527,000 gallons stormwater) x (3.785 liters/gallon) = 28,489,695 liters stormwater

(28,489,695 liters stormwater) x (1.4 mg. zinc/liter) = 39,885.6 grams zinc = 88 pounds zinc

The total amount of zinc discharged from all sources of your facility is:

250 pounds zinc from wastewater discharge
+ 88 pounds zinc from stormwater runoff
338 pounds zinc total water discharge

Round to 340 pounds of zinc on Form R.

The percentage of zinc discharged through stormwater is:

88/338 x 100 = 26%

Lawns: Heavy Soil
Flat, 2%   0.13-0.17
Average, 2-7%   0.18-0.22
Steep, 7%   0.25-0.35

Choose the most appropriate runoff coefficient for your site or calculate a weighted-average coefficient, which takes into account different types of land use at your facility:

Weighted-average runoff coefficient =
(Area 1 % of total)(C1) + (Area 2 % of total)(C2) + (Area 3 % of total)(C3) + ... + (Area i % of total)(Ci)

where C_i = runoff coefficient for a specific land use of Area i.

Section 6 Transfers of the Toxic Chemical in Wastes to Off-Site Locations

You must report in this section the total annual quantity of the toxic chemical in wastes sent to any off-site facility for the purposes of waste treatment, disposal, recycling, or energy recovery. **Note that beginning with reporting year 1991, off-site transfers for the purposes of recycling and energy recovery are required to be reported.** Report the total amount of the toxic chemical transferred off-site after any on-site waste treatment, recycling, or removal is completed. Do not report transfers of listed mineral acids if they have been neutralized to a pH of 6 or above prior to discharge to a Publicly Owned Treatment Works (POTW).

If you do not discharge wastewater containing the reported toxic chemical to a POTW, enter not applicable, NA, in the box for the POTW's name in Section 6.1.B._. If you do not ship or transfer wastes containing the reported toxic chemical to other off-site locations, enter not applicable, NA, in the box for the off-site location's EPA Identification Number in Section 6.2._.

Important: Beginning with the 1991 reporting year, you must number the boxes for reporting the information for each POTW or other off-site location in Sections 6.1 and 6.2. In the upper left hand corner of each box, the section number is either 6.1.B._ or 6.2._.

If you report a transfer of the listed toxic chemical to one or more POTW, number the boxes in Section 6.1.B as 6.1.B.1, 6.1.B.2, etc. If you transfer the listed toxic chemical to more than two POTWs, photocopy page 5 of Form R as many times as necessary and then number the boxes consecutively for each POTW. At the bottom of page 5 you will find instructions for indicating the total number of page 5s that you are submitting as part of Form R, as well as indicating the sequence of those pages. For

> **Example 10: Stormwater Runoff**
>
> Your facility is located in a semi-arid region of the United States which has an annual precipitation (including snowfall) of 12 inches of rain. (Snowfall should be converted to the equivalent inches of rain; assume one foot of snow is equivalent to one inch of rain.) The total area covered by your facility is 42 acres (about 170,000 square meters or 1,829,520 square feet). The area of your facility is 50 percent unimproved area, 10 percent asphaltic streets, and 40 percent concrete pavement.
>
> The total stormwater runoff from your facility is therefore calculated as follows:
>
> | Land Use | % Total Area | Runoff Coefficient |
> |---|---|---|
> | Unimproved area | 50 | 0.20 |
> | Asphaltic streets | 10 | 0.85 |
> | Concrete pavement | 40 | 0.90 |
>
> Weighted-average runoff coefficient = (50%) x (0.20) + (10%) x (0.85) + (40%) x (0.90) = 0.545
>
> (Rainfall) x (land area) x (conversion factor) x (runoff coefficient) = stormwater runoff
>
> (1 foot) x (1,829,520 ft$_2$) x (7.48 gal/ft$_3$) x (0.545) = 7,458,221 gallons/year
>
> Total stormwater runoff = 7.45 million gallons/year

example, your facility transfers the reported toxic chemical in wastewaters to three POTWs. You would photocopy page 5 once, indicate at the bottom of each page 5 that there are a total of two page 5s and then indicate the first and second page 5. The boxes for the two POTWs on the first page 5 would be numbered 6.1.B.1 and 6.1.B.2, while the box for the third POTW on the second page 5 would be numbered 6.1.B.3.

If you report a transfer of the listed toxic chemical to one or more other off-site locations, number the boxes in Section 6.2 as 6.2.1, 6.2.2, etc. If you transfer the listed toxic chemical to more than two other off-site locations, photocopy page 6 of Form R as many times as necessary and then number the boxes consecutively for each off-site location. At the bottom of page 6 you will find instructions for indicating the total number of page 6s that you are submitting as part of Form R as well as indicating the sequence of those pages. For example, your facility transfers the reported toxic chemical to three other off-site locations. You would photocopy page 6 once, indicate at the bottom of each page 6 that there are a total of two page 6s and then indicate the first and second page 6. The boxes for the two off-site locations on the first page 6 would be numbered 6.2.1 and 6.2.2, while the box for the third off-site location on the second page 6 would be numbered 6.2.3.

6.1 Discharges to Publicly Owned Treatment Works (POTW)

In Section 6.1.A, estimate the quantity of the reported toxic chemical transferred to all POTWs and the basis upon which the estimate was made. In Section 6.1.B, enter the name and address for each POTW to which your facility discharges wastewater containing the reported toxic chemical.

If you do not discharge wastewater containing the reported toxic chemical to a POTW, enter not applicable, NA, in the box for the POTW's name in Section 6.1.B._.

6.1.A.1 Total Transfers

Enter the total amount, in pounds, of the reported toxic chemical that is contained in the wastewaters transferred to all POTWs. Do not enter the total poundage of the wastewaters. If the total amount transferred is less than 1,000 pounds, you may report a range by entering the appropriate range code. The following reporting range codes are to be used:

Code	Reporting Range (in pounds)
A	1-10
B	11-499
C	500-99

6.1.A.2 Basis of Estimate

You must identify the basis for your estimate of the total quantity of the reported toxic chemical in the wastewaters transferred to all POTWs. Enter one of the following letter codes that applies to the method by which the largest percentage of the estimate was derived.

M-   Estimate is based on monitoring data or measurements for the toxic chemical as transferred to an off-site facility.

C-   Estimate is based on mass balance calculations, such as calculation of the amount of the toxic chemical in streams entering and leaving process equipment.

E-   Estimate is based on published emission factors, such as those relating release quantity to throughput or equipment type (e.g., air emission factors).

O-   Estimate is based on other approaches such as engineering calculations (e.g., estimating volatilization using published mathematical formulas) or best engineering judgment. This would include applying an estimated removal efficiency to a waste stream, even if the composition of the stream before treatment was fully identified through monitoring data.

If you transfer a toxic chemical to more than one POTW, you should report the basis of estimate that was used to determine the largest percentage of the toxic chemical that was transferred.

6.2 Transfers to Other Off-Site Locations

In Section 6.2, enter the EPA Identification Number, name, and address for each off-site location to which your facility ships or transfers wastes containing the reported toxic chemical for the purposes of waste treatment, disposal, recycling, or energy recovery. Also estimate the quantity of the reported toxic chemical transferred and the basis upon which the estimate was made. If appropriate, you must report multiple activities (up to four) for each off-site location. For example, if your facility sends a reported toxic chemical in wastes to an off-site location where some of the toxic chemical is to be recycled while the remainder of the quantity transferred is to be treated, you must report both the waste treatment and recycle activities, along with the quantity associated with each activity.

If you do not ship or transfer wastes containing the reported toxic chemical to other off-site locations, enter not applicable, NA, in the box for the off-site location's EPA Identification Number in Section 6.2._. The EPA Identification Number (defined in 40 CFR 260.10 and therefore commonly referred to as the RCRA ID Number) may be found on the Uniform Hazardous Waste Manifest, which is required by RCRA regulations. If you ship or transfer wastes containing a toxic chemical and the off-site location does not have an EPA Identification Number (e.g., it does not accept RCRA hazardous wastes or the wastes in question are not classified as hazardous), enter NA in the box for the off-site location EPA Identification Number. If you ship or transfer the reported toxic chemical in wastes to another country, enter the Federal Information Processing Standards (FIPS) code for that country in the county field of the address for the off-site facility. The most commonly used FIPS codes are listed below.

The following is an abridged list of countires to which a U.S. facility might ship a listed toxic chemical. For a complete listing of FIPS codes, consult your local library.

Country	Code
Argentina	AR
Belgium	BE
Bolivia	BL
Brazil	BR
Canada	CA
Chile	CI
Columbia	CO
Costa Rica	CS
Cuba	CU
Ecuador	EC
El Salvador	ES
France	FR
Guatemala	GT
Honduras	HO
Ireland	EI
Italy	IT
Mexico	MX
Nicaragua	NU
Panama	PM
Paraguay	PA
Peru	PE

Country	Code
Portugal	PO
Spain	SP
Switzerland	SZ
United Kingdom	UK
Uruguay	UY
Venezuela	VE

You must distinguish between incineration, which is always considered waste treatment, and combustion where energy is actually recovered. When the reported toxic chemical has a significant heat of combustion value, and is transferred to an off-site location for combustion in an industrial kiln, furnace, or boiler, report the quantity as used for the purposes of energy recovery. However, toxic chemicals with little or no heat of combustion value (e.g., metals, chlorofluorocarbons) must be reported as treated.

6.2 column A Total Transfers

For each off-site location, enter the total amount, in pounds, of the toxic chemical that is contained in the waste transferred to that location. Do not enter the total poundage of the waste. If the total amount transferred is less than 1,000 pounds, you may report a range by entering the appropriate range code. The following reporting range codes are to be used:

Code	Reporting Range (in pounds)
A	1-10
B	11-499
C	500-999

If you transfer the toxic chemical in wastes to an off-site facility for distinct and multiple purposes, you must report those activities (up to four) for each off-site location, along with the quantity of the reported toxic chemical associated with each activity. For example, your facility transfers a total of 15,000 pounds of toluene to an off-site location that will use 5,000 pounds for the purposes of energy recovery, enter 7,500 pounds into a recovery process, and dispose of the remaining 2,500 pounds. These quantities and the associated activity codes must be reported separately in Section 6.2. (See Figure 4 for a hypothetical Section 6.2 completed for two off-site locations, one of which receives the transfer of 15,000 pounds of toluene as detailed.) If more than four activities are performed on distinct quantities at the off-site location, list the predominant four activities but still report all quantities sent to the off-site location.

Do not double or multiple count amounts transferred off-site. For example, when a reported toxic chemical is sent to an off-site facility for sequential activities and the specific quantities associated with each activity are unknown, report only a single quantity (the total quantity transferred to the off-site location) along with a single activity code. In such a case, report the activity applied to the majority of the reported toxic chemical sent off-site, not the ultimate disposition of the toxic chemical. For example, when a toxic chemical is first treated and then recovered with the majority of the toxic chemical being treated and only a fraction subsequently recovered, report the appropriate waste treatment activity along with the quantity.

6.2 column B Basis of Estimate

You must identify the basis for your estimates of the quantities of the reported toxic chemical in wastes transferred to each off-site location. Enter one of the following letter codes that applies to the method by which the largest percentage of the estimate was derived.

M - Estimate is based on monitoring data or measurements for the toxic chemical as transferred to an off-site facility.

C - Estimate is based on mass balance calculations, such as calculation of the amount of the toxic chemical in wastes entering and leaving process equipment.

E - Estimate is based on published emission factors, such as those relating release quantity to throughput or equipment type (e.g., air emission factors).

O - Estimate is based on other approaches such as engineering calculations (e.g., estimating volatilization using published mathematical formulas) or best engineering judgment. This would include applying an estimated removal efficiency to a treatment, even if the composition of the waste before treatment was fully identified through monitoring data.

6.2 column C Type of Waste Treatment/Disposal/Recycling/Energy Recovery

Enter one of the following codes to identify the type of waste treatment, disposal, recycling or energy recovery methods used by the off-site location for the reported toxic chemical. You must use more than one line and

Figure 4
Hypothetical Section 6.2 Completed for Two Off-site Locations

SECTION 6.2 TRANSFERS TO OTHER OFF-SITE LOCATIONS

6.2. _1_   Off-site EPA Identification Number (RCRA ID No.): COD 566162461
Off-Site Location Name: Acme Waste Services
Street Address: 5 Market Street
City: Releaseville   County: Hill
State: CO   Zip Code: 80461   Is location under control of reporting facility or parent company? Yes [ ] No [X]

A. Total Transfers (pounds/year) (enter range code or estimate)	B. Basis of Estimate (enter code)	C. Type of Waste Treatment/Disposal/ Recycling/Energy Recovery (enter code)
1. 5,000	1. O	1. M 56
2. 7,500	2. C	2. M 20
3. 2,500	3. O	3. M 72
4. NA	4.	4. M

This off-site location receives a transfer of 15,000 pounds of toluene (as discussed earlier) and will combust 5,000 pounds for the purposes of energy recovery, enter 7,500 pounds into a recovery process, and dispose of the remaining 2,500 pounds.

SECTION 6.2 TRANSFERS TO OTHER OFF-SITE LOCATIONS

6.2. _2_   Off-site EPA Identification Number (RCRA ID No.): COD 61772543
Off-Site Location Name: Combustion, Inc.
Street Address: 25 Facility Road
City: Dumfry   County: Burns
State: CO   Zip Code: 80500   Is location under control of reporting facility or parent company? Yes [ ] No [X]

A. Total Transfers (pounds/year) (enter range code or estimate)	B. Basis of Estimate (enter code)	C. Type of Waste Treatment/Disposal/ Recycling/Energy Recovery (enter code)
1. 12,500	1. O	1. M 54
2. NA	2.	2. M
3.	3.	3. M
4.	4.	4. M

This off-site location receives a transfer of 12,500 pounds of tetrachloroethylene (perchloroethylene) that is part of a waste that is combusted for the purposes of energy recovery in an industrial furnace. Note that the perchloroethylene is reported using code M54 to indicate that it is combusted in an energy recovery unit but it does not contribute to the heating value of the waste.

code for a single location when distinct quantities of the reported toxic chemical are subject to different waste treatment, disposal, recycling, or energy recovery methods. You may have this information in your copy of EPA Form SO, Item S of the Annual/Biennial Hazardous Waste Treatment, Storage, and Disposal Report (RCRA), or in your invoices from the waste service(s) or broker(s) receiving your wastes for the purposes of waste treatment, disposal, recycling, or energy recovery.

You must distinguish between incineration, which is waste treatment, and legitimate energy recovery. In order for you to claim that a reported toxic chemical sent off-site is used for the purposes of energy recovery and not for waste treatment, the toxic chemical must have a heating value high enough to sustain combustion and must be combusted in an energy recovery unit such as an industrial boiler, furnace, or kiln. In a situation where the reported toxic chemical is in a waste that is combusted in an energy recovery unit, but the toxic chemical does not have a heating value high enough to sustain combustion, use code M54, Incineration/Insignificant Fuel Value, to indicate that the toxic chemical was incinerated in an energy recovery unit but did not contribute to the heating value of the waste (see Figure 4 for an example).

Applicable codes for Part II, Section 6.2, column C are:

Disposal

M10	Storage Only
M71	Underground Injection
M72	Landfill/Disposal Surface Impoundment
M73	Land Treatment
M79	Other Land Disposal
M94	Transfer to Waste Broker—Disposal

Recycling

M20	Solvents/Organics Recovery
M24	Metals Recovery
M26	Other Reuse or Recovery
M28	Acid Regeneration
M93	Transfer to Waste Broker—Recycling

Waste Treatment

M40	Solidification/Stabilization
M50	Incineration/Thermal Treatment
M54	Incineration/Insignificant Fuel Value
M61	Wastewater Treatment (Excluding POTW)
M69	Other Waste Treatment
M95	Transfer to Waste Broker—Waste Treatment

Energy Recovery

M56	Energy Recovery
M92	Transfer to Waste Broker—Energy Recovery

Section 7   On-Site Waste Treatment, Energy Recovery and Recycling Methods

You must report in this section the methods of waste treatment, energy recovery, and recycling applied to the reported toxic chemical in wastes on-site. There are three separate sections for reporting such activities.

Section 7A   On-Site Waste Treatment Methods and Efficiency

In Section 7A, you must provide the following information if you treat the reported toxic chemical on-site:

(a) the general waste stream types containing the toxic chemical being reported;
(b) the waste treatment method(s) or sequence used on all waste streams containing the toxic chemical;
(c) the range of concentration of the toxic chemical in the influent to the waste treatment method;
(d) the efficiency of each waste treatment method or waste treatment sequence in removing the toxic chemical; and
(e) whether the waste treatment efficiency figure was based on actual operating data.

Use a separate line in Section 7A for each general waste stream type. Report only information about treatment of waste streams at your facility, not information about off-site waste treatment.

If you do not perform on-site treatment of waste streams containing the reported toxic chemical, check the Not Applicable (NA) box at the top of Section 7A.

7A column a   General Waste Stream

For each waste treatment method, indicate the type of waste stream containing the toxic chemical that is treated. Enter the letter code that corresponds to the general waste stream type:

A	Gaseous (gases, vapors, airborne particulates)
W	Wastewater (aqueous waste)
L	Liquid waste streams (non-aqueous waste)
S	Solid waste streams (including sludges and slurries)

If a waste is a mixture of water and organic liquid and the organic content is less than 50 percent, report it as a wastewater (W). Slurries and sludges containing water must be reported as solid waste if they contain appreciable amounts of dissolved solids, or solids that may settle, such that the viscosity or density of the waste is considerably different from that of process wastewater.

7A column b   Waste Treatment Method(s) Sequence

Enter the appropriate code from the list below for each on-site waste treatment method used on a waste stream containing the toxic chemical, regardless of whether the waste treatment method actually removes the specific toxic chemical being reported. Waste treatment methods must be reported for each type of waste stream being treated (i.e., gaseous waste streams, aqueous waste streams, liquid non-aqueous waste streams, and solids). Except for the air emission treatment codes, the waste treatment codes are not restricted to any medium.

Waste streams containing the toxic chemical may have a single source or may be aggregates of many sources. For example, process water from several pieces of equipment at your facility may be combined prior to waste treatment. Report waste treatment methods that apply to the aggregate waste stream, as well as waste treatment methods that apply to individual waste streams. If your facility treats various wastewater streams containing the toxic chemical in different ways, the different waste treatment methods must be listed separately.

If your facility has several pieces of equipment performing a similar service in a waste treatment sequence, you may combine the reporting for such equipment. It is not necessary to enter four codes to cover four scrubber units, for example, if all four are treating waste streams of similar character (e.g., sulfuric acid mist emissions), have similar influent concentrations, and have similar removal efficiencies. If, however, any of these parameters differs from one unit to the next, each scrubber must be listed separately.

If your facility performs more than eight sequential waste treatment methods on a single general waste stream, continue listing the methods in the next row and renumber appropriately those waste treatment method code boxes you used to continue the sequence. For example, if the general waste stream in box 7A.1a had nine treatment methods applied to it, the ninth method would be indicated in the first method box for row 7A.2a. The numeral "1" would be crossed out, and a "9" would be inserted.

Treatment applied to any other general waste stream types would then be listed in the next empty row. In the scenario above, for instance, the second general waste stream would be reported in row 7A.3a. See Figure 5 below for an example of a hypothetical Section 7A completed for a nine-step waste treatment process and a single waste treatment method.

If you need additional space to report under Section 7A, photocopy page 7 of Form R as many times as necessary. At the bottom of page 7 you will find instructions for indicating the total number of page 7s that you are submitting as part of Form R, as well as instructions for indicating the sequence of those pages.

Waste Treatment Codes

Air Emissions Treatment (applicable to gaseous waste streams only)

A01	Flare
A02	Condenser
A03	Scrubber
A04	Absorber
A05	Electrostatic Precipitator
A06	Mechanical Separation
A07	Other Air Emission Treatment

Biological Treatment

B11	Biological Treatment — Aerobic
B21	Biological Treatment — Anaerobic
B31	Biological Treatment — Facultative
B99	Biological Treatment — Other

Chemical Treatment

C01	Chemical Precipitation — Lime or Sodium Hydroxide
C02	Chemical Precipitation — Sulfide
C09	Chemical Precipitation — Other
C11	Neutralization
C21	Chromium Reduction
C31	Complexed Metals Treatment (other than pH Adjustment)
C41	Cyanide Oxidation — Alkaline Chlorination
C42	Cyanide Oxidation — Electrochemical
C43	Cyanide Oxidation — Other
C44	General Oxidation (including Disinfection) — Chlorination
C45	General Oxidation (including Disinfection) — Ozonation

Figure 5
Hypothetical Section 7A

SECTION 7A. ON-SITE WASTE TREATMENT METHODS AND EFFICIENCY

☐ Not Applicable (NA) - Check here if **no** on-site waste treatment is applied to any waste stream containing the toxic chemical or chemical category.

a. General Waste Stream (enter code)	b. Waste Treatment Method(s) Sequence (enter 3-character code(s))			c. Range of Influent Concentration	d. Waste Treatment Efficiency Estimate	e. Based on Operating Data?
7A.1a W	7A.1b 1: P12 3: P17 6: P21	2: P18 4: P61 7: B21	5: P42 8: P11	7A.1c NA	7A.1d %	7A.1e ☐ Yes ☐ No
7A.2a	7A.2b 1: C44 3: 6:	2: NA 4: 7:	5: 8:	7A.2c 1	7A.2d 99 %	7A.2e ☒ Yes ☐ No
7A.3a A	7A.3b 1: A01 3: 6:	2: NA 4: 7:	5: 8:	7A.3c 1	7A.3d 91 %	7A.3e ☒ Yes ☐ No

C46 General Oxidation (including Disinfection) — Other
C99 Other Chemical Treatment

Incineration/Thermal Treatment

F01 Liquid Injection
F11 Rotary Kiln with Liquid Injection Unit
F19 Other Rotary Kiln
F31 Two Stage
F41 Fixed Hearth
F42 Multiple Hearth
F51 Fluidized Bed
F61 Infra-Red
F71 Fume/Vapor
F81 Pyrolytic Destructor
F82 Wet Air Oxidation
F83 Thermal Drying/Dewatering
F99 Other Incineration/Thermal Treatment

Physical Treatment

P01 Equalization
P09 Other Blending
P11 Settling/Clarification
P12 Filtration
P13 Sludge Dewatering (non-thermal)
P14 Air Flotation
P15 Oil Skimming
P16 Emulsion Breaking — Thermal
P17 Emulsion Breaking — Chemical
P18 Emulsion Breaking — Other
P19 Other Liquid Phase Separation
P21 Adsorption — Carbon
P22 Adsorption — Ion Exchange (other than for recovery/reuse)
P23 Adsorption — Resin
P29 Adsorption — Other
P31 Reverse Osmosis (other than for recovery/reuse)
P41 Stripping — Air
P42 Stripping — Steam
P49 Stripping — Other
P51 Acid Leaching (other than for recovery/reuse)
P61 Solvent Extraction (other than for recovery/reuse)
P99 Other Physical Treatment

Solidification/Stabilization

G01 Cement Processes (including Silicates)
G09 Other Pozzolonic Processes (including Silicates)
G11 Asphaltic Processes
G21 Thermoplastic Techniques
G99 Other Solidification Processes

7A column c   Range of Influent Concentration

The form requires an indication of the range of concentration of the toxic chemical in the waste stream (i.e., the influent) as it typically enters the waste treatment step or sequence. The concentration is based on the amount or mass of the toxic chemical in the waste stream as compared to the total amount or mass of the waste stream. Enter in the space provided one of the following code numbers corresponding to the concentration of the toxic chemical in the influent:

1 = Greater than 1 percent
2 = 100 parts per million (0.01 percent) to 1 percent (10,000 parts per million)
3 = 1 part per million to 100 parts per million
4 = 1 part per billion to 1 part per million
5 = Less than 1 part per billion

Note: Parts per million (ppm) is:

o   milligrams/kilogram (mass/mass) for solids and liquids;

o   cubic centimeters/cubic meter (volume/volume) for gases;

o   milligrams/liter for solutions or dispersions of the chemical in water; and

o   milligrams of chemical/kilogram of air for particulates in air.

If you have particulate concentrations (at standard temperature and pressure) as grains/cubic foot of air, multiply by 1766.6 to convert to parts per million; if in milligrams/cubic meter, multiply by 0.773 to obtain parts per million. These conversion factors are for standard conditions of 0°C (32°F) and 760 mmHg atmospheric pressure.

7A column d Waste Treatment Efficiency Estimate

In the space provided, enter the number indicating the percentage of the toxic chemical removed from the waste stream through destruction, biological degradation, chemical conversion, or physical removal. The waste treatment efficiency (expressed as percent removal) represents the percentage of the toxic chemical destroyed or removed (based on amount or mass), not merely changes in volume or concentration of the toxic chemical in the waste stream. The efficiency, which can reflect the overall removal from sequential treatment methods applied to the general waste stream, refers only to the percent destruction, degradation, conversion, or removal of the listed toxic chemical from the waste stream, not the percent conversion or removal of other constituents in the waste stream. The efficiency also does not refer to the general efficiency of the treatment method for any waste stream. For some waste treatment methods, the percent removal will represent removal by several mechanisms, as in an aeration basin, where a toxic chemical may evaporate, be biodegraded, or be physically removed from the sludge.

Percent removal can be calculated as follows:

$$\frac{(I - E)}{I} \times 100, \text{ where}$$

I = amount of the toxic chemical in the influent waste stream (entering the waste treatment step or sequence) and

E = amount of the toxic chemical in the effluent waste stream (exiting the waste treatment step or sequence).

Calculate the amount of the toxic chemical in the influent waste stream by multiplying the concentration (by weight) of the toxic chemical in the waste stream by the total amount or weight of the waste stream. In most cases, the percent removal compares the treated effluent to the influent for the particular type of waste stream. For solidification of wastewater, the waste treatment efficiency can be reported as 100 percent if no volatile toxic chemicals were removed with the water or evaporated into the air. Percent removal does not apply to incineration because the waste stream, such as wastewater or liquids, may not exist in a comparable form after waste treatment and the purpose of incineration as a waste treatment is to destroy the toxic chemical by converting it to carbon dioxide and water. In cases where the toxic chemical is incinerated, the percent efficiency must be based on the amount of the toxic chemical destroyed or combusted, except for metals or metal compounds. In the cases where a metal or metal compound is incinerated, the efficiency is always zero for the parent metal.

Similarly, an efficiency of zero must be reported for any waste treatment method(s) (e.g., evaporation) that does not destroy, chemically convert, or physically remove the toxic chemical from the waste stream.

For metal compounds, the calculation of the reportable concentration and waste treatment efficiency must be based on the weight of the parent metal, not on the weight of the metal compounds. Metals are not destroyed, only physically removed or chemically converted from one form into another. The waste treatment efficiency reported must represent only physical removal of the parent metal from the waste stream (except for incineration), not the percent chemical conversion of the metal compound. If a listed waste treatment method converts but does not remove a metal (e.g., chromium reduction), the method must be reported with a waste treatment efficiency of zero.

Listed toxic chemicals that are strong mineral acids neutralized to a pH of 6 or above are considered treated at a 100 percent efficiency.

All data available at your facility must be used to calculate waste treatment efficiency and influent toxic chemical concentration. If data are lacking, estimates must be made using best engineering judgment or other methods.

7A column e Based on Operating Data?

This column requires you to indicate "Yes" or "No" to whether the waste treatment efficiency estimate is based on actual operating data. For example, you would check "Yes" if the estimate is based on monitoring of influent and effluent wastes under typical operating conditions.

If the efficiency estimate is based on published data for similar processes or on equipment supplier's literature, or if you otherwise estimated either the influent or effluent waste comparison or the flow rate, check "No."

Section 7B On-Site Energy Recovery Processes

In Section 7B, you must indicate the on-site energy recovery methods used on the reported toxic chemical. If you do not perform on-site energy recovery for the reported toxic chemical, check the Not Applicable (NA) box at the top of Section 7B.

APPENDIX 10—FORM R

> **Example 11: Reporting On-Site Energy Recovery**
>
> One waste stream generated by your facility contains, among other chemicals, toluene and cadmium. Threshold quantities are exceeded for both of these toxic chemicals, and you would, therefore, submit two separate Form R reports. This waste stream is sent to an on-site industrial furnace which uses the heat generated in a thermal hydrocarbon cracking process at your facility. Because toluene has a significant heat value (17,440 BTU/pound) and the energy is recovered in an industrial furnace, the code "U02" would be reported in Section 7B for the Form R submitted for toluene.
>
> However, as cadmium is a non-combustible metal and therefore does not contribute any heat value for energy recovery purposes, the combustion of cadmium in the industrial furnace is considered waste treatment, not energy recovery. You would report cadmium as entering a waste treatment step (i.e., incineration), in Section 7A, column b.

Only listed toxic chemicals that have a significant heating value and are combusted in an energy recovery unit such as an industrial furnace, kiln, or boiler, can be reported as combusted for energy recovery in this section. If a reported toxic chemical is incinerated on-site but does not contribute energy to the process (e.g., metals and chlorofluorocarbons), it must be considered waste treated on-site and reported in Section 7A. Energy recovery may take place only in one of the types of energy recovery equipment listed below.

Energy Recovery Codes

U01	Industrial Kiln
U02	Industrial Furnace
U03	Industrial Boiler
U09	Other Energy Recovery Methods

If your facility uses more than one on-site energy recovery method for the reported toxic chemical, list the methods used in descending order (greatest to least) based on the amount of the toxic chemical entering such methods.

Section 7C   On-Site Recycling Processes

In Section 7C, you must report the recycling methods used on the listed toxic chemical. If you do not conduct any on-site recycling of the reported toxic chemical, check the Not Applicable (NA) box at the top of Section 7C.

In this section, use the codes below to report only the recycling methods in place at your facility that are applied to the listed toxic chemical. Do not list any off-site recycling activities (Information about off-site recycling must be reported in Part II, Section 6, "Transfers of the Toxic Chemical in Wastes to Off-Site Locations,").

On-Site Recycling Codes

R11	Solvents/Organics Recovery — Batch Still Distillation
R12	Solvents/Organics Recovery — Thin-Film Evaporation
R13	Solvents/Organics Recovery — Fractionation
R14	Solvents/Organics Recovery — Solvent Extraction
R19	Solvents/Organics Recovery — Other
R21	Metals Recovery — Electrolytic
R22	Metals Recovery — Ion Exchange
R23	Metals Recovery — Acid Leaching
R24	Metals Recovery — Reverse Osmosis
R26	Metals Recovery — Solvent Extraction
R27	Metals Recovery — High Temperature
R28	Metals Recovery — Retorting
R29	Metals Recovery — Secondary Smelting
R30	Metals Recovery — Other
R40	Acid Regeneration
R99	Other Reuse or Recovery

If your facility uses more than one on-site recycling method for a toxic chemical, enter the codes in the space provided in descending order (greatest to least) of the volume of the reported toxic chemical recovered by each process. If your facility uses more than ten separate methods for recycling the reported toxic chemical on-site, then list the ten activities that recover the greatest amount of the toxic chemical (again, in descending order).

Section 8 Source Reduction and Recycling Activities

This Section includes the new data elements mandated by section 6607 of the Pollution Prevention Act of 1990 (PPA). Section 8 is now a required section of Form R and must be completed. This is the first reporting year these data are being collected. They are included in the Form R for reports due on or before July 1, 1992, covering source reduction and recycling activities in calendar year 1991. You are not required to amend previous year's submissions to include this information.

In Section 8, you must provide information about source reduction and recycling activities related to the toxic chemical for which releases are being reported. For all appropriate questions, report only the quantity, in pounds, of the reported toxic chemical. Do not include the weight of water, soil, or other waste constituents. When reporting on a metal compound, report only the amount of the parent metal as you do when estimating release amounts. All amounts must be reported in whole numbers and up to two significant figures can be provided.

Section 8.1 through 8.9 must be completed for each toxic chemical. Section 8.10 must be completed only if a source reduction activity was newly implemented specifically (in whole or in part) for the reported toxic chemical during the reporting year. Section 8.11 allows you to indicate if you have attached additional optional information on source reduction, recycling, or pollution control activities implemented at any time at your facility.

Sections 8.1 through 8.7 require reporting of quantities for the current reporting year, the prior year, and quantities anticipated in both the first year immediately following the reporting year and the second year following the reporting year (future estimates).

Column A: 1990 (Prior Year)

Quantities for Sections 8.1 through 8.7 must be reported for the year immediately preceding the reporting year in column A. For reports due July 1, 1992, the prior year is 1990. Information available at the facility that may be used to estimate the prior year's quantities include the prior year's Form R submission, supporting documentation, and recycling, energy recovery, or treatment operating logs or invoices.

EPA believes that such data should be available, especially in those cases where the facility has filed a Form R for the reported toxic chemical in the prior year. However, for the first year of reporting these data elements, 1991, prior year quantities are required only to the extent such information is available. In the event that sufficient data are not available, enter not applicable, "NA."

Column B: 1991 (Reporting Year)

Quantities for Sections 8.1 through 8.7 must be reported for the current reporting year in column B.

Columns C and D: 1992 and 1993 (Following Year and Second Year)

Quantities for Sections 8.1 through 8.7 must be estimated for 1992 and 1993. EPA expects reasonable future quantity estimates using a logical basis. Information available at the facility to estimate quantities of the chemical expected during these years include planned source reduction activities, market projections, expected contracts, anticipated new product lines, company growth projections, and production capacity figures. Not applicable, "NA", may not be entered for these data elements. Respondents should take into account protections available for trade secrets as provided in EPCRA Section 322 (42 USC 11042).

Example 12: Reporting Future Estimates

A pharmaceutical manufacturing facility uses a listed toxic chemical in the manufacture of a prescription drug. During the reporting year (1991), the company received approval from the Food and Drug Administration to begin marketing their product as an over-the-counter drug beginning in 1992. This approval is publicly known and does not constitute confidential business information. As a result of this expanded market, the company estimates that sales and subsequent production of this drug will increase their use of the reported toxic chemical by 30 percent per year for the two years following the reporting year. The facility treats the toxic chemical on-site and the quantity treated is directly proportional to production activity. The facility thus estimates the total quantity of the reported toxic chemical treated for the following year (1992) by adding 30 percent to the amount in column B (the amount for the current reporting year). The second year (1993) figure can be calculated by adding an additional 30 percent to the amount reported in Column C (the amount for the following year (1992) projection).

Relationship to Other Laws

The reporting categories for quantities recycled, treated, used for energy recovery, and disposed apply to completing Section 8 of Form R as well as to the rest of Form R. These categories are to be used only for TRI reporting. They are not intended for use in determining, under the Resource Conservation and Recovery Act (RCRA) Subtitle C regulations, whether a secondary material is a waste when recycled. These definitions also do not apply to the information that may be submitted in the Biennial Report required under RCRA. In addition, these definitions do not imply any future redefinition of RCRA terms and do not affect EPA's RCRA authority or authority under any other statute administered by EPA.

Differences in terminology and reporting requirements for toxic chemicals reported on Form R and for hazardous wastes regulated under RCRA occur because EPCRA and the PPA focus on specific chemicals, while the RCRA regulations and the Biennial Report focus on wastes, including mixtures. For example, a RCRA hazardous waste containing a section 313 toxic chemical is recycled to recover certain constituents of that waste, but not the toxic chemical reported under EPCRA section 313. The toxic chemical simply passes through the recycling process and remains in the residual from the recycling process. While the waste may be considered recycled under RCRA, the toxic chemical constituent would be considered to be treated for TRI purposes.

Quantities Reportable in Sections 8.1 - 8.7

8.1 Report releases pursuant to EPCRA Section 329(8) including "any spilling, leaking, pumping, pouring, emitting, emptying, discharging, injecting, escaping, leaching, dumping, or disposing [on-site or off-site] into the environment (including the abandonment of barrels, containers, and other closed receptacles)." Do not include any quantity treated on-site or off-site.

8.2 - 8.3 A toxic chemical or a mixture containing a toxic chemical that is used for energy recovery on-site or is sent off-site for energy recovery, unless it is a commercially available fuel. For the purposes of reporting on Form R, reportable on-site and off-site energy recovery is the combustion of a residual material containing a TRI toxic chemical when:

(a) The combustion unit is integrated into an energy recovery system (i.e., industrial furnaces, industrial kilns, and boilers); and

(b) The toxic chemical is combustible and has a heating value high enough to sustain combustion.

8.4 - 8.5 A toxic chemical or a mixture containing a toxic chemical that is recycled on-site or is sent off-site for recycling.

8.6 - 8.7 A toxic chemical or a mixture containing a toxic chemical that is treated on-site or is sent to a POTW or other off-site location for waste treatment.

A toxic chemical or a toxic chemical in a mixture that is a waste under RCRA must be reported in Sections 8.1 through 8.7.

Avoid Double-Counting in Sections 8.1 Through 8.8

Section 8 of Form R uses data collected to complete Part II, Sections 5 through 7. For this reason, Section 8 should be completed last.

Do not double- or multiple-count quantities in Sections 8.1 through 8.7. The quantities reported in each of those sections must be mutually exclusive. Do not multiple-count quantities entering sequential reportable activities. For example, 5,000 pounds of toxic chemical enters a treatment operation. Three thousand pounds of the toxic chemical exits the treatment operation and then enters a recycling operation. Five hundred pounds of the toxic chemical is in residues from the recycling operation which is subsequently sent off-site for disposal. These quantities would be reported as follows in Section 8:

Section 8.1:	500 pounds disposed
Section 8.4:	2,500 pounds recycled
Section 8.6:	2,000 pounds treated (5,000 that initially entered - 3,000 that subsequently entered recycling)

To report that 5,000 pounds were treated, 3,000 pounds were recycled, and that 500 pounds were sent off-site for disposal would result in over-counting the quantities of toxic chemical recycled, treated, and disposed by 3,500 pounds.

Do not include in Sections 8.1 through 8.7 any quantities of the toxic chemical released into the environment due to remedial actions; catastrophic events such as earthquakes, fires, or floods; or unanticipated one-time events not associated with the production process such as tank ruptures or reactor explosions. These quantities should be reported in Section 8.8 only. For example, 10,000 pounds of diaminoanisole sulfate is released due to a catastrophic event and is subsequently treated off-site.

The 10,000 pounds is reported in Section 8.8, but the amount subsequently treated off-site is not reported in Section 8.7.

8.8 Quantity Released to the Environment as a Result of Remedial Actions, Catastrophic Events, or One-Time Events Not Associated with Production Processes.

In Section 8.8, enter the total quantity of toxic chemical released directly into the environment or sent off-site for recycling, waste treatment, energy recovery, or disposal during the reporting year due to any of the following events:

(1) remedial actions,
(2) catastrophic events such as earthquakes, fires, or floods; or
(3) one-time events not associated with normal or routine production processes.

These quantities should not be included in Sections 8.1 through 8.7. The amount of toxic chemical released into the environment during remediation or transferred off-site is to be reported in Part II, Sections 5 and 6 as appropriate.

The purpose of this section is to separate quantities recycled, used for energy recovery, treated, or disposed that are associated with normal or routine production operations from those that are not. While all quantities released, recycled, treated, or disposed may ultimately be preventable, this section separates the quantities that are more likely to be reduced or eliminated by process-oriented source reduction activities from those releases that are largely unpredictable and are less amenable to such source reduction activities. For example, spills that occur as a routine part of production operations and could be reduced or eliminated by improved handling, loading, or unloading procedures are included in the quantities reported in Section 8.1 through 8.7 as appropriate. A total loss of containment resulting from a tank rupture caused by a tornado would be included in the quantity reported in Section 8.8.

Similarly, the amount of a toxic chemical spilled or cleaned up from normal operations during the reporting year would be included in the quantities reported in Sections 8.1 through 8.7. However, the quantity of the reported toxic chemical generated from a remedial action (e.g., RCRA corrective action) to clean up the environmental contamination resulting from past practices should be reported in Section 8.8 because they cannot currently be addressed by source reduction methods. A remedial action for purposes of Section 8.8 is a waste cleanup (including RCRA and CERCLA operations) within the facility boundary. Most remedial activities involve collecting and treating contaminated material.

Also, releases caused by catastrophic events are to be incorporated into the quantity reported in Section 8.8. Such releases may be caused by natural disasters (e.g., hurricanes and earthquakes) or by large scale accidents (e.g., fires and explosions). These amounts are not included in the quantity reported in Sections 8.1 through 8.7 because such releases are generally unanticipated and

---

Example 13: Quantity Released to the Environment as a Result of Remedial Actions, Catastrophic Events, or One-Time Events Not Associated with Production Processes.

A chemical manufacturer produces a toxic chemical in a reactor that operates at low pressure. The reactants and the toxic chemical product are piped in and out of the reactor at monitored and controlled temperatures. During normal operations, small amounts of fugitive emissions occur from the valves and flanges in the pipelines.

Due to a malfunction in the control panel (which is state-of-the-art and undergoes routine inspection and maintenance), the temperature and pressure in the reactor increase, the reactor ruptures, and the toxic chemical is released. Because the malfunction could not be anticipated and, therefore, could not be reasonably addressed by specific source reduction activities, the amount released is included in Section 8.8. In this case, much of the toxic chemical is released as a liquid and pools on the ground. It is estimated that 1,000 pounds of the toxic chemical pooled on the ground and was subsequently collected and sent off-site for treatment. In addition, it is estimated that another 200 pounds of the toxic chemical vaporized directly to the air from the rupture. The total amount reported in Section 8.8 is the 1,000 pounds that pooled on the ground (and subsequently sent off-site), plus the 200 pounds that vaporized into the air, a total of 1,200 pounds. The quantity sent off-site must also be reported in Section 6 (but not in Section 8.7) and the quantity that vaporized must be reported as a fugitive emission in Section 5 (but not in 8.1).

cannot be addressed by routine process-oriented accident prevention techniques.

By checking your documentation for calculating estimates made for Part II, Section 5, "Releases of the Toxic Chemical to the Environment," you may be able to identify release amounts from the above sources. Emergency notifications under CERCLA and EPCRA as well as accident histories required under the Clean Air Act may provide useful information. You should also check facility incident reports and maintenance records to identify one-time or catastrophic events.

Note that while the information reported in Section 8.8 represents only remedial, catastrophic, or one-time events not associated with production processes, Section 5 of Form R (releases to the environment) and Section 6 (off-site transfers), must include all releases and transfers as appropriate, regardless of whether they arise from catastrophic, remedial, or routine process operations.

8.9 Production Ratio or Activity Index

For Section 8.9, you must provide a ratio of reporting year production to prior year production, or provide an "activity index" based on a variable other than production that is the primary influence on the quantity of the reported toxic chemical recycled, used for energy recovery, treated, or disposed. The ratio or index must be reported to the nearest tenths or hundredths place (e.g., one or two digits to the right of the decimal point). If the manufacture or use of the reported toxic chemical began during the current reporting year, enter not applicable, "NA," as the production ratio or activity index.

It is important to realize that if your facility reports more than one reported toxic chemical, the production ratio or activity index may vary for different chemicals. For facilities that manufacture reported toxic chemicals, the quantities of the toxic chemical(s) produced in the current and prior years provide a good basis for the ratio because that is the primary business activity associated with the reported toxic chemical(s). In most cases, the production ratio or activity index must be based on some variable of production or activity rather than on toxic

---

Example 14: Determining a Production Ratio

Your facility's only use of toluene is as a paint carrier for a painting operation. You painted 12,000 refrigerators in the current reporting year and 10,000 refrigerators during the preceding year. The production ratio for toluene in this case is 1.2 (12,000/10,000) because the number of refrigerators produced is the primary factor determining the quantity of toluene to be reported in Sections 8.1 through 8.7.

A facility manufactures inorganic pigments, including titanium dioxide. Hydrochloric acid is produced as a waste byproduct during the production process. An appropriate production ratio for hydrochloric acid is the annual titanium dioxide production, not the amount of byproduct generated. If the facility produced 20,000 pounds of titanium dioxide during the reporting year and 26,000 pounds in the preceding year, the production ratio would be 0.77 (20,000/26,000).

---

chemical or material usage. Indices based on toxic chemical or material usage may reflect the effect of source reduction activities rather than changes in business activity. Toxic chemical or material usage is therefore not a basis to be used for the production ratio or activity index where the toxic chemical is "otherwise-used" (i.e., non-incorporative activities such as extraction solvents, metal degreasers, etc.).

While several methods are available to the facility for determining this data element, the production ratio or activity index must be based on the variable that most directly affects the quantities of the toxic chemical recycled, used for energy recovery, treated, or disposed. Examples of methods available include:

(1) Amount of toxic chemical manufactured in 1991 divided by the amount of toxic chemical manufactured in 1990; or

(2) Amount of product produced in 1991 divided by the amount of product produced in 1990.

Example 15: Determining an Activity Index

Your facility manufactures organic dyes in a batch process. Different colors of dyes are manufactured, and between color changes, all equipment must be thoroughly cleaned with solvent containing glycol ethers to reduce color carryover. During the preceding year, the facility produced 2,000 pounds of yellow dye in January, 9,000 pounds of green dye for February through September, 2,000 pounds of red dye in November, and another 2,000 pounds of yellow dye in December. This adds up to a total of 15,000 pounds and four color changeovers. During the reporting year, the facility produced 10,000 pounds of green dye during the first half of the year and 10,000 pounds of red dye in the second half. If your facility uses glycol ethers in this cleaning process only, an activity index of 0.5 (based on two color changeovers for the reporting year divided by four changeovers for the preceding year) is more appropriate than a production ratio of 1.33 (based on 20,000 pounds of dye produced in the current year divided by 15,000 pounds in the preceding year). In this case, an activity index, rather than a production ratio, better reflects the factors that influence the amount of solvent recycled, used for energy recovery, treated, or disposed.

A facility that manufactures thermoplastic composite parts for aircraft uses acetone as a wipe solvent to clean molds. The solvent is stored in 55-gallon drums and is transferred to 1-gallon dispensers. The molds are cleaned on an as-needed basis that is not necessarily a function of the parts production rate. Operators cleaned 5,200 molds during the reporting year, but only cleaned 2,000 molds in the previous year. An activity index of 2.6 (5,200/2,000) represents the activities involving acetone usage in the facility. If the molds were cleaned after 1,000 parts were manufactured, a production ratio would equal the activity index and either could be used as the basis for the index.

A facility manufactures surgical instruments and cleans the metal parts with 1,1,1-trichloroethane in a vapor degreaser. The degreasing unit is operated in a batch mode and the metal parts are cleaned according to an irregular schedule. The activity index can be based upon the total time the metal parts are in the degreasing operation. If the degreasing unit operated 3,900 hours during the reporting year and 3,000 hours the prior year, the activity index is 1.3 (3,900/3,000).

A pharmaceutical plant uses hydrochloric acid to regenerate deionization units that supply deionized water to several operations in the facility. During the reporting year, the facility noted that the units were recharged once per week. Records for the prior year indicate that the units were recharged four times per week. Provided that the reduction in recharges per week is not part of a planned source reduction program, an index of 0.25 (1/4) represents the activities that were the primary influence on the amount of hydrochloric acid recycled, used for energy recovery, treated, or disposed.

Example 16: "NA" is Entered as the Production Ratio or Activity Index

Your facility began production of a microwidget during this reporting year. Perchloroethylene is used as a cleaning solvent for this operation and this is the only use of the toxic chemical in your facility. You would enter not applicable, "NA," in Section 8.9 because you have no basis of comparison in the prior year for the purposes of developing the activity index.

> **Example 17: Determining the Production Ratio Based on a Weighted Average**
>
> At many facilities, a reported toxic chemical is used in more than one production process. In these cases, a production ratio or activity index can be estimated by weighting the production ratio for each process based on the respective contribution of each process to the quantity of the reported toxic chemical recycled, used for energy recovery, treated, or disposed.
>
> Your facility paints bicycles with paint containing toluene. Sixteen thousand bicycles were produced in the reporting year and 14,500 were produced in the prior year. There were no significant design modifications that changed the total surface area to be painted for each bike. The bicycle production ratio is 1.1 (16,000/14,5000). You estimate 12,500 pounds of toluene treated, recycled, used for energy recovery, or disposed as a result of bicycle production. Your facility also uses toluene as a solvent in a glue that is used to make components and add-on equipment for the bicycles. Thirteen thousand components were manufactured in the reporting year as compared to 15,000 during the prior year. The production ratio for the components using toluene is 0.87 (13,000/15,000). You estimate 1,000 pounds of toluene treated, recycled, used for energy recovery, or disposed as a result of components production. A production ratio can be calculated by weighting each of the production ratios based on the relative contribution each has to the quantities of toluene treated, recycled, used for energy recovery, or disposed during the reporting year (13,500 pounds). The production ratio is calculated as follows:
>
> Production ratio = (12,500/13,500 x 1.1) + (1,000/13,500 x 0.87) = 1.08

8.10 Did Your Facility Engage in any Source Reduction Activities for this Chemical during the Reporting Year?

If your facility engaged in any source reduction activity for the reported toxic chemical during the reporting year, report the activity that was implemented and the method used to identify the opportunity for the activity implemented. If your facility did not engage in any source reduction activity for the reported toxic chemical, enter not applicable, "NA," in Section 8.10.1 and answer Section 8.11.

Source reduction means any practice which:

- Reduces the amount of any hazardous substance, pollutant, or contaminant entering any waste stream or otherwise released into the environment (including fugitive emissions) prior to recycling, treatment, or disposal; and

- Reduces the hazards to public health and the environment associated with the release of such substances, pollutants, or contaminants.

The term includes equipment or technology modifications, process or procedure modifications, reformulation or redesign of products, substitution of raw materials, and improvements in housekeeping, maintenance, training, or inventory control.

The term source reduction does not include any practice which alters the physical, chemical, or biological characteristics or the volume of a hazardous substance, pollutant, or contaminant through a process or activity which itself is not integral to and necessary for the production of a product or the providing of a service.

Source reduction activities do not include recycling, treating, using for energy recovery, or disposing of a toxic chemical. Report in this section only the source reduction activities implemented to reduce or eliminate the quantities reported in Sections 8.1 through 8.7 — the focus of the section is only those activities that are applied to reduce routine or reasonably anticipated releases and quantities of the reported toxic chemical recycled, treated, used for energy recovery, or disposed. Do not report in this section any activities taken to reduce or eliminate the quantities reported in Section 8.8.

> **Example 18: Source Reduction**
>
> A facility assembles and paints furniture. Both the glue used to assemble the furniture and the paints contain listed toxic chemicals. By examining the gluing process, the facility discovered that a new drum of glue is opened at the beginning of each shift, whether the old drum is empty or not. By adding a mechanism that prevents the drum from being changed before it is empty, the need for disposal of the glue is eliminated at the source. As a result, this activity is considered source reduction. The painting process at this facility generates a solvent waste which is collected and recovered. The recovered solvent is used to clean the painting equipment. The recycling activity does not reduce the amount of toxic chemical recycled, and therefore is not considered a source reduction activity.

Source Reduction Activities

You must enter in the first column of Section 8.10, "Source Reduction Activities," the appropriate code(s) indicating the type of actions taken to reduce the amount of the reported toxic chemical released (as reported in Section 8.1), used for energy recovery (as reported in Section 8.2), recycled (as reported in Section 8.4-8.5), or treated (as reported in Section 8.6-8.7). The list of codes below includes many, but not all, of the codes provided in the RCRA biennial report. Remember that source reduction activities include only those actions or techniques that reduce or eliminate the amounts of the toxic chemical reported in Section 8.1 through 8.7. Actions taken to recycle, treat, or dispose of the toxic chemical are not considered source reduction activities.

Source Reduction Activity Codes:

Good Operating Practices

W13	Improved maintenance scheduling, recordkeeping, or procedures
W14	Changed production schedule to minimize equipment and feedstock changeovers
W19	Other changes in operating practices

Inventory Control

W21	Instituted procedures to ensure that materials do not stay in inventory beyond shelf-life
W22	Began to test outdated material — continue to use if still effective
W23	Eliminated shelf-life requirements for stable materials
W24	Instituted better labelling procedures
W25	Instituted clearinghouse to exchange materials that would otherwise be discarded
W29	Other changes in inventory control

Spill and Leak Prevention

W31	Improved storage or stacking procedures
W32	Improved procedures for loading, unloading, and transfer operations
W33	Installed overflow alarms or automatic shut-off valves
W35	Installed vapor recovery systems
W36	Implemented inspection or monitoring program of potential spill or leak sources
W39	Other spill and leak prevention

Raw Material Modifications

W41	Increased purity of raw materials
W42	Substituted raw materials
W49	Other raw material modifications

Process Modifications

W51	Instituted recirculation within a process
W52	Modified equipment, layout, or piping
W53	Use of a different process catalyst
W54	Instituted better controls on operating bulk containers to minimize discarding of empty containers
W55	Changed from small volume containers to bulk containers to minimize discarding of empty containers
W58	Other process modifications

Cleaning and Degreasing

W59	Modified stripping/cleaning equipment
W60	Changed to mechanical stripping/cleaning devices (from solvents or other materials)
W61	Changed to aqueous cleaners (from solvents or other materials)

W63 Modified containment procedures for cleaning units
W64 Improved draining procedures
W65 Redesigned parts racks to reduce dragout
W66 Modified or installed rinse systems
W67 Improved rinse equipment design
W68 Improved rinse equipment operation
W71 Other cleaning and degreasing modifications

Surface Preparation and Finishing

W72 Modified spray systems or equipment
W73 Substituted coating materials used
W74 Improved application techniques
W75 Changed from spray to other system
W78 Other surface preparation and finishing modifications

Product Modifications

W81 Changed product specifications
W82 Modified design or composition of product
W83 Modified packaging
W89 Other product modifications

In columns a through c of Section 8.10, the "Methods to Identify Activity", you must enter one or more of the following code(s) that correspond to those internal and external method(s) or information sources you used to identify the possibility for a source reduction activity implementation at your facility. If more than three methods were used to identify the source reduction activity, enter only the three codes that contributed most to the decision to implement the activity.

Methods to Identify Activity

T01 Internal pollution prevention opportunity audit(s)
T02 External pollution prevention opportunity audit(s)
T03 Materials balance audits
T04 Participative team management
T05 Employee recommendation (independent of a formal company program)
T06 Employee recommendation (under a formal company program)
T07 State government technical assistance program
T08 Federal government technical assistance program
T09 Trade association/industry technical assistance program
T10 Vendor assistance
T11 Other

8.11 **Is Additional Information on Source Reduction, Recycling, or Pollution Control Activities Included with this Report?**

Check "Yes" for this data element if you have attached to this report any additional optional information on source reduction, recycling, or pollution control activities you have implemented in the reporting year or in prior years for the reported toxic chemical. If you are not including additional information, check "No."

If you submit additional optional information, try to limit this information to one page that summarizes the source reduction, recycling, or pollution control activities. If there is a contact person at the facility, other than the technical or public contact provided in Part I, Section 4, the summary page should include that person's name and telephone number for individuals who wish to obtain further information about those activities. Also submit a copy of this additional information to the appropriate state agency as part of the Form R submittal to that agency.

(IMPORTANT: Type or print; read instructions before completing form)

Form Approved OMB Number: 2070-0093
Approval Expires: 11/92

Page 1 of 9

TRI FACILITY ID NUMBER

Toxic Chemical, Category, or Generic Name

♻ EPA
United States
Environmental Protection
Agency

FORM R TOXIC CHEMICAL RELEASE INVENTORY REPORTING FORM

Section 313 of the Emergency Planning and Community Right-to-Know Act of 1986, also known as Title III of the Superfund Amendments and Reauthorization Act

WHERE TO SEND COMPLETED FORMS:
1. EPCRA Reporting Center
P.O. Box 23779
Washington, DC 20026-3779
ATTN: TOXIC CHEMICAL RELEASE INVENTORY

2. APPROPRIATE STATE OFFICE
(See instructions in Appendix F)

Enter "X" here if this is a revision

For EPA use only

IMPORTANT: See instructions to determine when "Not Applicable (NA)" boxes should be checked.

PART I. FACILITY IDENTIFICATION INFORMATION

SECTION 1.

REPORTING YEAR

19 ___

SECTION 2. TRADE SECRET INFORMATION

2.1 Are you claiming the toxic chemical identified on page 3 trade secret?
☐ Yes (Answer question 2.2; Attach substantiation forms)
☐ No (Do not answer 2.2; Go to Section 3)

2.2 If yes in 2.1, is this copy: ☐ Sanitized ☐ Unsanitized

SECTION 3. CERTIFICATION (Important: Read and sign after completing all form sections.)

I hereby certify that I have reviewed the attached documents and that, to the best of my knowledge and belief, the submitted information is true and complete and that the amounts and values in this report are accurate based on reasonable estimates using data available to the preparers of this report.

Name and official title of owner/operator or senior management official

Signature

Date Signed

SECTION 4. FACILITY IDENTIFICATION

4.1
Facility or Establishment Name

TRI Facility ID Number

Street Address

City

County

State

Zip Code

Mailing Address (if different from street address)

City

State

Zip Code

PUT LABEL HERE

EPA Form 9350-1 (Rev. 5/14/92) - Previous editions are obsolete.

EPA FORM R
PART I. FACILITY IDENTIFICATION INFORMATION (CONTINUED)

TRI FACILITY ID NUMBER

Toxic Chemical, Category, or Generic Name

SECTION 4. FACILITY IDENTIFICATION (Continued)

4.2	This report contains information for: (Important: check only one)	a. ☐ An entire facility	b. ☐ Part of a facility

4.3	Technical Contact	Name	Telephone Number (include area code)

4.4	Public Contact	Name	Telephone Number (include area code)

4.5	SIC Code (4-digit)	a.	b.	c.	d.	e.	f.

4.6	Latitude and Longitude	Latitude			Longitude		
		Degrees	Minutes	Seconds	Degrees	Minutes	Seconds

4.7	Dun & Bradstreet Number(s) (9 digits)	a.
		b.

4.8	EPA Identification Number(s) (RCRA I.D. No.) (12 characters)	a.
		b.

4.9	Facility NPDES Permit Number(s) (9 characters)	a.
		b.

4.10	Underground Injection Well Code (UIC) I.D. Number(s) (12 digits)	a.
		b.

SECTION 5. PARENT COMPANY INFORMATION

5.1	Name of Parent Company ☐ NA

5.2	Parent Company's Dun & Bradstreet Number ☐ NA (9 digits)

EPA Form 9350-1 (Rev. 5/14/92) - Previous editions are obsolete.

Page 3 of 9

| ⬧EPA
United States
Environmental Protection
Agency | EPA FORM R
PART II. CHEMICAL-SPECIFIC INFORMATION | TRI FACILITY ID NUMBER

Toxic Chemical, Category, or Generic Name |

SECTION 1. TOXIC CHEMICAL IDENTITY
(Important: DO NOT complete this section if you complete Section 2 below.)

1.1 CAS Number (Important: Enter only one number exactly as it appears on the Section 313 list. Enter category code if reporting a chemical category.)

1.2 Toxic Chemical or Chemical Category Name (Important: Enter only one name exactly as it appears on the Section 313 list.)

1.3 Generic Chemical Name (Important: Complete only if Part I, Section 2.1 is checked "yes." Generic Name must be structurally descriptive.)

SECTION 2. MIXTURE COMPONENT IDENTITY
(Important: DO NOT complete this section if you complete Section 1 above.)

2.1 Generic Chemical Name Provided by Supplier (Important: Maximum of 70 characters, including numbers, letters, spaces, and punctuation.)

SECTION 3. ACTIVITIES AND USES OF THE TOXIC CHEMICAL AT THE FACILITY
(Important: Check all that apply.)

3.1 Manufacture the toxic chemical:
- a. ☐ Produce
- b. ☐ Import

If produce or import:
- c. ☐ For on-site use/processing
- d. ☐ For sale/distribution
- e. ☐ As a byproduct
- f. ☐ As an impurity

3.2 Process the toxic chemical:
- a. ☐ As a reactant
- b. ☐ As a formulation component
- c. ☐ As an article component
- d. ☐ Repackaging

3.3 Otherwise use the toxic chemical:
- a. ☐ As a chemical processing aid
- b. ☐ As a manufacturing aid
- c. ☐ Ancillary or other use

SECTION 4. MAXIMUM AMOUNT OF THE TOXIC CHEMICAL ON-SITE AT ANY TIME DURING THE CALENDAR YEAR

4.1 ☐ (Enter two-digit code from instruction package.)

EPA Form 9350-1(Rev. 5/14/92) - Previous editions are obsolete.

APPENDIX 10—FORM R 279

EPA United States Environmental Protection Agency

EPA FORM R
PART II. CHEMICAL-SPECIFIC INFORMATION (CONTINUED)

Page 4 of 9

TRI FACILITY ID NUMBER

Toxic Chemical, Category, or Generic Name

SECTION 5. RELEASES OF THE TOXIC CHEMICAL TO THE ENVIRONMENT ON-SITE

			A. Total Release (pounds/year) (enter range code from instructions or estimate)	B. Basis of Estimate (enter code)	C. % From Stormwater
5.1	Fugitive or non-point air emissions	☐ NA			
5.2	Stack or point air emissions	☐ NA			
5.3	Discharges to receiving streams or water bodies (enter one name per box)				
5.3.1	Stream or Water Body Name				
5.3.2	Stream or Water Body Name				
5.3.3	Stream or Water Body Name				
5.4	Underground injections on-site	☐ NA			
5.5	Releases to land on-site				
5.5.1	Landfill	☐ NA			
5.5.2	Land treatment/application farming	☐ NA			
5.5.3	Surface impoundment	☐ NA			
5.5.4	Other disposal	☐ NA			

☐ Check here only if additional Section 5.3 information is provided on page 5 of this form.

EPA Form 9350-1 (Rev. 5/14/92) - Previous editions are obsolete.

Range Codes: A = 1 - 10 pounds; B = 11 - 499 pounds; C = 500 - 999 pounds.

		Page 5 of 9
⊕EPA United States Environmental Protection Agency	**EPA FORM R** PART II. CHEMICAL-SPECIFIC INFORMATION (CONTINUED)	TRI FACILITY ID NUMBER Toxic Chemical, Category, or Generic Name

SECTION 5.3 ADDITIONAL INFORMATION ON RELEASES OF THE TOXIC CHEMICAL TO THE ENVIRONMENT ON-SITE

5.3	Discharges to receiving streams or water bodies (enter one name per box)	A. Total Release (pounds/year) (enter range code from instructions or estimate)	B. Basis of Estimate (enter code)	C. % From Stormwater
5.3.__	Stream or Water Body Name			
5.3.__	Stream or Water Body Name			
5.3.__	Stream or Water Body Name			

SECTION 6. TRANSFERS OF THE TOXIC CHEMICAL IN WASTES TO OFF-SITE LOCATIONS

6.1 DISCHARGES TO PUBLICLY OWNED TREATMENT WORKS (POTW)

6.1.A Total Quantity Transferred to POTWs and Basis of Estimate

6.1.A.1 Total Transfers (pounds/year) (enter range code or estimate)	6.1.A.2 Basis of Estimate (enter code)

6.1.B POTW Name and Location Information

6.1.B.__ POTW Name	6.1.B.__ POTW Name
Street Address	Street Address
City	County	City	County
State	Zip Code	State	Zip Code

If additional pages of Part II, Sections 5.3 and/or 6.1 are attached, indicate the total number of pages in this box ☐ and indicate which Part II, Sections 5.3/6.1 page this is, here. ☐
(example: 1, 2, 3, etc.)

EPA Form 9350-1 (Rev. 5/14/92) - Previous editions are obsolete.

Range Codes: A = 1 - 10 pounds; B = 11 - 499 pounds; C = 500 - 999 pounds.

APPENDIX 10—FORM R   281

Page 6 of 9

EPA FORM R
PART II. CHEMICAL-SPECIFIC INFORMATION (CONTINUED)

United States Environmental Protection Agency

TRI FACILITY ID NUMBER

Toxic Chemical, Category, or Generic Name

SECTION 6.2   TRANSFERS TO OTHER OFF-SITE LOCATIONS

6.2.___ Off-site EPA Identification Number (RCRA ID No.)

Off-Site Location Name

Street Address

City   County

State   Zip Code   Is location under control of reporting facility or parent company?   ☐ Yes   ☐ No

A. Total Transfers (pounds/year) (enter range code or estimate)	B. Basis of Estimate (enter code)	C. Type of Waste Treatment/Disposal/ Recycling/Energy Recovery (enter code)
1.	1.	1. M
2.	2.	2. M
3.	3.	3. M
4.	4.	4. M

SECTION 6.2   TRANSFERS TO OTHER OFF-SITE LOCATIONS

6.2.___ Off-site EPA Identification Number (RCRA ID No.)

Off-Site Location Name

Street Address

City   County

State   Zip Code   Is location under control of reporting facility or parent company?   ☐ Yes   ☐ No

A. Total Transfers (pounds/year) (enter range code or estimate)	B. Basis of Estimate (enter code)	C. Type of Waste Treatment/Disposal/ Recycling/Energy Recovery (enter code)
1.	1.	1. M
2.	2.	2. M
3.	3.	3. M
4.	4.	4. M

If additional pages of Part II, Section 6.2 are attached, indicate the total number of pages in this box ☐ and indicate which Part II, Section 6.2 page this is, here. ☐ (example: 1, 2, 3, etc.)

EPA Form 9350-1 (Rev. 5/14/92) - Previous editions are obsolete.

Range Codes: A = 1 - 10 pounds; B = 11 - 499 pounds; C = 500 - 999 pounds.

EPA FORM R
PART II. CHEMICAL-SPECIFIC INFORMATION (CONTINUED)

United States Environmental Protection Agency

Page 7 of 9

TRI FACILITY ID NUMBER

Toxic Chemical, Category, or Generic Name

SECTION 7A. ON-SITE WASTE TREATMENT METHODS AND EFFICIENCY

☐ **Not Applicable (NA)** - Check here if <u>no</u> on-site waste treatment is applied to any waste stream containing the toxic chemical or chemical category.

a. General Waste Stream (enter code)	b. Waste Treatment Method(s) Sequence [enter 3-character code(s)]	c. Range of Influent Concentration	d. Waste Treatment Efficiency Estimate	e. Based on Operating Data?
7A.1a	7A.1b 1 ___ 2 ___ 3 ___ 4 ___ 5 ___ 6 ___ 7 ___ 8 ___	7A.1c	7A.1d ___ %	7A.1e Yes ☐ No ☐
7A.2a	7A.2b 1 ___ 2 ___ 3 ___ 4 ___ 5 ___ 6 ___ 7 ___ 8 ___	7A.2c	7A.2d ___ %	7A.2e Yes ☐ No ☐
7A.3a	7A.3b 1 ___ 2 ___ 3 ___ 4 ___ 5 ___ 6 ___ 7 ___ 8 ___	7A.3c	7A.3d ___ %	7A.3e Yes ☐ No ☐
7A.4a	7A.4b 1 ___ 2 ___ 3 ___ 4 ___ 5 ___ 6 ___ 7 ___ 8 ___	7A.4c	7A.4d ___ %	7A.4e Yes ☐ No ☐
7A.5a	7A.5b 1 ___ 2 ___ 3 ___ 4 ___ 5 ___ 6 ___ 7 ___ 8 ___	7A.5c	7A.5d ___ %	7A.5e Yes ☐ No ☐

If additional copies of page 7 are attached, indicate the total number of pages in this box ☐ and indicate which page 7 this is, here. ☐ (example: 1, 2, 3, etc.)

EPA Form 9350-1 (Rev. 5/14/92) - Previous editions are obsolete.

EPA FORM R
PART II. CHEMICAL-SPECIFIC INFORMATION (CONTINUED)

United States Environmental Protection Agency

TRI FACILITY ID NUMBER

Toxic Chemical, Category, or Generic Name

SECTION 7B. ON-SITE ENERGY RECOVERY PROCESSES

☐ Not Applicable (NA) - Check here if no on-site energy recovery is applied to any waste stream containing the toxic chemical or chemical category.

Energy Recovery Methods [enter 3-character code(s)]

1	2	3	4

SECTION 7C. ON-SITE RECYCLING PROCESSES

☐ Not Applicable (NA) - Check here if no on-site recycling is applied to any waste stream containing the toxic chemical or chemical category.

Recycling Methods [enter 3-character code(s)]

1	2	3	4	5
6	7	8	9	10

EPA Form 9350-1 (Rev. 5/14/92) - Previous editions are obsolete.

		Page 9 of 9
♦EPA United States Environmental Protection Agency	**EPA FORM R** **PART II. CHEMICAL-SPECIFIC INFORMATION (CONTINUED)**	TRI FACILITY ID NUMBER Chemical, Category, or Generic Name

SECTION 8. SOURCE REDUCTION AND RECYCLING ACTIVITIES

	All quantity estimates can be reported using up to two significant figures.	Column A 1990 (pounds/year)	Column B 1991 (pounds/year)	Column C 1992 (pounds/year)	Column D 1993 (pounds/year)
8.1	Quantity released *				
8.2	Quantity used for energy recovery on-site				
8.3	Quantity used for energy recovery off-site				
8.4	Quantity recycled on-site				
8.5	Quantity recycled off-site				
8.6	Quantity treated on-site				
8.7	Quantity treated off-site				
8.8	Quantity released to the environment as a result of remedial actions, catastrophic events, or one-time events not associated with production processes (pounds/year)				
8.9	Production ratio or activity index				
8.10	Did your facility engage in any source reduction activities for this chemical during the reporting year? If not, enter "NA" in Section 8.10.1 and answer Section 8.11.				

	Source Reduction Activities [enter code(s)]	Methods to Identify Activity (enter codes)		
8.10.1		a.	b.	c.
8.10.2		a.	b.	c.
8.10.3		a.	b.	c.
8.10.4		a.	b.	c.
8.11	Is additional optional information on source reduction, recycling, or pollution control activities included with this report? (Check one box)	YES ☐	NO ☐	

* Report releases pursuant to EPCRA Section 329(8) including "any spilling, leaking, pumping, pouring, emitting, emptying, discharging, injecting, escaping, leaching, dumping, or disposing into the environment." Do not include any quantity treated on-site or off-site.

EPA Form 9350 - 1 (Rev. 5/14/92) - Previous editions are obsolete.

APPENDIX I. SECTION 313 RELATED MATERIALS

To receive a copy of any of the section 313 documents listed below, check the box(es) next to the desired document(s). There is no charge for any of these documents. Be sure to type your full mailing address in the space provided on this form. Send this request form to:

> Section 313 Document Distribution Center
> P.O. Box 12505
> Cincinnati, OH 45212

- ☐ **Section 313 Rule (40 CFR 372)**

  A reprint of the final section 313 rule as it appeared in the Federal Register (FR) February 16, 1988.

- ☐ **Comprehensive List of Chemicals Subject to Reporting Under the Act** (Title III List of Lists) (EPA 500-B-92-002)

  A consolidated list of specific chemicals covered by the Emergency Planning and Community Right-to-Know Act. The list contains the chemical name, CAS Registry Number, and reporting requirement(s) to which the chemical is subject.

- ☐ **The Emergency Planning and Community Right-to-Know Act: Section 313 Release Reporting Requirements** December 1991 (EPA 700-K-92-001)

  This brochure alerts businesses to their reporting obligations under section 313 and assists in determining whether their facility is required to report. The brochure contains the EPA regional contacts, the list of section 313 toxic chemicals and a description of the Standard Industrial Classification (SIC) codes subject to section 313.

- ☐ **Supplier Notification Requirements** (EPA 560/4-91-006)

  This pamphlet assists chemical suppliers who may be subject to the supplier notification requirements under section 313 of EPCRA. The pamphlet explains the supplier notification requirements, gives examples of situations which require notification, describes the trade secret provision, and contains a sample notification.

- ☐ **Trade Secrets Rule and Form** (FR Reprint)

  A reprint of the final rule that appeared in the *Federal Register* of July 29, 1988. This rule implements the trade secrets provision of the Emergency Planning and Community Right-to-Know Act (section 322). Includes a copy of the trade secret substantiation form.

Industry Specific Technical Guidance Documents

EPA has developed a group of smaller, individual guidance documents that target activities in industries who primarily process or otherwise use the listed toxic chemicals.

- ☐ **Electrodeposition of Organic Coatings** January 1988 (EPA 560/4-88-004c)

- ☐ **Electroplating Operations** January 1988 (EPA 560/4-88-004g)

- ☐ **Formulating Aqueous Solutions** March 1988 (EPA 560/4-88-004f)

- ☐ **Leather Tanning and Finishing Processes** February 1988 (EPA 560/4-88-004l)

- ☐ **Monofilament Fiber Manufacture** January 1988 (EPA 560/4-88-004a)

- ☐ **Paper Paperboard Production** February 1988 (EPA 560/4-88-004k)

- ☐ **Presswood & Laminated Wood Products Manufacturing** March 1988 (EPA 560/4-88-004i)

- ☐ **Printing Operations** January 1988 (EPA 560/4-88-004b)

- ☐ **Roller, Knife and Gravure Coating Operations** February 1988 (EPA 560/4/88/004j)

- ☐ **Rubber Production and Compounding** March 1988 (EPA 560/4-88-004q)

- ☐ **Semiconductor Manufacture** January 1988 (EPA 560/4-88-004e)

❏ **Spray Application of Organic Coatings** January 1988 (EPA 560/4-88-004d)

❏ **Textile Dyeing** February 1988 (EPA 560/4-88-004h)

❏ **Wood Preserving** February 1988 (EPA 560/4-88-004p)

Please type mailing address here (Do not attach business cards)

Name/Title _____

Company Name _____

Mail Stop _____

Street Address _____

P.O. Box _____

City/State/Zip Code _____

OTHER RELEVANT SECTION 313 MATERIALS

Toxics in the Community: National and Local Perspectives (EPA 560/4-91-014)

This report summarizes the third year of toxic release inventory data - where, how much, and which types of toxic chemicals are being released into the environment - and provides comparisons to the first two years' releases. Available from: Superintendent of Documents, Government Printing Office, Washington, DC 20402-9325, Stock number: 055-000-00387-4, $24.00.

Toxic Release Inventory — On-line Database

A computerized on-line database of the toxic release inventory data is available through the National Library of Medicine's (NLM) TOXNET on-line system 24 hours a day. Other NLM files on TOXNET can provide supporting information in such areas as health hazards and emergency handling of toxic chemicals. Information on accessing the TOXNET system is available from: TRI Representative, Specialized Information Services, National Library of Medicine, 8600 Rockville Pike, Bethesda, MD 20894, (301) 496-6531, up to $37.00 per hour.

Toxic Release Inventory 1987-1989 — Magnetic Tape

Contains the complete toxic release inventory for reporting years 1987-1989. Includes brief overviews of section 313 reporting requirements, a sample Form R, lists of regional and state section 313 contacts. Available from: National Technical Information Service, 5285 Port Royal Road, Springfield, VA 22161, (703) 487-4650.

    1987 Document Number: PB89-186068
    1600 (BPI) Density — $1,770.00
    6250 (BPI) Density — $890.00
    This tape is also available from the Government Printing Office (GPO-(6250)) -- $500.00.

    1988 Document Number: PB90-502030
    1600 (BPI) Density — $1,550.00
    6250 (BPI) Density — $1,100.00
    The Government Printing Office also has this tape available, GPO-(6250) -- $500.00.

    1989 Document Number: PB91-507509
    Both 1600 and 6250 (BPI) density -- $1,550.00

Toxic Release Inventory 1987-1988: Reporting Facilities Names and Addresses — Magnetic Tape

Contains the name, address, public contact, phone number, SIC code, Dun and Bradstreet number of each facility that reported under section 313 in reporting year 1987. Also includes, if applicable, parent company name and the parent company's Dun and Bradstreet number. Available from: National Technical Information Service, 5285 Port Royal Road, Springfield, VA 22161, (703) 487-4650.

    1987 Document Number: PB89-186118, $220.00
    (1600 and 6250 (BPI) density.)

    1988 Document Number: PB91-506816, $220.00
    (1600 and 6250 (BPI) density.)

Section 313 Roadmaps Database — Diskette

A database of sources of information on the toxic chemicals listed in section 313. The database, created in 1988 and updated in 1990 (a new update is scheduled in 1992), is intended to assist users of the toxic release inventory data in performing exposure and risk assessments of these toxic chemicals. The roadmaps system displays information, including the section 313 toxic chemicals' health and environmental effects, the applicability of federal, state, and local regulations, and monitoring data. Available from: National Technical Information Service, 5285 Port Royal Road, Springfield, VA 22161, (703) 487-4650, Document Number: PB90-501487, $195.00.

Comprehensive List of Chemicals Subject to Reporting Under the Act (Title III List of Lists)

Available as an IBM compatible disk from: The National Technical Information Service, 5285 Port Royal Road, Springfield, VA 22161, (703) 487-4650, Document Number: PB90-501479, $90.00.

The Toxic Release Inventory: Meeting the Challenge (April 1988)

This 19 minute videotape explains the toxic release reporting requirements for plant facility managers and others. State governments, local Chambers of Commerce, labor organizations, public interest groups, universities, and others may also find the video program useful and informative.
3/4 inch = $30.75; Beta = $22.95; VHS = $22.00.

To purchase, write or call:

>Color Film Corporation
>Video Division
>770 Connecticut Avenue
>Norwalk, CT 06854
>(800) 882-1120

Form R: A Better Understanding

Developed by EPA Region 3, this videotape reviews the Form R and explains how to correctly fill-out the Form R. Available from: National Technical Information Service, 5285 Port Royal Road, Springfield, VA 22161, (703) 487-4650, Document number: PB90-780446, $35.00.

Chemicals in Your Community, A Citizen's Guide to the Emergency Planning and Community Right-to-Know Act, September 1988 (OSWER-88-002)

This booklet is intended to provide a general overview of the EPCRA requirements and benefits for all audiences. Part I of the booklet describes the provisions of EPCRA and Part II describes more fully the authorities and responsibilities of the groups of people affected by the law. Available through written request at no charge from:

>Emergency Planning and Community
>Right-to-Know Information Service
>Mailcode: OS-120
>401 M Street, SW
>Washington, DC 20460

POLLUTION PREVENTION INFORMATION

An up-to-date source of information on pollution prevention is the Pollution Prevention Information Exchange System (PIES), the computerized information network of EPA's Pollution Prevention Information Clearinghouse (PPIC). PIES includes a directory of representatives from Federal, State, and local governments; current news on pollution prevention activities; program summaries for government agencies, public interest groups, academic institutions, trade associations, and industry; a data base of industry case studies; a calendar of conferences, training seminars, and workshops; a legislation data base; and specialized forums dedicated to various topics. Further information on using PIES can be obtained from the PPIC Technical Support Hotline, (703) 821-4800.

Documents containing general information about the PIES system and how to access them are listed below and can be obtained by writing to:

>PPIC
>c/o SAIC
>7600-A Leesburg Pike
>Falls Church, VA 22043

The list below includes some of the material available about the PIES system. Requests for these items should include the code number found to the left of each entry.

PPIC-1. PPIC General Information Package

PPIC-2. "PPIC: The Pollution Prevention Information Clearinghouse." U.S. EPA Office of Environmental Engineering and Technology Demonstration and Office of Pollution Prevention. April 1990. Brochure.

PPIC-3. "PIES: The Pollution Prevention Information Exchange System." U.S. EPA Office of Environmental Engineering and Technology Demonstration and Office of Pollution Prevention. May 1989. Brochure.

PPIC-4. "Pollution Prevention Information Exchange System (PIES) User Guide, Version 1.1." U.S. EPA Office of Environmental Engineering and Technology Demonstration and Office of Pollution Prevention (EPA/600/9-89/086). September 1989. 70 pp.

Additional information on source reduction, reuse, and recycling approaches to waste minimization is available through state programs that offer technical and/or financial assistance in the areas of waste minimization and treatment. These state contacts are listed in Appendix H.

Appendix 11

EPA Catalogue of Hazardous and Solid Waste Publications

RCRA Information Center
U.S. Environmental Protection Agency
Office of Solid Waste (OS-305)
401 M Street, S.W.
Washington, DC 20460
(202) 260-9327

HOW TO USE THIS CATALOGUE

The following publications are the responsibility of the U.S. Environmental Protection Agency's Office of Solid Waste (OSW). This is not a comprehensive list of all OSW documents, but rather a select list of those titles that are frequently requested. Some of these publications are no longer available from OSW, but can be purchased from the National Technical Information Service and/or the U.S. Government Printing Office.

The catalogue is organized alphabetically by title. On the right hand side of each entry you will find:

    OSW _____
    GPO _____
    NTS _____

These are the accession numbers by which a document is identified. If only the OSW number is filled in, this indicates the document is available by writing to:

    RCRA Docket Information Center (RIC)
    Office of Solid Waste (OS-305)
    U.S. EPA
    401 M Street, S.W.
    Washington, D.C. 20460
    or calling: The RCRA/Superfund Hotline
    (800) 424-9346
    (800) 553-7672 TDD (hearing impaired)
    (703) 920-9810 Washington, DC metro area
    (703) 486-3323 TDD (hearing impaired)

If the GPO number is filled in, this indicates the document must be purchased through GPO by writing to:

Superintendent of Documents
Government Printing Office
Washington, D.C. 20402
or calling: (202) 783-3238

If the NTIS number is filled in, this indicates the document must be purchased through NTIS by writing to:

National Technical Information Service
U.S. Department of Commerce
Springfield, VA 22161
or calling: (703) 487-4650
(800) 336-4700 RUSH ORDER

When ordering through NTIS, please use the PB number listed next to the title.

When the OSW number is included with the GPO and/or NTIS number, the OSW number only serves as an additional reference to assist you in identifying the document. The document must be purchased from GPO or NTIS, whichever is indicated.

Copies of the catalogue are available by writing the RCRA Docket Information Center or calling the RCRA/Superfund Hotline

NTIS CUSTOMER SERVICE INFORMATION

Telephone Orders
The NTIS telephone sales desk is available between 7:45 a.m. and 5:00 p.m. Eastern Time.

TELEX and Telecopier Orders
The NTIS TELEX number is 89-9405. Orders also may be sent to NTIS via Telecopier: Panafax MV3000. Dial NTIS, (703) 321-8547.

Method of Payment
Customers may pay for reports (and other NTIS products and services) by: (1) credit card - American Express, MasterCard, or VISA; (2) check or money order, on a U.S. bank, payable to NTIS; (3) an NTIS Deposit Account; or (4) by asking to be billed (add $7.50 per order) - United States, Canada, and Mexico only.

Handling Fee
A $3.00 handling fee per total order (not per item) applies to U.S. orders. In all other countries the handling fee is $4.00 per order.

Postage and Shipping
Orders are shipped First Class mail or equivalent except for addresses outside the United States. For air mail service to Canada and Mexico add $3.00 per printed report and $0.75 per microfiche copy.

Order Turnaround Time
Orders for technical reports are generally shipped within 3 to 8 days of receipt.

Rush and Express Order Service
NTIS Rush Order Service (add $10.00 per item) guarantees that an order will be processed through NTIS within 24 hours of its receipt. Rush orders receive immediate, individual attention. The items ordered are delivered by First Class mail unless express mail service is requested. Express service (add $20.00 per item) is for customers who need overnight delivery. Guidelines are the same as for Rush Order Service. Delivery is by overnight courier. To place a Rush or Express Order, call (800) 336-5700 (in Virginia, call (703) 487-4700.

Customer Service
If you have a problem with your order, contact a Customer Services Representative at (703) 487-4600.

OFFICE OF SOLID WASTE CATALOG OF GUIDANCE DOCUMENTS

Title of Document	Document Availability Date	Document No.
Adventures of the Garbage Gremlin: Recycle and Combat a Life of Grime (Comic Book)	08/15/90	OSW: 530/SW-90-024 GPO: NTS:
Alternate Concentration Limit Guidance; Part I: ACL Policy and Information Requirements	07/15/87	OSW: 530/SW-87-017 GPO: NTS: PB87-206 165
Alternate Concentration Limit Guidance; Part II: Based on 264.94(B) Criteria; Case Studies	05/15/88	OSW: 530/SW-87-031 GPO: NTS: PB88-214 267
Analysis of U.S. Municipal Waste Combustion Operating Practices	05/18/89	OSW: 530/SW-89-061 GPO: NTS: PB89-220 578
Assessment of Hazardous Waste Mismanagement Damage Case Histories	04/15/84	OSW: 530/SW-84-002 GPO: NTS: PB84-212 356
Background Document for First Third Wastes to Support 40 CFR 268 Land Disposal Restrictions; Final Rule: First Third Waste Volumes, Characteristics, and Required and Availability Treatment Capacity	08/15/88	OSW: 530/SW-88-049 GPO: NTS: PB88-246 145
Background Document for First Third Wastes to Support 40 CFR 268 Land Disposal Restrictions; Proposed Rule: First Third Waste Volumes, Characteristics, and Required and Availability Treatment Capacity; Part I	03/15/88	OSW: 530/SW-88-030A GPO: NTS: PB88-217 575
Background Document for First Third Wastes to Support 40 CFR 268 Land Disposal Restrictions; Proposed Rule: First Third Waste Volumes, Characteristics, and Required and Availability Treatment Capacity; Part II	05/15/88	OSW: 530/SW-88-030 GPO: NTS: PB88-213 368
Background Document for Second Third Wastes to Support 40 CFR 268 Land Disposal Restrictions; Final Rule Second Third Waste Volumes, Characteristics, and Required and Availability Treatment Capacity; Volume I	06/08/89	OSW: 530/SW-89-057A GPO: NTS: PB89-220 552
Background Document for Second Third	06/08/89	OSW: 530/SW-89-057B

Wastes to Support 40 CFR 268 Land Disposal Restrictions; Final Rule Second Third Waste Volumes, Characteristics, and Required and Availability Treatment Capacity; Volume II		GPO: NTS: PB89-220 560
Background Document for Second Third Wastes to Support 40 CFR 268 Land Disposal Restrictions; Proposed Rule: Second Third Waste Volumes, Characteristics, and Required and Availability Treatment Capacity	12/15/88	OSW: 530/SW-89-034 GPO: NTS: PB89-179 535
Background Document for Solvents to Support 40 CFR Part 268 Land Disposal Restrictions; Volume I	11/15/86	OSW: 530/SW-86-060 GPO: NTS: PB87-146 361/AS
Background Document for Solvents to Support 40 CFR Part 268 Land Disposal Restrictions; Volume II	11/15/86	OSW: 530/SW-86-061 GPO: NTS: PB87-146 379
Background Document for Solvents to Support 40 CFR Part 268 Land Disposal Restrictions: Solvent Waste Volumes and Characteristics, Required Treatment and Recycling Capacity, and Available Treatment and Recycling	11/15/86	OSW: GPO: NTS: PB87-163 481/AS
Background Document for the Groundwater Screening Procedure to Support 40 CFR 268 Land Disposal Restrictions	01/15/86	OSW: 530/SW-86-047 GPO: NTS: PB87-101 606
Background Document for Third Third Wastes to Support 40 CFR Part 268 Land Disposal Restrictions; Final Rule: Third Third Waste Volumes, Characteristics, and Required and Available Treatment Capacity; (Complete Set)	05/08/90	OSW: 530/SW-90-062 GPO: NTS: PB90-234 675
Background Document for Third Third Wastes to Support 40 CFR Part 268 Land Disposal Restrictions; Final Rule: Third Third Waste Volumes, Characteristics, and Required and Available Treatment Capacity; Volume I: Executive Summary, Chapter 1 and Chapter 2	05/08/90	OSW: 530/SW-90-062A GPO: NTS: PB90-234 683
Background Document for Third Third Wastes to Support 40 CFR Part 268 Land Disposal Restrictions; Final Rule: Third Third Waste Volumes, Characteristics, and Required and Available Treatment Capacity; Volume II: Chapter 3	05/08/90	OSW: 530/SW-90-062B GPO: NTS: PB90-234 691

Background Document for Third Third Wastes to Support 40 CFR Part 268 Land Disposal Restrictions; Final Rule: Third Third Waste Volumes, Characteristics, and Required and Available Treatment Capacity; Volume III: Chapter 4 and Appendix A - Appendix I	05/08/90	OSW: 530/SW-90-062C GPO: NTS: PB90-234 709
Background Document for Third Third Wastes to Support 40 CFR Part 268 Land Disposal Restrictions; Final Rule: Third Third Waste Volumes, Characteristics, and Required and Available Treatment Capacity; Volume IV: Chapter 4 and Appendix J - Appendix M	05/08/90	OSW: 530/SW-90-062D GPO: NTS: PB90-234 717
Background Document for Third Third Wastes to Support 40 CFR Part 268 Land Disposal Restrictions; Proposed Rule: Third Third Waste Volumes, Characteristics, and Required and Available Treatment Capacity (Complete Set)	11/15/89	OSW: 530/SW-90-033 GPO: NTS: PB90-187 113
Background Document for Third Third Wastes to Support 40 CFR Part 268 Land Disposal Restrictions; Proposed Rule: Third Third Waste Volumes, Characteristics, and Required and Available Treatment Capacity; Volume I: Executive Summary, Chapter 1, Chapter 2	11/15/89	OSW: 530/SW-90-033A GPO: NTS: PB90-187 121
Background Document for Third Third Wastes to Support 40 CFR Part 268 Land Disposal Restrictions; Proposed Rule: Third Third Waste Volumes, Characteristics, and Required and Available Treatment Capacity; Volume II: Chapter 3	11/15/89	OSW: 530/SW-90-033B GPO: NTS: PB90-187 139
Background Document for Third Third Wastes to Support 40 CFR Part 268 Land Disposal Restrictions; Proposed Rule: Third Third Waste Volumes, Characteristics, and Required and Available Treatment Capacity; Volume III: Chapter 4, Appendices	11/15/89	OSW: 530/SW-90-033C GPO: NTS: PB90-187 147
Background Document on Bottom Liner Performance in Double-Lined Landfills and Surface Impoundments	04/15/87	OSW: 530/SW-87-013 GPO: NTS: PB87-182 291

Background Document on Proposed Liner and Leak Detection Rule	05/15/87	OSW: 530/SW-87-015 GPO: NTS: PB87-191 383
Background Document on Development and Use of Reference Doses; Part I: Data Needs and Apportionment; Part II: Considerations Related to the Development of Protocols for Toxicity Studies	12/20/85	OSW: 530/SW-86-048 GPO: NTS: PB87-107 173/AS
Background Documentation for Minimum Content Standards	03/06/87	OSW: 530/SW-88-046 GPO: NTS: PB88-242 052
Background Information Document for the Development of Regulations to Control the Burning of Hazardous Wastes in Boilers and Industrial Furnaces (Complete Set)	01/15/87	OSW: 530/SW-87-014 GPO: NTS: PB87-173 811/AS
Background Information Document for the Development of Regulations to Control the Burning of Hazardous Wastes in Boilers and Industrial Furnaces; Volume I: Industrial Boilers	01/15/87	OSW: 530/SW-87-014A GPO: NTS: PB87-173 829
Background Information Document for the Development of Regulations to Control the Burning of Hazardous Wastes in Boilers and Industrial Furnaces; Volume II: Industrial Furnaces	01/15/87	OSW: 530/SW-87-014B GPO: NTS: PB87-173 837
Background Information Document for the Development of Regulations to Control the Burning of Hazardous Wastes in Boilers and Industrial Furnaces; Volume III: Risk Assessment	02/15/87	OSW: 530/SW-87-014C GPO: NTS: PB87-173 845/AS
Batch-Type Adsorption Procedures for Estimating Soil Attenuation of Chemicals; Draft Technical Resource Document for Public Comment	01/15/87	OSW: 530/SW-87-006 GPO: NTS: PB87-146 155/AS
Be an Environmentally Alert Consumer (Brochure)	04/15/90	OSW: 530/SW-90-034A GPO: NTS:
Best Demonstrated Available Technology (BDAT) Background Document for Barium Wastes D005 and P013 (Final)	05/08/90	OSW: 530/SW-90-059T GPO: NTS: PB90-234 204
Best Demonstrated Available	05/08/90	OSW: 530/SW-90-059B

Technology (BDAT) Background Document for Characteristic Ignitable Wastes (D001), Characteristic Corrosive Wastes (D002), Characteristic Reactive Wastes (D003), and P and U Wastes Containing Reactive Listing Constituents (Final)

GPO:
NTS: PB90-234 022

Best Demonstrated Available Tech. (BDAT) Background Document for Chromium Wastes D007 and U032 (Final)

05/08/90

OSW: 530/SW-90-059V
GPO:
NTS: PB90-234 220

Best Demonstrated Available Tech. (BDAT) Background Document for Cyanide Wastes (F006, F007-F012, F019, and Various P and U Codes) (Final)

06/15/89

OSW: 530/SW-89-048K
GPO:
NTS: PB89-221 485

Best Demonstrated Available Tech. (BDAT) Background Document for Cyanide Wastes; Addendum for F019 (Final)

05/08/90

OSW: 530/SW-90-059N
GPO:
NTS: PB90-234 147

Best Demonstrated Available Tech. (BDAT) Background Document for D006 Cadmium Wastes (Final)

05/08/90

OSW: 530/SW-90-059U
GPO:
NTS: PB90-234 212

Best Demonstrated Available Tech. (BDAT) Background Document for D008 and P and U Lead Wastes (Final)

05/08/90

OSW: 530/SW-90-059W
GPO:
NTS: PB90-234 238

Best Demonstrated Available Tech. (BDAT) Background Document for Distillation Bottoms from the Production of Aniline K083 (Final)

05/08/90

OSW: 530/SW-90-060J
GPO:
NTS: PB90-234 378

Best Demonstrated Available Tech. (BDAT) Background Document for Distillation Bottoms from the Production of Nitrobenzene by the Nitration of Benzene, K025 (Final)

05/08/90

OSW: 530/SW-90-060K
GPO:
NTS: PB90-234 386

Best Demonstrated Available Tech. (BDAT) Background Document for F001-F005 Spent Solvents; Volume 1

11/15/86

OSW: 530/SW-86-056A
GPO:
NTS: PB87-120 267/AS

Best Demonstrated Available Tech. (BDAT) Background Document for F001-F005 Spent Solvents; Volume 2

11/15/86

OSW: 530/SW-86-056B
GPO:
NTS: PB87-120 275/AS

Best Demonstrated Available Tech. (BDAT) Background Document for F001-F005 Spent Solvents; Volume 3

11/15/86

OSW: 530/SW-86-056C
GPO:
NTS: PB87-120 283

Best Demonstrated Available Tech. (BDAT) Background Document for F001-F005 Spent Solvents; Amendment to

08/15/88

OSW: 530/SW-88-031R
GPO:
NTS: PB89-142 525

Volumes 1 and 2 (Final)

Best Demonstrated Available Tech. 05/08/90 OSW: 530/SW-90-059P
(BDAT) Background Document for F002 GPO:
(1, 1, 2-Trichloroethane) and F005 (Benzene, NTS: PB90-234 162
2-Ethoxyethanol, and 2-Nitropropane);
Amendment (Final)

Best Demonstrated Available Tech. 08/08/88 OSW: 530/SW-88-031L
(BDAT) Background Document for F006 GPO:
(Final) NTS: PB89-142 467

Best Demonstrated Available Tech. 05/08/90 OSW: 530/SW-90-059M
(BDAT) Background Document for F006; GPO:
Addendum (Final) NTS: PB90-234 139

Best Demonstrated Available Tech. 05/08/90 OSW: 530/SW-90-060P
(BDAT) Background Document for GPO:
Halogenated Pesticide and Chlorobenzene NTS: PB90-234 436
Wastes (K032-K034, K041, K042, K085, K097,
K098, K105, and D012-D017) (Final)

Best Demonstrated Available Tech. 05/08/90 OSW: 530/SW-90-059Y
(BDAT) Background Document for GPO:
Inorganic Pigment Wastes, K002-K008 NTS: PB90-234 253
(Final)

Best Demonstrated Available Tech. 08/15/88 OSW: 530/SW-88-031O
(BDAT) Background Document for K001 GPO:
(Final) NTS: PB89-142 491

Best Demonstrated Available Tech. 05/-8/90 OSW: 530/SW-90-059C
(BDAT) Background Document for K001 GPO:
(Addendum) and U051 (Creosote) NTS: PB90-234 030

Best Demonstrated Available Tech. 06/15/89 OSW: SW-89-048I
(BDAT) Background Document for K009 GPO:
and K010 (Final) NTS: PB89-221 469

Best Demonstrated Available Tech. 06/15/89 OSW: 530/SW-89-048J
(BDAT) Background Document for K011, GPO:
K013, and K014 (Final) NTS: PB89-221 477

Best Demonstrated Available Tech. 05/08/90 OSW: 530/SW-90-059O
(BDAT) Background Document for K011, GPO:
K013, and K014; Addendum for Acrylonitrile NTS: PB90-234 154
Wastes (Final)

Best Demonstrated Available Tech. 08/08/88 OSW: 530/SW-88-031A
(BDAT) Background Document for K015 GPO:
(Final) NTS: PB89-142 350

Best Demonstrated Available Tech. 05/08/90 OSW: 530/SW-90-059D
(BDAT) Background Document for K015; GPO:

Addendum (Final)		NTS: PB90-234 048
Best Demonstrated Available Tech. (BDAT) Background Document for K016, K018, K019, K020, and K030 (Final)	08/15/88	OSW: 530/SW-88-031B GPO: NTS: PB89-142 368
Best Demonstrated Available Tech. (BDAT) Background Document for K021 (Final)	05/08/90	OSW: 530/SW-90-059F GPO: NTS: PB90-234 063
Best Demonstrated Available Tech. (BDAT) Background Document for K022 (Non-CBI Version) (Final)	08/15/88	OSW: 530/SW-88-031Q GPO: NTS: PB89-142 517
Best Demonstrated Available Tech. (BDAT Background Document for K022; Amendment (Final)	05/08/90	OSW: 530/SW-90-060I GPO: NTS: PB90-234 360
Best Demonstrated Available Tech. (BDAT) Background Document for K024 (Final)	08/15/88	OSW: 530/SW-88-031H GPO: NTS: PB89-142 426
Best Demonstrated Available Tech. (BDAT) Background Document for K031, K084, K101, K102, Characteristic Arsenic Wastes (D004), Characteristic Selenium Wastes (D010), and P and U Wastes Containing Arsenic and Selenium Listing Constituents (Final)	05/08/90	OSW: 530/SW-90-059A GPO: NTS: PB90-234 014
Best Demonstrated Available Tech. (BDAT) Background Document for K037 (Final)	08/15/88	OSW: 530/SW-88-031I GPO: NTS: PB89-142 434
Best Demonstrated Available Tech. (BDAT) Background Document for K037; Amendment (Final)	05/08/90	OSW: 530/SW-90-060O GPO: NTS: PB90-234 428
Best Demonstrated Available Tech. (BDAT) Background Document for K043 (Final)	06/15/89	OSW: 530/SW-89-048L GPO: NTS: PB89-221 493
Best Demonstrated Available Tech. (BDAT) Background Document for K046 Nonreactive Subcategory (Final)	08/15/88	OSW: 530/SW-88-031J GPO: NTS: PB89-142 442
Best Demonstrated Available Tech. (BDAT) Background Document for K046; Addendum (Final)	05/08/90	OSW: 530/SW-90-059J GPO: NTS: PB90-234 105
Best Demonstrated Available Tech. (BDAT) Background Document for K048, K049, K050, K051, and K052 (Final)	08/15/88	OSW: 530/SW-88-031C GPO: NTS: PB89-142 376

Best Demonstrated Available Tech. (BDAT) Background Document for K048, K049, K050, K051, and K052; Amendment (Final)	05/08/90	OSW: 530/SW-90-060R GPO: NTS: PB90-234 451
Best Demonstrated Available Tech. (BDAT) Background Document for K060 (Final)	05/08/90	OSW: 530/SW-90-059H GPO: NTS: PB90-234 089
Best Demonstrated Available Tech. (BDAT) Background Document for K061 (Final)	08/15/88	OSW: 530/SW-88-031D GPO: NTS: PB89-142 384
Best Demonstrated Available Tech. (BDAT) Background Document for K061; Addendum (Final)	05/08/90	OSW: 530/SW-90-059I GPO: NTS: PB90-234 097
Best Demonstrated Available Tech. (BDAT) Background Document for K062 (Final)	08/15/88	OSW: 530/SW-88-031E GPO: NTS: PB89-142 392
Best Demonstrated Available Tech. (BDAT) Background Document for K071 (Final)	08/08/88	OSW: 530/SW-88-031F GPO: NTS: PB89-142 400
Best Demonstrated Available Tech. (BDAT) Background Document for K073 (Final)	05/08/90	OSW: 530/SW-90-059E GPO: NTS: PB90-234 055
Best Demonstrated Available Tech. (BDAT) Background Document for K086 Solvent Wash (Final)	08/15/88	OSW: 530/SW-88-031N GPO: NTS: PB89-142 483
Best Demonstrated Available Tech. (BDAT) Background Document for K086 (Ink Formulation Equipment Cleaning Wastes); Addendum (Final)	05/08/90	OSW: 530/SW-90-059G GPO: NTS: PB90-234 071
Best Demonstrated Available Tech. (BDAT) Background Document for K087 (Final)	08/08/88	OSW: 530/SW-88-031M GPO: NTS: PB89-142 475
Best Demonstrated Available Tech. (BDAT) Background Document for K099 (Final)	08/15/88	OSW: 530/SW-88-031S GPO: NTS: PB89-142 533
Best Demonstrated Available Tech. (BDAT) Background Document for K101 and K102 Low Arsenic Subcategory (Final)	08/15/88	OSW: 530/SW-88-031K GPO: NTS: PB89-142 459
Best Demonstrated Available Tech. (BDAT) Background Document for K103 and K104 (Final)	08/15/88	OSW: 530/SW-88-031G GPO: NTS: PB89-142 418

Best Demonstrated Available Tech. 05/08/90 OSW: 530/SW-90-059Q
(BDAT) Background Document for Mercury-   GPO:
Containing Wastes (D009, K106, P065, P092, NTS: PB90-234 170
and U151) (Final)

Best Demonstrated Available Tech. 05/08/90 OSW: 530/SW-90-060N
(BDAT) Background Document for   GPO:
Organophosphorus Wastes, K036   NTS: PB90-234 410
Nonwastewaters; Amendment (Final)

Best Demonstrated Available Tech. 06/15/89 OSW: 530/SW-89-048G
(BDAT) Background Document for   GPO:
Organophosphorus Wastes, K038-K040,   NTS: PB89-221 444
and Various P and U Codes (Final)

Best Demonstrated Available Tech. 05/08/90 OSW: 530/SW-90-059R
(BDAT) Background Document for P and U   GPO:
Thallium Wastes (Final)   NTS: PB90-234 188

Best Demonstrated Available Tech. 06/15/89 OSW: 530/SW-89-048H
(BDAT) Background Document for   GPO:
Phthalate Wastes, K023, K093, K094, and   NTS: PB89-221 451
Various P and U Codes (Final)

Best Demonstrated Available Tech. 05/08/90 OSW: 530/SW-90-059X
(BDAT) Background Document for Silver-   GPO:
Containing Wastes D011, P099, P104 (Final) NTS: PB90-234 246

Best Demonstrated Available Tech. 05/08/90 OSW: 530/SW-90-060L
(BDAT) Background Document for Stripping GPO:
Still Tails from the Production of Methyl NTS: PB90-234 394
Ethyl Pyridine, K026 (Final)

Best Demonstrated Available Tech. 05/08/90 OSW: 530/SW-90-060F
(BDAT) Background Document for U and P   GPO:
Wastes, Multisource Leachate (F039),   NTS: PB90-234 337
Volume A: Wastewater Forms of Organic
U and P Wastes and Multisource Leachate
(F039) for which there are Concentration-
Based Treatment Standards

Best Demonstrated Available Tech. 05/08/90 OSW: 530/SW-90-060G
(BDAT) Background Document for U and P   GPO:
Wastes, Multisource Leachate (F039),   NTS: PB90-234 345
Volume B: U and P Wastewaters and
Nonwastewaters with Methods of Treatment
as Treatment Standards

Best Demonstrated Available Tech. 05/08/90 OSW: 530/SW-90-060H
(BDAT) Background Document for U and P   GPO:
Wastes, Multisource Leachate (F039),   NTS: PB90-234 352
Volume C: Nonwastewater Forms of Organic
U and P Wastes and Multisource Leachate
(F039) for which there are Concentration-

Based Treatment Standards

Best Demonstrated Available Tech. 05/08/90 OSW: 530/SW-90-060C
(BDAT) Background Document for U and P GPO:
Wastes, Multisource Leachate (F039), NTS: PB90-234 303
Volume D: Reactive U and P Wastewaters
and Nonwastewaters

Best Demonstrated Available Tech. 05/08/90 OSW: 530/SW-90-060S
(BDAT) Background Document for U and P GPO:
Wastes, Multisource Leachate (F039), NTS: PB90-234 469
Volume E: U and P Gaseous Wastes

Best Demonstrated Available Tech. 05/08/90 OSW: 530/SW-90-059S
(BDAT) Background Document for Vanadium- GPO:
Containing Wastes P119 and P120 (Final) NTS: PB90-234 196

Best Demonstrated Available Tech. 05/15/89 OSW: 530/SW-89-048M
(BDAT) Background Document for Wastes GPO:
from the Production of Chlorinated NTS: PB89-221 501
Aliphatic Hydrocarbons, F024 (Final)

Best Demonstrated Available Tech. 05/08/90 OSW: 530/SW-90-060Q
(BDAT) Background Document for Wastes GPO:
from the Production of Chlorinated NTS: PB90-234 444
Aliphatic Hydrocarbons, F024;
Amendment (Final)

Best Demonstrated Available Tech. 05/08/90 OSW: 530/SW-90-060A
(BDAT) Background Document for Wastes GPO:
from the Production of Chlorinated NTS: PB90-234 287
Aliphatics F025 (Final)

Best Demonstrated Available Tech. 05/15/89 OSW: 530/SW-89-048O
(BDAT) Background Document for Wastes GPO:
from the Production of Dinitrotoluene, NTS: PB89-221 527
Toluenediamine, and Toluene
Diisocyanate, K027, K111-K116, U221,
and U223 (Final)

Best Demonstrated Available Tech. 05/08/90 OSW: 530/SW-90-060D
(BDAT) Background Document for Wastes GPO:
from the Production of Epichlorohydrin, NTS: PB90-234 311
K017 (Final)

Best Demonstrated Available Tech. 05/25/89 OSW: 530/SW-89-048N
(BDAT) Background Document for Wastes GPO:
from the Production of 1,1,1-Trichloroethane, NTS: PB89-221 519
K028, K029, K095, and K096 (Final)

Best Demonstrated Available Tech. 05/08/90 OSW: 530/SW-90-060E
(BDAT) Background Document for Wastes GPO:
from the Production of 1,1,1-Trichloroethane, NTS: PB90-234 329
K028, K029, K095, and K096; Amendment

(Final)

Best Demonstrated Available Tech. (BDAT) Background Document for Wastewater Treatment Sludges Generated in the Production of Creosote K035 (Final)	05/08/90	OSW: 530/SW-90-060M GPO: NTS: PB90-234 402
Best Demonstrated Available Tech. (BDAT) Background Documents for First Third Wastes; Final Rule (Complete Set)	08/15/88	OSW: 530/SW-88-031 GPO: NTS: PB89-142 343
Best Demonstrated Available Tech. (BDAT) Background Documents for Second Third Wastes; Final Rule (Complete Set)	06/15/89	OSW: 530/SW-89-048 GPO: NTS: PB89-221 402
Best Demonstrated Available Tech. (BDAT) Background Documents for Third Third Wastes; Final Rule (First Part of Complete Set)	05/08/90	OSW: 530/SW-90-059 GPO: NTS: PB90-234 006
Best Demonstrated Available Tech. (BDAT) Background Documents for Third Third Wastes; Final Rule (Second Part of Complete Set)	05/08/90	OSW: 530/SW-90-060 GPO: NTS: PB90-234 279
Bibliographic Series: Waste Minimization: Hazardous and Non-Hazardous Solid Waste (1980 to Present)	09/15/87	OSW: IMSD/87-007 GPO: NTS: PB88-163 787/AS
Bibliography of Municipal Solid Waste Management Alternatives	08/15/89	OSW: 530/SW-89-055 GPO: NTS:
Census of State and Territorial Subtitle D Non-Hazardous Waste Programs	10/15/86	OSW: 530/SW-86-039 GPO: NTS: PB87-108 080/AS
Characterization of Municipal Solid Waste in the United States, 1960-2000	07/25/86	OSW: GPO: NTS: PB87-178 323
Characterization of Municipal Solid Waste in the United States: 1990 Update	06/15/90	OSW: 530/SW-90-042 GPO: NTS: PB90-215 112
Characterization of Municipal Solid Waste in the United States: 1990 Update; Executive Summary	06/15/90	OSW: 530/SW-90-042A GPO: NTS:
Characterization of Municipal Waste Combustion Ash, Ash Extracts, and	03/15/90	OSW: 530/SW-90-029A GPO:

Leachates		NTS: PB90-187 154
Characterization of Municipal Waste Combustion Ash, Ash Extracts, and Leachates; Executive Summary	03/15/90	OSW: 530/SW-90-029B GPO: NTS:
Characterization of MWC Ashes and Leachates from MSW Landfills, Monofills, and Co-Disposal Sites (Complete Set)	10/15/87	OSW: 530/SW-87-028 GPO: NTS: PB88-127 931
Characterization of MWC Ashes and Leachates from MSW Landfills, Monofills, and Co-Disposal Sites; Volume I: Summary	10/15/87	OSW: 530/SW-87-028A GPO: NTS: PB88-127 949
Characterization of MWC Ashes and Leachates from MSW Landfills, Monofills, and Co-Disposal Sites; Volume II: Leachate Baseline Report; Determination of Municipal Landfill Leachate Characteristics	10/15/87	OSW: 530/SW-87-028B GPO: NTS: PB88-127 956
Characterization of MWC Ashes and Leachates from MSW Landfills, Monofills, and Co-Disposal Sites; Volume III: Addendum to Characterization of Municipal Landfill Leachates; A Literature Review	10/15/87	OSW: 530/SW-87-028C GPO: NTS: PB88-127 964
Characterization of MWC Ashes and Leachates from MSW Landfills, Monofills, and Co-Disposal Sites; Volume IV: Characterization of Municipal Waste Combustion Residues and their Leachates; A Literature Review	10/15/87	OSW: 530/SW-87-028D GPO: NTS: PB88-127 972
Characterization of MWC Ashes and Leachates from MSW Landfills, Monofills, and Co-Disposal Sites; Volume V: Characterization of Municipal Waste Combustor Residues	10/15/87	OSW: 530/SW-87-028E GPO: NTS: PB88-127 980
Characterization of MWC Ashes and Leachates from MSW Landfills, Monofills, and Co-Disposal Sites; Volume VI: Characterization of Leachates from Municipal Waste Disposal Sites and Co-Disposal Sites	10/15/87	OSW: 530/SW-87-028F GPO: NTS: PB88-127 998
Characterization of MWC Ashes and Leachates from MSW Landfills, Monofills, and Co-Disposal Sites;	10/15/87	OSW: 530/SW-87-028G GPO: NTS: PB88-128 004

Volume VII: Addendum to Monofill Report

Characterization of Products Containing Lead and Cadmium in Municipal Solid Waste in the United States, 1970 to 2000	01/15/89	OSW: 530/SW-89-015A GPO: NTS: PB89-151 039
Characterization of Products Containing Lead and Cadmium in Municipal Solid Waste in the United States, 1970 to 2000; Executive Summary	01/15/89	OSW: 530/SW-89-015C GPO: NTS:
Characterization of Products Containing Lead and Cadmium in Municipal Solid Waste in the United States, 1970 to 2000; Executive Summary and Chapter 1: Lead and Cadmium in Municipal Solid Waste; Overview and Summary	01/15/89	OSW: 530/SW-89-015B GPO: NTS:
Chemical, Physical, and Biological Properties of Compounds Present at Hazardous Waste Sites (Final Report)	09/27/85	OSW: 530/SW-89-010 GPO: NTS: PB89-132 203
Closure of Hazardous Waste Surface Impoundments	09/15/82	OSW: SW-873 GPO: NTS: PB87-155 537/AS
Closure/Post-Closure and Financial Responsibility Requirements for Hazardous Waste Treatment, Storage, and Disposal Facilities; Final Rule; Background Document	04/15/86	OSW: 530/SW-86-009 GPO: NTS: PB86-210 671
Closure/Post-Closure Interim Status Standards (40 CFR 265, Subpart G): Standards Applicable to Owners and Operators of Hazardous Waste Treatment, Storage, and Disposal Facilities under RCRA, Subtitle C, Section 3004	01/15/84	OSW: SW-912 GPO: NTS: PB87-156 683/AS
Combined NRC/EPA Siting Guidelines for Disposal of Commercial Mixed Low-Level Radioactive and Hazardous Waste	06/29/87	OSW: 530/SW-87-029 GPO: NTS:
Commercial Treatment/Recovery Data Set	05/08/90	OSW: 530/SW-90-078 GPO: NTS: PB90-259 789
Commercial Treatment/Recovery TSDR Survey Data Set	06/15/89	OSW: 530/SW-89-058 GPO: NTS: PB89-220 545
Compendium of ORD and OSWER Documents	05/15/88	OSW: 530/SW-88-010

Relevant to RCRA Corrective Action		GPO: NTS:
Compilation of Persons Who Design, Test, Inspect, and Install Storage Tank Systems	02/29/88	OSW: 530/SW-88-019 GPO: NTS: PB88-197 611
Composition and Management of Used Oil Generated in the United States	11/15/84	OSW: 530/SW-84-013 GPO: NTS: PB85-180 297
Composition and Management of Used Oil Generated in the United States; Appendix	11/15/84	OSW: 530/SW-84-013A GPO: NTS: PB88-111 752
Construction Quality Assurance for Hazardous Waste Land Disposal Facilities; Technical Guidance Document	10/15/86	OSW: 530/SW-86-031 GPO: NTS: PB87-132 825
Corrective Measures for Releases to Ground Water from Solid Waste Management Units; Draft Final Report	08/15/85	OSW: 530/SW-88-020 GPO: NTS: PB88-185 251
Corrective Measures for Releases to Soil from Solid Waste Management Units; Draft Final Report	08/15/88	OSW: 530/SW-88-022 GPO: NTS: PB88-185 277
Corrective Measures for Releases to Surface Waters, Draft Final Report	08/15/85	OSW: 530/SW-90-085 GPO: NTS: PB91-102 046
Corrosivity Characteristic (40 CFR 261.22); Identifcation and Listing of Hazardous Waste Under RCRA, Subtitle C, Section 3001	05/01/80	OSW: GPO: NTS: PB81-184 319
Criteria for Classification of Solid Waste Disposal Facilities and Practices; Notification Requirements (40 CFR Part 257) (Draft)	07/15/88	OSW: 530/SW-88-044 GPO: NTS: PB88-242 508
Criteria for Identifying Areas of Vulnerable Hydrogeology under RCRA (Complete Set)	07/15/86	OSW: 530/SW-86-022 GPO: NTS: PB86-224 946
Criteria for Identifying Areas of Vulnerable Hydrogeology: A RCRA Statutory Interpretative Guidance	07/15/86	OSW: 530/SW-86-022 GPO: NTS: PB86-224 953
Criteria for Identifying Areas of Vulnerable Hydrogeology under RCRA; Appendix A: Technical Methods for Evaluating Hydrogeologic Parameters	07/15/86	OSW: 530/SW-86-022A GPO: NTS: PB86-224 961/AS

Criteria for Identifying Areas of Vulnerable Hydrogeology under RCRA; Appendix B: Groundwater Flow Net/Flow Line Construction and Analysis	07/15/86	OSW: 530/SW-86-022B GPO: NTS: PB86-224 979/AS
Criteria for Identifying Areas of Vulnerable Hydrogeology under RCRA; Appendix C: Technical Methods for Calculating Time of Travel in the Unsaturated Zone	07/15/86	OSW: 530/SW-86-022C GPO: NTS: PB86-224 987/AS
Criteria for Identifying Areas of Vulnerable Hydrogeology under RCRA; Appendix D: Development of Vulnerability Criteria Based on Risk Assessments and Theoretical Modeling	07/15/86	OSW: 530/SW-86-022D GPO: NTS: PB86-224 995/AS
Criteria for Identifying Characteristics of Hazardous Waste (40 CFR 261.10); Criteria for Listing Hazardous Waste (40 CFR 261.11); Petitions to Amend Part 261 to Exclude a Waste Produced at a Particular Facility (40 CFR 260.22); Identification and Listing of Hazardous Waste Under RCRA, Subtitle C, Section 3001	04/01/80	OSW: GPO: NTS: PB81-184 962
Criteria for Municipal Solid Waste Landfills: Case Studies on Ground Water and Surface Water Contamination from Municipal Solid Waste Landfills	07/15/88	OSW: 530/SW-88-040 GPO: NTS: PB88-242 466
Criteria for Municipal Solid Waste Landfills: Closure and Post-Closure Care and Financial Responsibility Requirements (Subpart C, Sections 258.30-258.32); Draft Background Document	07/15/88	OSW: 530/SW-88-041 GPO: NTS: PB88-242 474
Criteria for Municipal Solid Waste Landfills: Design Criteria (Subpart D); Draft Background Document	07/15/88	OSW: 530/SW-88-042 GPO: NTS: PB88-242 482
Criteria for Municipal Solid Waste Landfills: Draft Regulatory Impact Analysis to Subtitle D	08/15/88	OSW: 530/SW-88-045 GPO: NTS: PB88-242 516
Criteria for Municipal Solid Waste Landfills: Ground-Water Monitoring and Corrective Action (Subpart E)	07/15/88	OSW: 530/SW-88-043 GPO: NTS: PB88-242 490
Criteria for Municipal Solid Waste Landfills: Location Restrictions	07/15/88	OSW: 530/SW-88-036 GPO:

APPENDIX 11—EPA PUBLICATIONS

(Subpart B); Draft Background Document		NTS: PB88-242 425
Criteria for Municipal Solid Waste Landfills: Operating Criteria (Subpart C); Draft Background Document	07/15/88	OSW: 530/SW-88-037 GPO: NTS: PB88-242 433
Criteria for Municipal Solid Waste Landfills: Summary of Data on Municipal Solid Waste Landfill Leachate Characteristics	07/15/88	OSW: 530/SW-88-038 GPO: NTS: PB88-242 441
Criteria for Municipal Solid Waste Landfills: Updated Review of Selected Provisions of Solid Waste Regulations	07/15/88	OSW: 530/SW-88-039 GPO: NTS: PB88-242 458
Decision-Makers Guide to Solid Waste Management	11/15/89	OSW: 530/SW-89-072 GPO: NTS:
Decision-Makers Guide to Solid Waste Management (Promotional Brochure)	09/15/90	OSW: 530/SW-89-073 GPO: NTS:
Definitions and General Provisions Under RCRA, Subtitle C; Definitions and Provisions for Confidentiality (40 CFR 260, Subparts A and B)	04/15/80	OSW: GPO: NTS: PB81-181 489
Degree of Hazard as an Approach to Defining and Managing Hazardous Wastes; Identification and Listings of Hazardous Waste Under RCRA, Subtitle C, Section 3001	04/01/80	OSW: GPO: NTS: PB81-188 161
Design and Development of a Hazardous Waste Reactivity Testing Protocol	02/01/84	OSW: 600/2-84-057 GPO: NTS: PB84-158 807
Design, Construction, and Evaluation of Clay Liners for Waste Management Facilities	11/15/88	OSW: 530/SW-86-007F GPO: NTS: PB89-181 937
Directory of Commercial Hazardous Waste Management Facilities	08/15/87	OSW: 530/SW-87-024 GPO: NTS: PB88-109 699
Disposal Tips for Home Health Care (Patient Flyer)	01/15/90	OSW: 530/SW-90-014B GPO: NTS:
Disposal Tips for Home Health Care (Professional Brochure)	01/15/90	OSW: 530/SW-90-014A GPO: NTS:

Documentation for Development of Toxicity and Volumes Scores for the Purpose of Scheduling Hazardous Waste	03/15/85	OSW: 530/SW-85-031 GPO: NTS: PB87-154 936/AS
Does Your Business Produce Hazardous Waste? Many Small Businesses Do	04/15/90	OSW: 530/SW-90-027 GPO: NTS:
Chemical Manufacturers		OSW: 530/SW-90-027H
Cleaning and Cosmetics		OSW: 530/SW-90-027Q
Construction		OSW: 530/SW-90-027J
Drycleaning and Laundry		OSW: 530/SW-90-027B
Educational/Vocational		OSW: 530/SW-90-027L
Equipment Repair		OSW: 530/SW-90-027D
Formulators		OSW: 530/SW-90-027P
Furniture/Wood Refinishing		OSW: 530/SW-90-027C
Laboratories		OSW: 530/SW-90-027M
Leather/Leather Products		OSW: 530/SW-90-027R
Metal Manufacturing		OSW: 530/SW-90-027N
Motor Freight Terminals/Railroad Transport		OSW: 530/SW-90-027K
Pesticide End-Users		OSW: 530/SW-90-027I
Printing and Allied Industry		OSW: 530/SW-90-027G
Pulp and Paper Industry		OSW: 530/SW-90-027O
Textile Manufacturing		OSW: 530/SW-90-027E
Uniform Hazardous Waste Manifest Instructions		OSW: 530/SW-90-027S
Vehicle Maintenance		OSW: 530/SW-90-027A
Wood Preserving		OSW: 530/SW-90-027F
Emissions Testing of a Precalciner Cement Kiln at Louisville, Nebraska	12/15/90	OSW: 530/SW-91-016 GPO: NTS: PB91-130 195
Emissions Testing of a Wet Cement Kiln at Hannibal, Missouri; Draft Final Report	12/15/90	OSW: 530/SW-91-017 GPO: NTS: PB91-130-203
Engineering Handbook for Hazardous Waste Incineration	09/01/81	OSW: SW-889 GPO: NTS: PB81-248 163
Environmental Fact Sheet: About the Municipal Solid Waste Stream...	08/15/90	OSW: 530/SW-90-042B GPO: NTS:
Environmental Fact Sheet: Agency Releases Report to Congress on Special Wastes from Mineral Processing	07/15/90	OSW: 530/SW-90-070A GPO: NTS:
Environmental Fact Sheet: Amendment to Requirements for Hazardous Waste Incinerator Permits	01/15/89	OSW: 530/SW-89-004 GPO: NTS:
Environmental Fact Sheet: America's War on Waste	10/15/89	OSW: 530/SW-90-002 GPO:

APPENDIX 11—EPA PUBLICATIONS

		NTS:	
Environmental Fact Sheet: Application of Land Disposal Restrictions to CERCLA Remedial Actions	10/15/89	OSW: GPO: NTS:	530/SW-90-003
Environmental Fact Sheet: Changes to Interim Status Facilities; Modifications to Hazardous Waste Permits; Procedures for Post-Closure Permitting	03/15/89	OSW: GPO: NTS:	530/SW-89-028
Environmental Fact Sheet: Delay of Closure Period for Hazardous Waste Facilities	08/15/89	OSW: GPO: NTS:	530/SW-89-007
Environmental Fact Sheet: Delisting Regulation Amendment	06/15/89	OSW: GPO: NTS:	530/SW-89-052
Environmental Fact Sheet: EPA Proposes Modifications to the Definition of Wastewater Treatment Unit	07/15/90	OSW: GPO: NTS:	530/SW-90-030
Environmental Fact Sheet: EPA's Final Conditional No-Migration Determination for DOE's Waste Isolation Pilot Plant (WIPP)	11/15/90	OSW: GPO: NTS:	530/SW-91-026
Environmental Fact Sheet: Final Rule for Third Third Scheduled Wastes Completes Statutory Requirements for Land Disposal Restrictions	05/08/90	OSW: GPO: NTS:	530/SW-90-046
Environmental Fact Sheet: Final Rule to Eliminate Mineral Processing Wastes from the Bevill Amendment	08/15/89	OSW: GPO: NTS:	530/SW-89-062
Environmental Fact Sheet: Final Rule to Identify the Status of Twenty Mineral Processing Wastes Conditionally Retained Within the Bevill Amendment	01/15/90	OSW: GPO: NTS:	530/SW-90-013
Environmental Fact Sheet: Hazardous Waste Boilers and Industrial Furnaces Now Under Strict RCRA Regulations	12/15/90	OSW: GPO: NTS:	530/SW-91-014
Environmental Fact Sheet: Hazardous Waste Management System; Final Codification Rule Correction	04/15/90	OSW: GPO: NTS:	530/SW-90-026
Environmental Fact Sheet: Interim Final Rule Suspending Application of the Toxicity Characteristic for Used Chlorofluorocarbon Refrigerants Being Reclaimed	02/15/91	OSW: GPO: NTS:	530/SW-91-028

Environmental Fact Sheet: Land Disposal Restrictions - Second Third (Final)	06/15/89	OSW: 530/SW-89-046 GPO: NTS:
Environmental Fact Sheet: Listing of Wastes from Primary Treatment of Oily Wastewaters	11/15/90	OSW: 530/SW-90-043 GPO: NTS:
Environmental Fact Sheet: Milestone! Fifth Rulemaking Finalizes Land Disposal Restrictions	05/08/90	OSW: 530/SW-90-048 GPO: NTS:
Environmental Fact Sheet: New Treatment Standards Proposed for K061 High Zinc Subcategory Wastes	04/15/91	OSW: 530/SW-91-032 GPO: NTS:
Environmental Fact Sheet: Nineteen Eighty-Five National Biennial Report of Hazardous Waste Generators and Treatment, Storage, and Disposal Facilities	03/15/89	OSW: 530/SW-89-033C GPO: NTS:
Environmental Fact Sheet: No Land Disposal Standards Amendment	05/01/89	OSW: 530/SW-89-044 GPO: NTS:
Environmental Fact Sheet: Notification of Ground-water Monitoring Policy for Delisting Petitions	10/15/89	OSW: 530/SW-89-064 GPO: NTS:
Environmental Fact Sheet: Plastics: The Facts About Production, Use, and Disposal	02/15/90	OSW: 530/SW-90-017A GPO: NTS:
Environmental Fact Sheet: Plastics: The Facts on Source Reduction	02/15/90	OSW: 530/SW-90-017C GPO: NTS:
Environmental Fact Sheet: Proposed Amendments to Interim Status Standards for Downgradient Monitoring Well Locations at Hazardous Waste Facilities	01/15/91	OSW: 530/SW-91-036 GPO: NTS:
Environmental Fact Sheet: Proposed Rulemaking on Corrective Action for Solid Waste Management Units at Hazardous Waste Management Facilities	07/15/90	OSW: 530/SW-90-067 GPO: NTS:
Environmental Fact Sheet: Report Analyzes Municipal Waste Combustion Ash, Ash Extracts, and Leachates	03/15/90	OSW: 530/SW-90-029C GPO: NTS:
Environmental Fact Sheet: SWICH: EPA's National Solid Waste Information	01/15/91	OSW: 530/SW-91-025 GPO:

APPENDIX 11—EPA PUBLICATIONS 311

Clearinghouse		NTS:	
Environmental Fact Sheet: The Facts About Plastics in the Marine Environment	02/15/90	OSW: GPO: NTS:	530/SW-90-017B
Environmental Fact Sheet: The Facts on Degradable Plastics	02/15/90	OSW: GPO: NTS:	530/SW-90-017D
Environmental Fact Sheet: The Facts on Recycling Plastics	02/15/90	OSW: GPO: NTS:	530/SW-90-017E
Environmental Fact Sheet: Tighter Controls Proposed for Hazardous Waste Incinerators	03/15/90	OSW: GPO: NTS:	530/SW-90-015
Environmental Fact Sheet: Toxicity Characteristic Rule Finalized	03/15/90	OSW: GPO: NTS:	530/SW-89-045
Environmental Fact Sheet: Wood Preserving Wastes Listed as Hazardous	11/15/90	OSW: GPO: NTS:	530/SW-91-008
Environmental Fact Sheet: Yard Waste Composting	01/15/91	OSW: GPO: NTS:	530/SW-91-009
EP Toxicity Characteristic (40 CFR 261.24); Identification and Listing of Hazardous Waste Under RCRA, Subtitle C, Section 3001	05/01/80	OSW: GPO: NTS:	PB81-185 027
EPA Activities Under RCRA - Annual Report to the President and Congress, Fiscal Year 1978	03/15/79	OSW: GPO: NTS:	SW-755
EPA Activities Under RCRA of 1976 - Annual Report to the President and Congress, Fiscal Year 1977	02/01/78	OSW: GPO: NTS:	SW-663 PB88-197 603
EPA Guide for Infectious Waste Management	05/15/86	OSW: GPO: NTS:	530/SW-86-014 PB86-199 130
Equivalency of State Financial Responsibility Mechanisms	09/01/82	OSW: GPO: NTS:	PB87-157 475
Evaluating Cover Systems for Solid and Hazardous Waste	09/15/82	OSW: GPO: NTS:	SW-867 PB87-154 894

Evaluation Guidelines for Toxic Air Emissions from Land Disposal Facilities	08/01/81	OSW: GPO: NTS: PB87-157 418/AS
Fate and Transport of Hazardous Constituents; Identification and Listing of Hazardous Waste Under RCRA, Subtitle C, Section 3001	05/01/80	OSW: GPO: NTS: PB81-190 027
Final Draft Guidance for Subpart G of the Interim Status Standards for Owners and Operators of Hazardous Waste Treatment, Storage, and Disposal Facilities	11/15/81	OSW: GPO: NTS: PB87-193 397
Final Interim Status Standards for Surface Impoundments (40 CFR 265.220); Standards Applicable to Owners and Operators of Hazardous Waste Treatment, Storage, and Disposal Facilities Under RCRA, Subtitle C, Section 3004	04/15/80	OSW: GPO: NTS: PB81-185 001
Financial Assurance for Closure and Post-Closure Care; Requirements for Owners and Operators of Hazardous Waste Treatment, Storage, and Disposal Facilities; A Guidance Manual	05/15/82	OSW: SW-955 GPO: NTS: PB82-237 595
Financial Requirements (40 CFR 264 and 265, Subpart H); Standards Applicable to Owners and Operators of Hazardous Waste Treatment, Storage, and Disposal Facilities Under RCRA, Subtitle C, Section 3004	01/15/81	OSW: GPO: NTS: PB81-189 326
Financial Requirements; Interim Status Standards (40 CFR 265, Subpart H); Final Draft Guidance	01/15/84	OSW: SW-913 GPO: NTS: PB88-197 595
First Report to Congress: Resource Recovery and Source Reduction	01/15/74	OSW: SW-352 GPO: NTS: PB255 139/8
Forty CFR Parts 190-259	07/01/90	OSW: GPO: 869-011-00146-7 NTS:
Forty CFR Parts 260-299	07/01/90	OSW: GPO: 869-011-00147-5 NTS:
Fourth Report to Congress: Resource	08/01/77	OSW: SW-600

Recovery and Waste Reduction		GPO: NTS: PB88-197 579
General and Interim Status Standards for Tanks (40 CFR 264 and 265, Subpart J); Interim Status Standards for Chemical, Physical, and Biological Treatment (40 CFR 265, Subpart Q); Standards Applicable to Owners and Operators of Hazardous Waste Treatment, Storage, and Disposal Facilities Under RCRA, Subtitle C, Section 3004	12/15/80	OSW: GPO: NTS: PB81-190 050
General Facility Standards for Location of Facilities (40 CFR 264, Subpart B, Section 264.18); Standards Applicable to Owners and Operators of Hazardous Waste Treatment, Storage, and Disposal Facilities Under RCRA, Subtitle C, Section 3004	12/15/80	OSW: GPO: NTS: PB81-189 755
General Issues Concerning Interim Status Standards (40 CFR 265); Standards Applicable to Owners and Operators of Hazardous Waste Treatment, Storage, and Disposal Facilities Under RCRA, Subtitle C, Section 3004	04/15/80	OSW: GPO: NTS: PB81-181 414
General Waste Analysis (40 CFR 264.13); Interim Status Standards for General Waste Analysis (40 CFR 265.13); Standards Applicable to Owners and Operators of Hazardous Waste Treatment, Storage, and Disposal Facilities Under RCRA, Subtitle C, Section 3004	04/29/80	OSW: GPO: NTS: PB81-181 406
Generic Quality Assurance Project Plan for Land Disposal Restrictions Program (BDAT)	03/15/87	OSW: 530/SW-87-011 GPO: NTS: PB88-170 766
Groundwater Monitoring (40 CFR 265, Subpart F); Standards Applicable to Owners and Operators of Hazardous Waste Treatment, Storage, and Disposal Facilities Under RCRA, Subtitle C, Section 3004	05/20/80	OSW: GPO: NTS: PB81-189 797
Groundwater Monitoring Guidance for Owners and Operators of Interim Status Facilities	03/15/83	OSW: SW-963 GPO: NTS: PB83-209 445

314   RCRA REGULATORY COMPLIANCE GUIDE

Guidance Document for Subpart F: Air Emission Monitoring; Land Disposal Toxic Air Emissions Evaluation Guideline	12/15/80	OSW: GPO: NTS: PB87-155 578/AS
Guidance Document: Seismic Considerations in Hazardous Waste Management Facilities	12/29/80	OSW: GPO: NTS: PB87-155 552/AS
Guidance for Facility Management Planning; Draft	07/01/85	OSW: GPO: NTS: PB87-188 090
Guidance for Implementing the RCRA Dioxin Listing Rule	08/01/85	OSW: GPO: NTS: PB87-202 040
Guidance for Permit Writers: Facilities Storing Hazardous Waste in Containers	11/02/82	OSW: GPO: NTS: PB88-105 689
Guidance for the Analysis of Refinery Wastes	07/05/85	OSW: GPO: NTS: PB87-154 910/AS
Guidance Manual for Cost Estimates for Closure and Post-Closure Plans (Subparts G and H); Volume I: Treatment and Storage Facilities	11/15/86	OSW: 530/SW-87-009A GPO: NTS: PB87-158 994/AS
Guidance Manual for Cost Estimates for Closure and Post-Closure Plans (Subparts G and H); Volume II: Land Disposal Facilities	11/15/86	OSW: 530/SW-87-009B GPO: NTS: PB87-159 000/AS
Guidance Manual for Cost Estimates for Closure and Post-Closure Plans (Subparts G and H); Volume III: Unit Costs	11/15/86	OSW: 530/SW-87-009C GPO: NTS: PB87-159 018/AS
Guidance Manual for Cost Estimates for Closure and Post-Closure Plans (Subparts G and H); Volume IV: Documentation	11/15/86	OSW: 530/SW-87-009D GPO: NTS: PB87-159 026/AS
Guidance Manual for Hazardous Waste Incinerator Permits; Final Report	07/15/83	OSW: SW-966 GPO: NTS: PB86-100 577
Guidance Manual for Research, Development, and Demonstration Permits (40 CFR Section 270.65)	07/15/86	OSW: 530/SW-86-008 GPO: NTS: PB86-229 192/AS
Guidance Manual for the Class-	01/15/81	OSW: SW-199C

ification of Solid Waste Disposal Facilities		GPO: NTS: PB81-218 505
Guidance Manual on Hazardous Waste Land Treatment Closure/Post-Closure (40 CFR Part 265)	04/14/87	OSW: GPO: NTS: PB87-183 695
Guidance Manual on the RCRA Regulation of Recycled Hazardous Wastes	03/15/86	OSW: 530/SW-86-015 GPO: NTS: PB86-208 584/AS
Guidance on Implementation of the Minimum Technological Requirements of HSWA of 1984, Respecting Liners and Leachate Collection Systems; Reauthorization Statutory Interpretation #5D	05/24/85	OSW: 530/SW-85-012 GPO: NTS: PB87-163 242/AS
Guidance on Issuing Permits to Facilities Required to Analyze Groundwater for Appendix VIII Constituents	02/15/86	OSW: GPO: NTS: PB87-188 082
Guidance on Public Involvement in the RCRA Permitting Program	01/15/86	OSW: OSWER9500.00-1A GPO: NTS:
Guidance on the Definition and Identification of Commercial Mixed Low-Level Radioactive and Hazardous Waste and Answers to Anticipated Questions	10/04/89	OSW: 530/SW-90-016 GPO: NTS:
Guide for Preparing RCRA Permit Applications for Existing Facilities	01/15/82	OSW: GPO: NTS: PB87-193 371
Guide to Energy from Municipal Waste for Small Communities	10/01/82	OSW: SW-958 GPO: NTS: PB88-138 177
Guide to the Disposal of Chemically Stabilized and Solidified Waste	09/15/82	OSW: SW-872 GPO: NTS: PB87-154 902/AS
Handle with Care: How to Throw Out Used Insulin Syringes and Lancets at Home; A Booklet for Young People with Diabetes and their Families	10/15/90	OSW: 530/SW-90-089 GPO: NTS:
Hazardous Waste from Discarding of Commercial Chemical Products and the Containers and Spill Residues Thereof (40 CFR 261.33); Identification and	04/01/81	OSW: 530/SW-89-005 GPO: NTS: PB89-126 460

Listing of Hazardous Waste Under RCRA,
Subtitle C, Section 3001

Title	Date	OSW	GPO	NTS
Hazardous Waste Incineration Permitting Study	08/15/86			PB87-202 420
Hazardous Waste Incineration: Questions and Answers	04/15/88	530/SW-88-018		
Hazardous Waste Land Treatment	04/15/83	SW-874		PB89-179 014
Hazardous Waste Tanks Failure Model; Description of Methodology and Appendices A, B, C, D, and E	01/13/86	530/SW-86-012		PB86-192 937/AS
Hazardous Waste Tanks Risk Analysis	06/15/86	530/SW-86-011		PB86-212 289/AS
Health and Environmental Effects Profiles (40 CFR 261); Identification and Listing of Hazardous Waste Under RCRA, Subtitle C, Section 3001	10/15/81			PB81-190 019
Heuristic Routing for Solid Waste Collection Vehicles	01/01/74	SW-356		
Hospital Waste Combustion Study: Data Gathering Phase; Final Report	12/15/88	450/3-88-017		PB89-148 308
Household Hazardous Waste: Bibliography of Useful References and List of State Experts	03/15/88	530/SW-88-014		
How to Set Up a Local Program to Recycle Used Oil	05/15/89	530/SW-89-039A		
Hydrologic Evaluation of Landfill Performance (HELP) Model; Volume 1: User's Guide for Version 1; Second Edition	06/15/84	530/SW-84-009		PB85-100 840
Hydrologic Evaluation of Landfill Performance (HELP) Model; Volume 2: Documentation for Version 1	06/15/84	530/SW-84-010		PB85-100 832

Hydrologic Simulation on Solid Waste Disposal Sites	09/15/82	OSW: SW-868 GPO: NTS:
Ignitability Characteristic (40 CFR 261.21); Identification and Listing of Hazardous Waste Under RCRA, Subtitle C, Section 3001	05/02/80	OSW: GPO: NTS: PB81-187 890
Incineration Standards (40 CFR 264 and 265, Subpart O); Standards Applicable to Owners and Operators of Hazardous Waste Treatment, Storage, and Disposal Facilities Under RCRA, Subtitle C, Section 3004	12/15/80	OSW: GPO: NTS: PB81-190 092
Infectious Waste (40 CFR 250.14); Identification and Listing of Hazardous Waste Under RCRA, Subtitle C, Section 3001	12/15/78	OSW: 530/SW-88-055 GPO: NTS: PB89-102 594
Inter-Industry Collaborative Study of Toxicity Characteristic Leaching Procedures; Addendum to Compilation of Phase IA and Phase II Data	09/15/86	OSW: GPO: NTS: PB87-155 545/AS
Interim National Criteria for a Quality Hazardous Waste Management Program Under RCRA	05/01/84	OSW: 530/SW-84-006 GPO: NTS:
Interim Status Standards and General Status Standards for Closure and Post-Closure Care (40 CFR 264 and 265, Subpart G); Standards Applicable to Owners and Operators of Hazardous Waste Treatment, Storage, and Disposal Facilities Under RCRA, Subtitle C, Section 3004	12/15/80	OSW: GPO: NTS: PB81-189 763
Interim Status Standards for Land Treatment Facilities (40 CFR 265, Subpart M); Standards Applicable to Owners and Operators of Hazardous Waste Treatment, Storage, and Disposal Facilities Under RCRA, Subtitle C, Section 3004	04/15/80	OSW: GPO: NTS: PB81-190 068
Interim Status Standards for Landfills (40 CFR 265, Subpart N); Standards Applicable to Owners and Operators of Hazardous Waste Treatment, Storage, and Disposal Facilities Under RCRA, Subtitle C, Section 3004	02/15/81	OSW: GPO: NTS: PB81-189 789

Interim Status Standards for Thermal Treatment Processes Other Than Incineration and for Open Burning (40 CFR 265, Subpart P); Standards Applicable to Owners and Operators of Hazardous Waste Treatment, Storage, and Disposal Facilities Under RCRA, Subtitle C, Section 3004	04/5/80	OSW: GPO: NTS: PB81-189 771
Interim Status Surface Impoundments; Retrofitting Variances; Guidance Document	06/15/86	OSW: 530/SW-86-017 GPO: NTS: PB86-212 263/AS
Inventory of Open Dumps	06/15/85	OSW: SW-964 GPO: NTS: PB91-181 594
Joint NRC/EPA Guidance on a Conceptual Design Approach for Commercial Mixed Low-Level Radioactive and Hazardous Waste Disposal Facilities	08/03/87	OSW: 530/SW-87-027 GPO: NTS:
Land Disposal Restrictions Summary; Volume I: Solvents and Dioxins	05/15/87	OSW: 530/SW-87-019A GPO: NTS:
Land Disposal Restrictions Summary; Volume II: California List Wastes	10/15/87	OSW: 530/SW-88-012 GPO: NTS:
Landfill and Surface Impoundment Performance Evaluation	04/15/83	OSW: SW-869R GPO: NTS: PB91-181 586
Let's Reduce and Recycle: A Curriculum for Solid Waste Awareness	08/15/90	OSW: 530/SW-90-005 GPO: NTS:
Liability Coverage; Requirements for Owners and Operators of Hazardous Waste Treatment, Storage, and Disposal Facilities; A Guidance Manual	11/15/82	OSW: SW-961 GPO: NTS: PB83-144 675
Liner Location Risk and Cost Analysis Model	01/15/85	OSW: GPO: NTS: PB87-157 756/AS
Liner Location Risk and Cost Analysis Model; Appendices	01/15/85	OSW: GPO: NTS: PB87-157 210/AS
Lining of Waste Impoundment and Disposal Facilities	03/01/83	OSW: SW-870 GPO: NTS: PB86-192 796/AS

List of Municipal Waste Landfills	12/15/86	OSW: GPO: NTS: PB87-178 331
Listing of Hazardous Waste; Finalization of July 16. 1980, Hazardous Waste List (40 CFR 261.31 and 261.32); Identification and Listing of Hazardous Waste Under RCRA, Subtitle C, Section 3001	01/12/81	OSW: GPO: NTS: PB81-190 076
Listing of Hazardous Waste (40 CFR 261.31 and 261.32); Identification and Listing of Hazardous Waste Under RCRA, Subtitle C, Section 3001	05/02/80	OSW: GPO: NTS: PB81-190 035
Location of Mines and Factors Affecting Exposure	06/30/86	OSW: 530/SW-86-023 GPO: NTS: PB86-219 409/AS
Location Standards for RCRA Hazardous Waste Facilities; Regulatory Development Plan	12/15/85	OSW: GPO: NTS: PB87-162 954/AS
Low-Level Mixed Waste: A RCRA Perspective for NRC Licensees	08/15/90	OSW: 530/SW-90-057 GPO: NTS:
Management of Hazardous Waste Leachate	09/15/82	OSW: SW-871R GPO: NTS: PB91-181 578
Managing and Tracking Medical Wastes: A Guide to the Federal Program for Generators	09/15/89	OSW: 530/SW-89-021 GPO: NTS:
Managing and Tracking Medical Wastes: A Guide to the Federal Program for Transporters	09/15/89	OSW: 530/SW-89-022 GPO: NTS:
Managing and Tracking Medical Wastes: A Guide to the Federal Program for Treatment, Destruction, and Disposal Facilities	09/15/89	OSW: 530/SW-89-023 GPO: NTS:
Manifest System, Recordkeeping, and Reporting (40 CFR 264 and 265, Subpart E); Standards Applicable to Owners and Operators of Hazardous Waste Treatment, Storage, and Disposal Facilities Under RCRA, Subtitle C, Section 3004	04/15/80	OSW: GPO: NTS: PB81-190 043

Measurements of Particulates, Metals and Organics at a Hazardous Waste Incinerator	11/15/88	OSW: 530/SW-89-067 GPO: NTS: PB89-230 668
Medical Waste Management in the United States: First Interim Report to Congress	05/15/90	OSW: 530/SW-90-051A GPO: NTS: PB90-219 874
Medical Waste Management in the United States: First Interim Report to Congress, Executive Summary	05/15/90	OSW: 530/SW-90-051B GPO: NTS:
Medical Waste Management in the United States: Second Interim Report to Congress	12/15/90	OSW: 530/SW-90-087A GPO: NTS: PB91-130-187
Medical Waste Management in the United States: Second Interim Report to Congress; Executive Summary	12/15/90	OSW: 530/SW-90-087B GPO: NTS:
Medical Waste Tracking Act of 1988	11/01/88	OSW: 530/SW-89-008 GPO: NTS:
Memorandum to Monica Chatmon-McEaddy of Office of Solid Waste Regarding Final Treatment Standards for K069 Nonwastewaters in the Calcium Sulfate/Sodium Subcategory and Wastewater Forms of K069	05/08/90	OSW: 530/SW-90-059K GPO: NTS: PB90-234 113
Memorandum to Monica Chatmon-McEaddy of Office of Solid Waste Regarding Final Treatment Standards for Nonwastewater and Wastewater Forms of K100	05/08/90	OSW: 530/SW-90-059L GPO: NTS: PB90-234 121
Memorandum to the Docket Regarding Final Treatment Standards for Nonwastewater and Wastewater Forms of K044, K045, and K047	05/08/90	OSW: 530/SW-90-060B GPO: NTS: PB90-234 295
Metals Control Efficiency Test at a Dry Scrubber and Baghouse Equipped Hazardous Waste Incinerator	09/15/90	OSW: 530/SW-91-004 GPO: NTS: PB91-101 865
Methodology for Developing Best Demonstrated Available Technology (BDAT) Treatment Standards	05/08/89	OSW: 530/SW-89-048B GPO: NTS: PB89-221 428
Methods Manual for Compliance with the BIF Regulations: Burning Hazardous Waste in Boilers and Industrial Furnaces	12/15/90	OSW: 530/SW-91-010 GPO: NTS: PB91-120-006
Methods for the Storage and Retrieval of RCRA Groundwater Monitoring Data on STORET; User's Manual	10/01/85	OSW: GPO: NTS: PB87-154 928/AS

Minimum Technology Guidance on Double Liner Systems for Landfills and Surface Impoundments; Design, Construction, and Operation	05/24/85	OSW: 530/SW-85-014 GPO: NTS: PB87-151 072/AS
Minimum Technology Guidance on Single Liner Systems for Landfills, Surface Impoundments, and Waste Piles; Design, Construction, and Operation (Draft)	05/24/85	OSW: 530/SW-85-013 GPO: NTS: PB87-173 159
Mobile Incineration: An Analysis of the Industry	06/30/89	OSW: 530/SW-90-076 GPO: NTS: PB90-255 449
Model RCRA Permit for Hazardous Waste Management Facilities	09/15/88	OSW: 530/SW-90-049 GPO: NTS: PB90-210 998
Modifying RCRA Permits	09/15/89	OSW: 530/SW-89-050 GPO: NTS:
Multimaterial Source Separation in Marblehead and Somerville, Massachusetts; Volume II: Collection and Marketing	12/15/79	OSW: SW-822 GPO: NTS:
Multimaterial Source Separation in Marblehead and Somerville, Massachusetts; Volume III: Materials and Refuse; Composition of Source Separated Materials and Refuse	12/15/79	OSW: SW-823 GPO: NTS:
Multimaterial Source Separation in Marblehead and Somerville, Massachusetts; Volume IV: Energy Use and Savings from Source Separated Materials and Other Solid Waste Management Alternatives for Marblehead	12/15/79	OSW: SW-824 GPO: NTS:
Multimaterial Source Separation in Marblehead and Somerville, Massachusetts; Volume V: Citizen Attitudes Toward Source Separation	12/15/79	OSW: SW-825 GPO: NTS:
Municipal Waste Combustion Ash and Leachate Characterization; Monofill Baseline Year; Woodburn Monofill; Woodburn, Oregon	08/15/89	OSW: 530/SW-89-074 GPO: NTS: PB90-104 746
Municipal Waste Combustion Study (Complete Set)	09/15/87	OSW: 530/SW-87-021 GPO: NTS: PB87-206 066

Municipal Waste Combustion Study: Assessment of Health Risks Associated With Exposure to Municipal Waste Combustion Emissions	09/15/87	OSW: 530/SW-87-021G GPO: NTS: PB87-206 132
Municipal Waste Combustion Study: Characterization of the Municipal Waste Combustion Industry	06/15/87	OSW: 530/SW-87-021H GPO: NTS: PB87-206 140
Municipal Waste Combustion Study: Control of Organic Emissions	06/15/87	OSW: 530/SW-87-021C GPO: NTS: PB87-206 090
Municipal Waste Combustion Study: Costs of Flue Gas Cleaning Technologies	06/15/87	OSW: 530/SW-87-021E GPO: NTS: PB87-206 116
Municipal Waste Combustion Study: Emissions Data Base for Municipal Waste Combustors	06/15/87	OSW: 530/SW-87-021B GPO: NTS: PB87-206 082
Municipal Waste Combustion Study: Flue Gas Cleaning Technology	06/15/87	OSW: 530/SW-87-021D GPO: NTS: PB87-206 108
Municipal Waste Combustion Study: Recycling of Solid Waste	06/15/87	OSW: 530/SW-87-021I GPO:
Municipal Waste Combustion Study: Report to Congress	06/15/87	OSW: 530/SW-87-021A GPO: NTS: PB87-206 074
Municipal Waste Combustion Study: Sampling and Analysis	06/15/87	OSW: 530/SW-87-021F GPO: NTS: PB87-206 124
National Criteria for a Quality Hazardous Waste Management Program	06/01/86	OSW: 530/SW-89-011 GPO: NTS:
National Dioxin Study	08/15/87	OSW: 530/SW-87-025 GPO: NTS: PB88-192 687
National RCRA Corrective Action Strategy	10/01/86	OSW: 530/SW-86-045 GPO: NTS:
National Small Quantity Hazardous Waste Generator Survey; Final Report	02/15/85	OSW: GPO: NTS: PB85-180 438
National Survey of Hazardous Waste	04/20/84	OSW: 530/SW-84-005

Generators and Treatment, Storage, and Disposal Facilities Regulated Under RCRA in 1981		GPO: NTS: PB86-197 837/AS
National Survey of Solid Waste (Municipal) Landfill Facilities	09/15/88	OSW: 530/SW-88-034 GPO: NTS: PB89-118 525
The Nation's Hazardous Waste Management Program at a Crossroads: The RCRA Implementation Study	07/15/90	OSW: 530/SW-90-069 GPO: NTS:
Native American Network: A RCRA Information Exchange	09/15/90	OSW: 530/SW-90-079 GPO: NTS:
Native American Network: A RCRA Information Exchange (Second Issue)	04/15/91	OSW: 530/SW-91-001 GPO: NTS:
New RCRA: A Fact Book	01/01/85	OSW: 530/SW-85-035 GPO: NTS:
New Rule for Wood Preserving Wastes	12/15/90	OSW: 530/SW-91-012 GPO: NTS:
The New Toxicity Characteristic Rule: Information and Tips for Generators	04/15/90	OSW: 530/SW-90-028 GPO: NTS:
Nineteen Eighty-Five National Biennial Report of Hazardous Waste Generators and Treatment, Storage, and Disposal Facilities Regulated Under RCRA; Volume I: Summary	03/15/89	OSW: 530/SW-89-033A GPO: NTS: PB89-187 645
Nineteen Eighty-Six National Screening Survey of Hazardous Waste Treatment, Storage, Disposal, and Recycling Facilities	09/01/88	OSW: 530/SW-88-035 GPO: NTS: PB89-106 058
Ninety-Day Subchronic Oral Toxicity in Rats Test Materials: Pyridine; Volume I	12/15/87	OSW: 530/SW-88-016A GPO: NTS: PB88-176 136
Ninety-Day Subchronic Oral Toxicity in Rats Test Materials: Pyridine; Volume II	12/15/87	OSW: 530/SW-88-016B GPO: NTS: PB88-176 144
No Migration Variances to the Hazardous Waste Land Disposal Prohibitions; A Guidance Manual for	03/15/90	OSW: 530/SW-90-045 GPO: NTS: PB90-204 736

Petitioners

Title	Date	Identifiers
Office Paper Recycling: An Implementation Manual	01/15/90	OSW: 530/SW-90-001 GPO: NTS: PB90-199 431
Once There Lived a Wicked Dragon (Coloring Book)	01/15/74	OSW: SW-335 GPO: NTS:
Performance Test on a Spray Dryer, Fabric Filter, and Wet Scrubber; Draft Test Report	10/15/89	OSW: 530/SW-90-008 GPO: NTS: PB90-120 544
Permit Applicants' Guidance Manual for Exposure Information Requirements Under RCRA, Section 3019	07/03/85	OSW: GPO: NTS: PB87-193 694
Permit Applicants' Guidance Manual for Hazardous Waste Land Treatment, Storage, and Disposal Facilities; Final Draft	05/15/84	OSW: 530/SW-84-004 GPO: NTS: PB89-115 695
Permit Applicants' Guidance Manual for the General Facility Standards of 40 CFR 264	10/15/83	OSW: SW-968 GPO: NTS: PB87-151 064/AS
Permit Guidance Manual on Hazardous Waste Land Treatment Demonstrations	07/15/86	OSW: 530/SW-86-032 GPO: NTS: PB86-229 184/AS
Permit Guidance Manual on Unsaturated Zone Monitoring for Hazardous Waste Land Treatment Units	10/15/86	OSW: 530/SW-86-040 GPO: NTS: PB87-215 463/AS
Permit Writers' Guidance Manual for Hazardous Waste Land Treatment, Storage, and Disposal Facilities; Phase 1: Criteria for Location Acceptability and Existing Applicable Regulations	02/15/85	OSW: 530/SW-85-024 GPO: NTS: PB86-125 580/AS
Permit Writers' Guidance Manual for Hazardous Waste Tank Standards	07/15/86	OSW: 530/SW-89-003 GPO: NTS: PB89-126 478
Permitting Hazardous Waste Incinerators	04/15/88	OSW: 530/SW-88-024 GPO: NTS:
Permitting of Land Disposal Facilities: Groundwater and Air Emission Monitoring	07/31/81	OSW: GPO: NTS: PB81-246 431

Permitting of Land Disposal Facilities: Groundwater Protection Standard	07/31/81	OSW: GPO: NTS: PB81-246 423
Permitting of Land Disposal Facilities: Information Requirements for Permitting Discharges; General Standards Applicable to Owners and Operators of Hazardous Waste Treatment, Storage, and Disposal Facilities	07/31/81	OSW: GPO: NTS: PB81-246 415
Permitting of Land Disposal Facilities: Land Treatmen	07/31/81	OSW: GPO: NTS: PB81-246 381
Permitting of Land Disposal Facilities: Landfills	07/31/81	OSW: GPO: NTS: PB81-246 399
Permitting of Land Disposal Facilities: Overview; Background Document	07/31/81	OSW: GPO: NTS: PB81-246 357
Permitting of Land Disposal Facilities: Performance Standards for Land Disposal Facilities	07/31/81	OSW: GPO: NTS: PB81-246 449
Permitting of Land Disposal Facilities: Surface Impoundments	07/31/81	OSW: GPO: NTS: PB81-246 365
Permitting of Land Disposal Facilities: Underground Injection; Background Document	07/31/81	OSW: GPO: NTS: PB81-246 407
Permitting of Land Disposal Facilities: Waste Piles	07/31/81	OSW: GPO: NTS: PB81-246 373
Petitions to Delist Hazardous Wastes; A Guidance Manual	04/01/85	OSW: 530/SW-85-003 GPO: NTS: PB85-194 488
Pilot-Scale ESP and Hydro-Sonic Scrubber Parametric Tests for Particulate, Metals and HCl Emissions; John Zink Company Research Facility; Tulsa, Oklahoma; Draft Test Report	06/15/89	OSW: 530/SW-90-009 GPO: NTS: PB90-129 362
Plans, Recordkeeping, Variances, and Demonstrations for Hazardous Waste Treatment, Storage, and Disposal Facilities; A Guidance Manual	01/01/81	OSW: SW-921 GPO: NTS: PB87-155 503/AS

Post-Closure Liability Trust Fund Simulation Model; Volume I: Model Overview and Results; Volume II: Graphs and Tables of Model Results; Volume III: Model Description	05/01/85	OSW: GPO: NTS: PB86-212 479/AS
Procedural Guidance for Reviewing Exposure Information Under RCRA, Section 3019	09/26/86	OSW: GPO: NTS: PB87-193 702
Procedures for Modeling Flow Through Clay Liners to Determine Required Liner Thickness	04/15/84	OSW: 530/SW-84-001 GPO: NTS: PB87-191 029
Procedures Manual for Groundwater Monitoring at Solid Waste Disposal Facilities	08/15/77	OSW: SW-611 GPO: NTS: PB84-174 820
Proceedings of the First International Symposium on Oil and Gas Exploration and Production Waste Management Practices, September 10-13, 1990, New Orleans, Louisiana	09/13/90	OSW: 530/SW-91-030 GPO: NTS: PB91-160 549
Proceedings of the Fourth National Conference on Household Hazardous Waste Management	02/15/90	OSW: 530/SW-89-042D GPO: NTS: PB90-163 189
Procurement Guidelines for Government Agencies	12/15/90	OSW: 530/SW-91-011 GPO: NTS:
Prohibition on the Disposal of Bulk Liquid Hazardous Waste in Landfills; Statutory Interpretive Guidance	06/11/86	OSW: 530/SW-86-016 GPO: NTS: PB86-212 271/AS
Promoting Source Reduction and Recyclability in the Marketplace	09/15/89	OSW: 530/SW-89-066 GPO: NTS: PB90-163 122
Proposed Additions to Standards for Hazardous Waste Incineration (40 CFR 264.342 and 264.343); Standards Applicable to Owners and Operators of Hazardous Waste Treatment, Storage, and Disposal Facilities Under RCRA, Subtitle C, Section 3004	12/15/80	OSW: GPO: NTS: PB81-193 021
Proposed Sampling and Analytical Methodologies for Addition to Test Methods for Evaluating Solid Waste: Physical/Chemical Methods (SW-846, Second Edition	01/01/84	OSW: GPO: NTS: PB85-103 026

Protocol for Evaluating Interim Status Closure/Post-Closure Plans	08/15/86	OSW: GPO: NTS: PB87-178 315
Quantities of Cyanide-Bearing and Acid-Generating Wastes Generated by the Mining and Beneficiating Industries, and the Potentials for Contaminant Release	06/27/86	OSW: 530/SW-86-025 GPO: NTS: PB86-219 391/AS
Questions and Answers on Land Disposal Restrictions for Solvents and Dioxins	05/15/87	OSW: 530/SW-87-020 GPO: NTS:
Questions and Answers on the Land Disposal Restrictions Program (Brochure)	11/15/86	OSW: 530/SW-86-057 GPO: NTS:
Questions and Answers Regarding the July 14, 1986, Hazardous Waste Tank System Regulatory Amendments	10/02/87	OSW: 530/SW-87-012 GPO: NTS:
Rat Oral Subchronic Toxicity Study; Final Report; Compound: Isobutyl Alcohol	07/30/85	OSW: 530/SW-88-015 GPO: NTS: PB88-176 177
RCRA Corrective Action Interim Measures Guidance; Interim Final	06/15/88	OSW: 530/SW-88-029 GPO: NTS:
RCRA Corrective Action Plan; Interim Final	06/15/88	OSW: 530/SW-88-028 GPO: NTS:
RCRA Facility Assessment Guidance	10/09/86	OSW: 530/SW-86-053 GPO: NTS: PB87-107 769
RCRA Facility Investigation (RFI) Guidance; Interim Final; Volume I: Development of an RFI Work Plan and General Considerations for RCRA Facility Investigations; Volume II: Soil, Ground Water, and Subsurface Gas Releases; Volume III: Air and Surface Water Releases; Volume IV: Case Study Examples	05/15/89	OSW: 530/SW-89-031 GPO: NTS: PB89-200 299
RCRA Final Authorization Guidance Manual	06/10/83	OSW: SW-862 GPO: NTS: PB87-155 057/AS
RCRA Groundwater Monitoring Compliance Order Guidance	08/01/85	OSW: GPO:

		NTS: PB87-193 710
RCRA Groundwater Monitoring Technical Enforcement Guidance Document (TEGD) (Final)	09/15/86	OSW: 530/SW-86-055 GPO: NTS: PB87-107 751/AS
RCRA Guidance Document: Land Treatment	10/06/82	OSW: GPO: NTS: PB87-155 065/AS
RCRA Guidance Document: Landfill Design, Liner Systems and Final Cover; Draft	07/01/82	OSW: GPO: NTS: PB87-157 657/AS
RCRA Guidance Document: Surface Impoundments, Liner Systems, Final Cover, and Free Board Control; Draft	11/15/86	OSW: GPO: NTS: PB87-157 665/AS
RCRA Guidance Manual for Subpart G Closure and Post-Closure Care Standards and Subpart H Cost Estimating Requirements	01/15/87	OSW: 530/SW-87-010 GPO: NTS: PB87-158 978/AS
RCRA Implementation Plan (RIP): Fiscal Year 1987	05/19/86	OSW: OSWER9420.00-1 GPO: NTS: PB87-157 673/AS
RCRA Implementation Plan (RIP): Fiscal Year 1988	03/31/87	OSW: 530/SW-91-039 GPO: NTS: PB91-156 679
RCRA Implementation Plan (RIP): Fiscal Year 1989	04/05/88	OSW: 530/SW-91-040 GPO: NTS: PB91-156 687
RCRA Implementation Plan (RIP): Fiscal Year 1990	04/15/89	OSW: 530/SW-91-041 GPO: NTS: PB91-156 695
RCRA Implementation Plan (RIP): Fiscal Year 1991	04/15/89	OSW: 530/SW-91-042 GPO: NTS:
RCRA Information Center (RIC): How to Find Hazardous and Solid Waste Information	08/15/89	OSW: 530/SW-87-023 GPO: NTS:
RCRA Inspection Manual	03/15/88	OSW: 530/SW-89-068 GPO: NTS: PB89-208 69/AS
RCRA Liability Coverage for Bodily Injury and Property Damage Survey Results	09/15/89	OSW: 530/SW-89-043 GPO: NTS: PB90-113 945

RCRA Orientation Manual: 1990 Edition	08/01/90	OSW: 530/SW-90-036 GPO: 055-000-00364-5 NTS:
RCRA Permit Quality Protocol; Draft	09/15/88	OSW: 530/SW-90-050 GPO: NTS: PB90-211 004
RCRA Personnel Training Guidance for Owners or Operators of Hazardous Waste Management Facilities	09/01/80	OSW: SW-915 GPO: NTS: PB87-193 348
Reactivity Characteristic (40 CFR 261.23); Identification and Listing of Hazardous Waste Under RCRA, Subtitle C, Section 3001	05/01/80	OSW: GPO: NTS: PB81-184 988
Recycle (Brochure)	10/15/88	OSW: 530/SW-88-050 GPO: NTS:
Recycle Today! Educational Materials for Grades K-12 (Brochure)	04/15/90	OSW: 530/SW-90-025 GPO: NTS:
Recycling in Federal Agencies	10/15/90	OSW: 530/SW-90-082 GPO: NTS:
Recycling Used Oil: For Service Stations and Other Vehicle-Service Facilities	06/15/89	OSW: 530/SW-89-039D GPO: NTS:
Recycling Used Oil: Ten Steps to Change Your Oil	06/15/89	OSW: 530/SW-89-039C GPO: NTS:
Recycling Used Oil: What Can You Do?	06/15/89	OSW: 530/SW-89-039B GPO: NTS:
Recycling Works: State and Local Solutions to Solid Waste Management Problems	01/15/89	OSW: 530/SW-89-014 GPO: NTS:
Regional Guidance Manual for Selected Interim Status Requirements (Draft)	09/15/80	OSW: GPO: NTS: PB87-194 130
Regulatory Analysis of RCRA Regulations	04/30/80	OSW: GPO: NTS: PB81-181 471

Regulatory Impact Analysis for the Proposed Rulemaking on Corrective Action for Solid Waste Management Units	06/25/90	OSW: 530/SW-90-081 GPO: NTS: PB91-102 061
Report on Minimum Criteria to Assure Data Quality	12/12/89	OSW: 530/SW-90-021 GPO: NTS:
Report to Congress on Special Wastes from Mineral Processing: Summary and Findings	07/15/90	OSW: 530/SW-90-070B GPO: NTS:
Report to Congress on Special Wastes from Mineral Processing: Summary and Findings; Methods and Analyses; Appendices	07/15/90	OSW: 530/SW-90-070C GPO: NTS: PB90-258 492
Report to Congress on the Discharge of Hazardous Wastes to Publicly Owned Treatment Works (The Domestic Sewage Study)	02/07/86	OSW: 530/SW-86-004 GPO: NTS: PB86-184 017/AS
Report to Congress on the Minimization of Hazardous Wastes (Complete Set)	10/15/86	OSW: 530/SW-86-033 GPO: NTS: PB87-114 328
Report to Congress on the Minimization of Hazardous Wastes	10/15/86	OSW: 530/SW-86-033A GPO: NTS: PB87-114 336/AS
Report to Congress on the Minimization of Hazardous Wastes; Appendices	10/15/86	OSW: 530/SW-86-033B GPO: NTS: PB87-114 344
Report to Congress on Waste Minimization: Issues and Options; Volume I	10/15/86	OSW: 530/SW-86-041 GPO: NTS: PB87-114 351/AS
Report to Congress on Waste Minimization: Issues and Options; Volume II	10/15/86	OSW: 530/SW-86-042 GPO: NTS: PB87-114 369/AS
Report to Congress on Waste Minimization: Issues and Options; Volume III	10/15/86	OSW: 530/SW-86-043 GPO: NTS: PB87-114 377/AS
Report to Congress: EPA Activities and Accomplishments Under the Resource Conservation and Recovery Act Fiscal Years 1980-1985	07/15/86	OSW: 530/SW-86-027 GPO: NTS: PB86-232 154
Report to Congress: EPA Activities and Accomplishments Under the Resource Conservation and Recovery Act: Fourth	12/15/87	OSW: 530/SW-88-007 GPO: NTS:

Quarter Fiscal Year 1986 Through
Fiscal Year 1987

Title	Date	Identifiers
Report to Congress: Management of Hazardous Wastes from Educational Institutions	04/15/89	OSW: 530/SW-89-040 GPO: NTS: PB89-187 629
Report to Congress: Management of Hazardous Wastes from Educational Institutions; Executive Summary	04/15/89	OSW: 530/SW-89-040A GPO: NTS:
Report to Congress: Management of Wastes from the Exploration, Development, and Production of Crude Oil, Natural Gas, and Geothermal Energy (Complete Set)	12/15/87	OSW: 530/SW-88-003 GPO: NTS: PB88-146 212
Report to Congress: Management of Wastes from the Exploration, Development, and Production of Crude Oil, Natural Gas, and Geothermal Energy; Executive Summaries	12/15/87	OSW: 530/SW-88-003D GPO: NTS: PB88-146 253
Report to Congress: Management of Wastes from the Exploration, Development, and Production of Crude Oil, Natural Gas, and Geothermal Energy; Volume 1: Oil and Gas	12/15/87	OSW: 530/SW-88-003A GPO: NTS: PB88-146 220
Report to Congress: Management of Wastes from the Exploration, Development, and Production of Crude Oil, Natural Gas, and Geothermal Energy; Volume 2: Geothermal Energy	12/15/87	OSW: 530/SW-88-003B GPO: NTS: PB88-146 238
Report to Congress: Management of Wastes from the Exploration, Development, and Production of Crude Oil, Natural Gas, and Geothermal Energy; Volume 3: Appendices	12/15/87	OSW: 530/SW-88-003C GPO: NTS: PB88-146 246
Report to Congress: Methods to Manage and Control Plastic Wastes	02/15/90	OSW: 530/SW-89-051 GPO: NTS: PB90-163 106
Report to Congress: Methods to Manage and Control Plastic Wastes; Executive Summary	02/15/90	OSW: 530/SW-89-051A GPO: NTS:
Report to Congress: Solid Waste Disposal in the United States; Volume I	10/15/88	OSW: 530/SW-88-011 GPO: NTS: PB89-110 381
Report to Congress: Solid Waste	10/15/88	OSW: 530/SW-88-011B

Disposal in the United States; Volume II		GPO: NTS: PB89-110 399
Report to Congress: Solid Waste Disposal in the United States; Executive Summary	10/15/88	OSW: 530/SW-88-011A GPO: NTS:
Report to Congress: Wastes from the Combustion of Coal by Electric Utility Power Plants	02/15/88	OSW: 530/SW-88-002 GPO: NTS: PB88-177 977
Report to Congress: Wastes from the Combustion of Coal by Electric Utility Power Plants; Appendices	02/15/88	OSW: 530/SW-88-002A GPO: NTS: PB88-177 985
Report to Congress: Wastes from the Combustion of Coal by Electric Utility Power Plants; Executive Summary	02/15/88	OSW: 530/SW-88 002A.1 GPO: NTS:
Report to Congress: Wastes from the Extraction and Beneficiation of Metallic Ores, Phosphate Rock, Asbestos, Overburden from Uranium Mining, and Oil Shale	12/01/85	OSW: 530/SW-85-033 GPO: NTS: PB88-162 631
Report to the Congress of the United States on the Post-Closure Liability Trust Fund Under Section 3001(A)(2)(II) of the Comprehensive Environmental Response, Compensation, and Liability Act of 1980	05/15/85	OSW: GPO: NTS: PB86-210 176/AS
Residential Paper Recovery: A Community Action Program	01/01/76	OSW: SW-553 GPO: NTS:
Resource Conservation and Recovery Act Regulations Affecting Generators and Transporters (40 CFR 262 and 263); Explanation of Revisions in Reporting Burden Estimates	03/25/80	OSW: GPO: NTS: PB87-155 776/AS
Resource Recovery Plant Implementation; Guides for Municipal Officials: Accounting Format	01/15/76	OSW: SW-493 GPO: NTS: PB259 143/6
Resource Recovery Plant Implementation; Guides for Municipal Officials: Financing	01/15/75	OSW: SW-471 GPO: NTS: PB256 461/5
Resource Recovery Plant Implementation; Guides for Municipal Officials: Further Assistance	01/15/75	OSW: SW-470 GPO: NTS: PB256 460/7

APPENDIX 11—EPA PUBLICATIONS

Title	Date	OSW / GPO / NTS
Resource Recovery Plant Implementation; Guides for Municipal Officials: Interim Report	01/15/75	OSW: SW-480 GPO: NTS: PB259 139/4
Resource Recovery Plant Implementation; Guides for Municipal Officials: Marketing	01/15/76	OSW: SW-499 GPO: NTS: PB259 141/0
Resource Recovery Plant Implementation; Guides for Municipal Officials: Planning and Overview	01/15/76	OSW: SW-533 GPO: NTS: PB88-197 561
Resource Recovery Plant Implementation; Guides for Municipal Officials: Procurement	01/01/76	OSW: SW-495 GPO: NTS: PB259 140/2
Resource Recovery Plant Implementation; Guides for Municipal Officials: Risks and Contracts	01/15/76	OSW: SW-496 GPO: NTS: PB259 142/8
Resource Recovery Plant Implementation; Guides for Municipal Officials: Technologies	01/15/76	OSW: SW-550 GPO: NTS: PB259 144/4
Response to Comments Background Document for the Second Third Land Disposal Restrictions Proposed Rule; Volume 1: Treatment Standards Related Comments	06/08/89	OSW: 530/SW-89-048D GPO: NTS: PB89-221 535
Response to Comments Background Document for the Second Third Land Disposal Restrictions Proposed Rule; Volume 2: Capacity Related Comments	06/08/89	OSW: 530/SW-89-048E GPO: NTS: PB89-221 543
Response to Comments Background Document for the Second Third Land Disposal Restrictions Proposed Rule; Volume 3: Policy Related Comments	06/08/89	OSW: 530/SW-89-048C GPO: NTS: PB89-221 436
Response to Comments Background Document for the Third Third Land Disposal Restrictions Proposed Rule; Volume 1: BDAT Related Comments (Complete Set)	05/08/90	OSW: 530/SW-90-061 GPO: NTS: PB90-234 477
Response to Comments Background Document for the Third Third Land Disposal Restrictions Proposed Rule; Volume 1-A-1: BDAT Related Comments; General BDAT Issues	05/08/90	OSW: 530/SW-90-061A GPO: NTS: PB90-234 485
Response to Comments Background Document for the Third Third Land Disposal Restrictions Proposed Rule; Volume 1-A-2: BDAT Related Comments; General BDAT Issues	05/08/90	OSW: 530/SW-90-061B GPO: NTS: PB90-234 493

Response to Comments Background   05/08/90   OSW: 530/SW-90-061C
Document for the Third Third Land Disposal   GPO:
Restrictions Proposed Rule; Volume 1-A-3:   NTS: PB90-234 501
BDAT Related Comments; General BDAT Issues

Response to Comments Background   05/08/90   OSW: 530/SW-90-061D
Document for the Third Third Land Disposal   GPO:
Restrictions Proposed Rule; Volume 1-B: NTS: PB90-234 519
BDAT Related Comments; D001: Characteristic
Ignitable Wastes

Response to Comments Background   05/08/90   OSW: 530/SW-90-061E
Document for the Third Third Land Disposal   GPO:
Restrictions Proposed Rule; Volume 1-C: NTS: PB90-234 527
BDAT Related Comments; D002: Characteristic
Corrosive Wastes

Response to Comments Background   05/08/90   OSW: 530/SW-90-061F
Document for the Third Third Land Disposal   GPO:
Restrictions Proposed Rule; Volume 1-D: NTS: PB90-234 535
BDAT Related Comments; D003: Characteristic
Reactive Wastes and P and U Wastes
Containing Reactive Listing Constituents

Response to Comments Background   05/08/90   OSW: 530/SW-90-061G
Document for the Third Third Land Disposal   GPO:
Restrictions Proposed Rule; Volume 1-E: NTS: PB90-234 543
BDAT Related Comments; D004: Characteristic
Wastes for Arsenic and K, P, and U Wastes
Containing Arsenic; and D010: Characteristic
Wastes for Selenium

Response to Comments Background   05/08/90   OSW: 530/SW-90-061H
Document for the Third Third Land Disposal   GPO:
Restrictions Proposed Rule; Volume 1-F: NTS: PB90-234 550
BDAT Related Comments; D005: Characteristic
Wastes for Barium and P013; and D006:
Characteristic Wastes for Cadmium

Response to Comments Background   05/08/90   OSW: 530/SW-90-061I
Document for the Third Third Land Disposal   GPO:
Restrictions Proposed Rule; Volume 1-G: NTS: PB90-234 568
BDAT Related Comments; D007: Characteristic
Wastes for Chromium

Response to Comments Background   05/08/90   OSW: 530/SW-90-061J
Document for the Third Third Land Disposal   GPO:
Restrictions Proposed Rule; Volume 1-H: NTS: PB90-234 576
BDAT Related Comments; D008: Characteristic
Wastes for Lead and P and U Wastes
Containing Lead

Response to Comments Background   05/08/90   OSW: 530/SW-90-061K
Document for the Third Third Land Disposal   GPO:

Restrictions Proposed Rule; Volume 1-I: BDAT Related Comments; D009: Characteristic Wastes for Mercury and P and U Wastes Containing Mercury; K045, K046, and K047: Wastes from the Manufacturing and Processing of Explosives; D011: Characteristic Wastes for Silver; P119, P120: Vanadium Containing Wastes; and P and U Wastes Containing Thallium

NTS: PB90-234 584

Response to Comments Background Document for the Third Third Land Disposal Restrictions Proposed Rule; Volume 1-J: BDAT Related Comments; Mixed Radioactive Hazardous Wastes; Polynuclear Aromatic U Wastes; Halogenated Aliphatic U Wastes; Non-Halogenated Aromatic U Wastes; F002-F005: Solvents; F006: Electroplating Wastewater Treatment Sludges; and F019: Aluminum Conversion Coating Treatment Sludges

05/08/90

OSW: 530/SW-90-061L
GPO:
NTS: PB90-234 592

Response to Comments Background Document for the Third Third Land Disposal Restrictions Proposed Rule; Volume 1-K BDAT Related Comments; F025: Wastes from the Production of Chlorinated Aliphatics; K002-K008: Inorganic Pigments; K011, K013, K014: Acrylonitriles; K015: Still Bottoms from Benzal Chloride; K046: Wastewater Treatment Sludges from Manufacturing, Formulating, and Loading of Lead-Based Initiating Compounds; K061: Electric Arc Furnace Dust; and K069: Emission Control Dust/Sludge from Secondary Lead Smelting

05/08/90

OSW: 530/SW-90-061M
GPO:
NTS: PB90-234 600

Response to Comments Background for the Third Third Land Disposal Restrictions Proposed Rule; Volume 1-L: BDAT Related Comments; K071 and K106: Mercury Cell Process Wastes; K086: Residues from Ink Production, Wastes Containing Cyanide

05/08/90

OSW: 530/SW-90-061N
GPO:
NTS: PB90-234 618

Response to Comments Background Document for the Third Third Land Disposal Restrictions Proposed Rule; Volume 1-M: BDAT Related Comments; K017: Wastes from the Production of Epichlorohydrin; K028, K029, K095, and K096: Production of 1,1,1-Trichloroethane Wastewaters Containing BDAT Constituents; K022: Wastes from the Production of Phenol/Acetone; K025: Distillation Bottoms from the Production of

05/08/90

OSW: 530/SW-90-061O
GPO:
NTS: PB90-234 626

Nitrobenzene; K035: Wastewater Treatment
Sludges from the Production of Creosote;
K026: Stripping Still Tails from the
Production of Methyl Ethyl Pyridine; K083:
Distillation Bottoms from the Production of
Aniline Oxygenated Hydrocarbons and Hetero-
cyclic U and P Wastes; F024: Production of
Chlorinated Aliphatic Hydrocarbons

Response to Comments Background   05/08/90   OSW: 530/SW-90-061P
Document for the Third Third Land Disposal   GPO:
Restrictions Proposed Rule; Volume 1-N: NTS: PB90-234 634
BDAT Related Comments; Halogenated
Organic, Pharmaceutical, Brominated
Organic, Organo-Sulfur Compounds, and
Organo-Nitrogen Compound Wastes; and
Halogenated Pesticide and Chlorobenzene,
Halogenated Phenolic, and Phenolic Wastes

Response to Comments Background   05/08/90   OSW: 530/SW-90-061Q
Document for the Third Third Land Disposal   GPO:
Restrictions Proposed Rule; Volume 1-O: NTS: PB90-234 642
BDAT Related Comments; K048-K052:
Petroleum Refining Industry Wastes;
K036: Organophosphorous Wastes
(Nonwastewaters); K037: Wastewater
Treatment Sludges from the Production
of Disulfoton

Response to Comments Background   05/08/90   OSW: 530/SW-90-061R
Document for the Third Third Land Disposal   GPO:
Restrictions Proposed Rule; Volume 1-P: NTS: PB90-234 659
BDAT Related Comments; Leachates

Response to Comments Background   05/08/90   OSW: 530/SW-90-061S
Document for the Third Third Land Disposal   GPO:
Restrictions Proposed Rule; Volume 1-Q: NTS: PB90-234 667
BDAT Related Comments; Gases

Response to Comments Background   05/08/90   OSW: 530/SW-90-063
Document for the Third Third Land Disposal   GPO:
Restrictions Proposed Rule; Volume 2:   NTS: PB90-234 725
Capacity Related Comments (Complete Set)

Response to Comments Background   05/08/90   OSW: 530/SW-90-063A
Document for the Third Third Land Disposal   GPO:
Restrictions Proposed Rule; Volume 2-1:   NTS: PB90-234 733
Capacity Related Comments

Response to Comments Background   05/08/90   OSW: 530/SW-90-063B
Document for the Third Third Land Disposal   GPO:
Restrictions Proposed Rule; Volume 2-2:   NTS: PB90-234 741
Capacity Related Comments

Response to Comments Background Document for the Third Third land Disposal Restrictions Proposed Rule; Volume 2-3: Capacity Related Comments	05/08/90	OSW: 530/SW-90-063C GPO: NTS: PB90-234 758
Response to Comments Background Document for the Third Third Land Disposal Restrictions Proposed Rule; Volume 3: Policy Related Comments	05/08/90	OSW: 530/SW-90-064 GPO: NTS: PB90-234 766
Restrictions on the Placement of Nonhazardous Liquids in Hazardous Waste Landfills	04/15/86	OSW: 530/SW-86-013 GPO: NTS: PB86-215 043/AS
Reusable News (Fall 1990)	09/15/90	OSW: 530/SW-90-056 GPO: NTS:
Reusable News (Spring 1990)	05/15/90	OSW: 530/SW-90-039 GPO: NTS:
Reusable News (Summer 1990)	06/15/90	OSW: 530/SW-90-055 GPO: NTS:
Reusable News (Winter 1990)	01/15/90	OSW: 530/SW-90-018 GPO: NTS:
Reusable News (Winter 1991)	01/15/91	OSW: 530/SW-91-120 GPO: NTS:
Ride the Wave of the Future: Recycle Today! (Poster)	04/15/90	OSW: 530/SW-90-010 GPO: NTS:
Rules for Hazardous Waste Tank Systems	01/15/88	OSW: 530/SW-88-004 GPO: NTS:
School Recycling Programs: A Handbook for Educators	08/15/90	OSW: 530/SW-90-023 GPO: NTS:
Second Report to Congress: Resource Recovery and Source Reduction	01/15/74	OSW: SW-353 GPO: NTS: PB253 406/3
Sites for our Solid Waste: A Guidebook for Effective Public Involvement	04/15/90	OSW: 530/SW-90-019 GPO: NTS:

Title	Date	ID
Siting our Solid Waste: Making Public Involvement Work (Brochure)	04/15/90	OSW: 530/SW-90-020 GPO: NTS:
Soil Properties, Classification, and Hydraulic Conductivity Testing	03/15/84	OSW: SW-925 GPO: NTS: PB87-155-784/AS
Soiliner Model	04/15/86	OSW: 530/SW-86-006 GPO: NTS: PB87-100 038
Solid Waste Dilemma: An Agenda for Action; Background Document	08/15/88	OSW: 530/SW-88-054A GPO: NTS: PB88-251 137
Solid Waste Dilemma: An Agenda for Action; Appendices A-C	08/15/88	OSW: 530/SW-88-054B GPO: NTS: PB88-251 145
Solid Waste Dilemma: An Agenda for Action (Final Report of the Municipal Waste Task Force)	02/15/89	OSW: 530/SW-89-019 GPO: NTS:
The Solid Waste Disposal Act as Amended by the Hazardous and Solid Waste Amendments (The Resource Conservation and Recovery Act)	01/01/87	OSW: 530/SW-85-022 GPO: NTS:
Solid Waste Leaching Procedure; Technical Resource Document for Public Comment	03/15/84	OSW: SW-924 GPO: NTS: PB87-152 054/AS
Solving the Hazardous Waste Problem: EPA's RCRA Program	11/15/86	OSW: 530/SW-86-037 GPO: NTS:
Source Separation Collection and Processing Equipment; A User's Guide	01/15/80	OSW: SW-842 GPO: NTS: PB81-158 297
Source Separation: The Community Awareness Program in Somerville and Marblehead, Massachusetts	11/15/76	OSW: SW-551 GPO: NTS: PB260 654/9
Standards for Inspection (40 CFR 264.15) and Interim Status Standards for Inspection (40 CFR 265.15); Standards Applicable to Owners and Operators of Hazardous Waste Treatment, Storage, and Disposal Facilities Under RCRA, Subtitle C, Section 3004	04/15/80	OSW: GPO: NTS: PB81-190 001

Standards for Personnel Training (40 CFR 264.16); Interim Status Standards for Personnel Training (40 CFR 265.16); Standards Applicable to Owners and Operators of Hazardous Waste Treatment, Storage, and Disposal Facilities Under RCRA, Subtitle C, Section 3004	04/15/80	OSW: GPO: NTS: PB81-181 380
Standards for Preparedness and Prevention (40 CFR 264 and 265, Subpart C); Standards for Contingency Plan and Emergency Procedures (40 CFR 264 and 265, Subpart D); Standards Applicable to Owners and Operators of Hazardous Waste Treatment, Storage, and Disposal Facilities Under RCRA, Subtitle C, Section 3004	04/15/80	OSW: GPO: NTS: PB81-181 372
Standards for Security (40 CFR 264.14); Interim Status Standards for Security (40 CFR 265.14); Standards Applicable to Owners and Operators of Hazardous Waste Treatment, Storage, and Disposal Facilities Under RCRA, Subtitle C, Section 3004	04/15/80	OSW: GPO: NTS: PB81-181 398
State Authorization Manual; Volume I	10/15/90	OSW: 530/SW-91-018A GPO: NTS: PB91-130 211
State Authorization Manual; Volume II	10/15/90	OSW: 530/SW-91-018B GPO: NTS: PB91-130 229
State Decision-Makers Guide for Hazardous Waste Management	01/01/77	OSW: SW-612 GPO: NTS:
Statistical Analysis of Ground-Water Monitoring Data at RCRA Facilities; Interim Final Guidance	04/15/89	OSW: 530/SW-89-026 GPO: NTS: PB89-151 047
Statistical Analysis of Mining Waste Data	06/30/86	OSW: 530/SW-86-024 GPO: NTS: PB86-219 383/AS
Subchronic Toxicity of Meta-Cresol in Sprague Dawley Rats	03/25/88	OSW: 530/SW-88-026 GPO: NTS: PB88-195 284
Subchronic Toxicity of Ortho-Cresol in Sprague Dawley Rats	03/21/88	OSW: 530/SW-88-027 GPO: NTS: PB88-197 496

Subchronic Toxicity of Para-Cresol in Sprague Dawley Rats; MBA Chemical No. 25	04/04/88	OSW: 530/SW-88-025 GPO: NTS: PB88-195 292
Subtitle D Study: Phase I Report	10/15/86	OSW: 530/SW-86-054 GPO: NTS: PB87-116 810/AS
Summary of Appropriate Analytical Methods for Appendix IX; Parts I and II	07/15/87	OSW: GPO: NTS: PB87-230 371
Summary of Comments on Mining Waste Report to Congress	05/09/86	OSW: 530/SW-86-030 GPO: NTS: PB86-222 486/AS
Summary of the First National Conference on Household Hazardous Waste Collection Programs	11/18/86	OSW: 530/SW-89-042A GPO: NTS: PB89-179 501
Summary of the Second National Conference on Household Hazardous Waste Collection Programs	11/02/87	OSW: 530/SW-89-042B GPO: NTS: PB89-179 519
Summary of the Third National Conference on Household Hazardous Waste Collection Programs	11/02/88	OSW: 530/SW-89-042C GPO: NTS: PB89-179 527
Summary Report of Capacity at Commercial Facilities; Volume I	01/31/89	OSW: 530/SW-89-035A GPO: NTS: PB89-179 022
Summary Report of Capacity at Commercial Facilities; Volume II	01/31/89	OSW: 530/SW-89-035B GPO: NTS: PB89-179 030
Summary Report of Capacity at Limited Commercial and Company Captive Facilities; Draft; Volume I	02/15/89	OSW: 530/SW-89-036A GPO: NTS: PB89-179 048
Summary Report of Capacity at Limited Commercial and Company Captive Facilities; Draft; Volume II	02/15/89	OSW: 530/SW-89-036B GPO: NTS: PB89-179 055
Surface Water Screening Procedure; Background Document	12/15/85	OSW: 530/SW-86-050 GPO: NTS: PB87-101 614
Survey of Household Hazardous Waste and Related Collection Programs	10/15/86	OSW: 530/SW-86-038 GPO: NTS:
Technical Background Document and	04/15/89	OSW: 530/SW-90-080

Response to Comments: Method 1311 - Toxicity Characteristic Leaching Procedure		GPO: NTS: PB91-102 053
Technical Evaluation of the Combustion System of the Marine Shale Processors, Inc. Facility in Amelia, Louisiana	10/15/90	OSW: 530/SW-90-086 GPO: NTS: PB91-111 492
Technical Guidance Document: Fabrication of Polyethylene FML Field Seams	09/15/89	OSW: 530/SW-89-069 GPO: NTS:
Technical Guidance Document: Final Covers on Hazardous Waste Landfills and Surface Impoundments	07/15/89	OSW: 530/SW-89-047 GPO: NTS: PB89-233 489
Technical Guidance for Corrective Measures: Determining Appropriate Technology and Response for Air Releases; Draft Final Report	03/15/85	OSW: 530/SW-88-021 GPO: NTS: PB88-185 269
Technical Guidance for Corrective Measures: Subsurface Gas	03/28/85	OSW: 530/SW-88-023 GPO: NTS: PB88-185 285
Technical Report: Exploration, Development, and Production of Crude Oil and Natural Gas; Field Sampling and Analytical Results	01/31/87	OSW: 530/SW-87-005 GPO: NTS: PB87-165 403
Technical Report: Exploration, Development, and Production of Crude Oil and Natural Gas; Appendix A: Analytical Results	01/31/87	OSW: 530/SW-87-005A GPO: NTS: PB87-165 411
Technical Report: Exploration, Development, and Production of Crude Oil and Natural Gas; Appendix B: Sampling Strategy	01/31/87	OSW: 530/SW-87-005B GPO: NTS: PB87-165 429
Technical Report: Exploration, Development, and Production of Crude Oil and Natural Gas; Appendix C: Sampling Reports; Volumes 1 and 2	01/31/87	OSW: 530/SW-87-005C GPO: NTS: PB87-165 437
Technical Report: Exploration, Development, and Production of Crude Oil and Natural Gas; Appendix D: Analytical Methods	01/31/87	OSW: 530/SW-87-005D GPO: NTS: PB87-165 445
Technical Report: Exploration, Development, and Production of Crude Oil and Natural Gas; Appendix E: Role	01/31/87	OSW: 530/SW-87-005E GPO: NTS: PB87-165 452

and Function of EPA Sample Control
Center

Technical Report: Exploration, Development, and Production of Crude Oil and Natural Gas; Appendix F: List of Analytes	01/31/87	OSW: GPO: NTS:	530/SW-87-005F PB87-165 460
Technical Report: Exploration, Development, and Production of Crude Oil and Natural Gas; Appendix G: Sampling Plan and Sampling Quality Assurance/Quality Control	01/31/87	OSW: GPO: NTS:	530/SW-87-005G PB87-165 478
Technical Resource Document for Obtaining Variances from the Secondary Containment Requirement of Hazardous Waste Tank Systems; Volume I	02/15/87	OSW: GPO: NTS:	530/SW-87-002A PB87-158 655/AS
Technical Resource Document for Obtaining Variances from the Secondary Containment Requirement of Hazardous Waste Tank Systems; Volume II	02/15/87	OSW: GPO: NTS:	530/SW-87-002B PB87-158 663
Technical Resource Document for the Storage and Treatment of Hazardous Waste in Tank Systems	8/22/86	OSW: GPO: NTS:	530/SW-86-044 PB87-134 391/AS
Technical Studies Supporting the Mining Waste Regulatory Determination	06/30/86	OSW: GPO: NTS:	530/SW-86-026 PB86-219 417/AS
Teratologic Evaluation of 2, 3, 4, 6-Tetrachlorophenol Administered to CD Rats on Gestational Days 6 Through 15; Final Report	08/21/87	OSW: GPO: NTS:	530/SW-88-017A PB88-176 151
Teratologic Evaluation of 2, 3, 4, 6-Tetrachlorophenol Administered to CD Rats on Gestational Days 6 Through 15; Final Report; Appendices I-IX	08/21/87	OSW: GPO: NTS:	530/SW-88-017B PB88-176 169
Test Method Equivalency Petitions; A Guidance Manual	02/15/87	OSW: GPO: NTS:	530/SW-87-008 PB87-178 349
Test Methods for Evaluating Solid Waste: Physical/Chemical Methods; Second Edition	07/15/82	OSW: GPO: NTS:	SW-846 PB87-120 291/AS
Test Methods for Evaluating Solid Waste: Physical/Chemical Methods; Third Edition; Volumes IA, IB, IC, and II	11/15/86	OSW: GPO: NTS:	SW-846 955-001-00000-1 PB88-239 223

APPENDIX 11—EPA PUBLICATIONS 343

Title	Date	IDs
Test Methods for Evaluating Solid Waste: Physical/Chemical Methods; Third Edition: Proposed Update Package	01/23/89	OSW: SW-846.3-1 GPO: 955-001-00000-1 NTS: PB89-148 076
Third Report to Congress: Resource Recovery and Source Reduction	01/15/75	OSW: SW-448 GPO: NTS: PB88-174 677
Toxicity Characteristic Regulatory Impact Analysis	03/15/90	OSW: 530/SW-90-088 GPO: NTS: PB91-101 873
Tracking Medical Wastes (Brochure)	05/15/89	OSW: 530/SW-89-020 GPO: NTS:
Treatment Technology Background Document; Third Third; Final	01/15/91	OSW: 530/SW-90-059Z GPO: NTS: PB91-160 556
Trial Burn Observation Guide	03/15/89	OSW: 530/SW-89-027 GPO: NTS: PB89-179 543
Understanding the Small Quantity Generator Hazardous Waste Rules: A Handbook for Small Business	09/15/86	OSW: 530/SW-86-019 GPO: NTS:
U.S. Department of Transportation Hazardous Materials Regulations as They Apply to the U.S. Environmental Protection Agency's Hazardous Waste Regulations	11/01/81	OSW: SW-935 GPO: NTS: PB82-182 361
Use of the Water Balance Method for Predicting Leachate Generation from Solid Waste Disposal Sites	10/15/75	OSW: SW-168 GPO: NTS: PB87-194 643
Used Oil Recycling (First Newsletter)	03/15/88	OSW: 530/SW-88-047 GPO: NTS:
Used Oil Recycling (Second Newsletter)	11/15/88	OSW: 530/SW-89-006 GPO: NTS:
Used Oil Recycling (Third Newsletter)	09/15/90	OSW: 530/SW-90-068 GPO: NTS:
Variable Rates in Solid Waste: Handbook for Solid Waste Officials; Volume I: Executive Summary	06/15/90	OSW: 530/SW-90-084A GPO: NTS:

Variable Rates in Solid Waste: Handbook for Solid Waste Officials; Volume II: Detailed Manual	06/15/90	OSW: GPO: NTS:	530/SW-90-084B PB90-272 063
Waste Analysis Plans; A Guidance Manual	10/15/84	OSW: GPO: NTS:	530/SW-84-012
Waste Minimization: Environmental Quality with Economic Benefits	10/15/87	OSW: GPO: NTS:	530/SW-90-044
Waste Minimization; Executive Summary	10/15/86	OSW: GPO: NTS:	530/SW-86-041A
Waste Minimization in Metals Parts Cleaning	08/15/89	OSW: GPO: NTS:	530/SW-89-049
Yard Waste Composting; A Study of Eight Programs	04/15/89	OSW: GPO: NTS:	530/SW-89-038 PB90-163 114

List of Regulations and Statutes Referenced in Text

Code of Federal Regulations

29 CFR Section 1910.1200. Ch. 7, n. 2
29 CFR Section 1910.1200(e). Ch. 7, n. 6
29 CFR Section 1910.1200(f). Ch. 7, n. 5
29 CFR Section 1910.1200(h)(1). Ch. 7, n. 3
29 CFR Section 1910.1200(h)(2). Ch. 7, n. 7

33 CFR Section 153.203. Ch. 6, n. 19

40 CFR Part 261. Ch. 3, ns. 1, 4
40 CFR Part 262. Ch. 4, n. 5
40 CFR Part 264. Ch. 4, n. 7
40 CFR Part 302. Ch. 6, n. 18
40 CFR Part 763. Ch. 3, n. 15
40 CFR Section 116.4. Ch. 6, n. 17
40 CFR Section 117.3. Ch. 6, n. 18
40 CFR Section 117.22. Ch. 6, n. 20
40 CFR Section 261. Ch. 8, n. 10
40 CFR Section 261.2(c)(1). Ch. 2, n. 5
40 CFR Section 261.2(c)(2). Ch. 2, n. 5
40 CFR Section 261.2(c)(3). Ch. 2, n. 5
40 CFR Section 261.2(c)(4). Ch. 2, n. 5
40 CFR Section 261.3(a)(2)(iii). Ch. 2, n. 12
40 CFR Section 261.3(a)(2)(iv). Ch. 3, n. 7
40 CFR Section 261.3(b)(2). Ch. 3, n. 7
40 CFR Section 261.3(c). Ch. 2, n. 13
40 CFR Section 261.3(c)(1). Ch. 3, n. 7
40 CFR Section 261.3(c)(2)(i). Ch. 3, n. 6
40 CFR Section 261.3(d)(2). Ch. 3, n. 7
40 CFR Section 261.4. Ch. 2, n. 14
40 CFR Section 261.5. Ch. 4, n. 4
40 CFR Section 261.6(c)(1). Ch. 2, n. 17
40 CFR Section 261.21. Ch. 2, n. 8
40 CFR Section 261.22. Ch. 2, n. 9
40 CFR Section 261.23. Ch. 2, n. 10
40 CFR Section 261.24. Ch. 2, n. 11
40 CFR Section 261.30(b). Ch. 3, n. 8
40 CFR Section 261.32. Ch. 2, ns. 7, 27; Ch. 8, n. 12
40 CFR Section 261.33. Ch. 2, ns. 7, 27.

40 CFR Section 262.11. Ch. 8, n. 10
40 CFR Section 262.11. Ch. 2, n. 15
40 CFR Section 262.11(b), (c). Ch. 3, n. 2
40 CFR Section 262.12. Ch. 2, n. 16
40 CFR Section 262.20 - 262.22. Ch. 2, n. 26
40 CFR Section 262.30. Ch. 2, n. 27
40 CFR Section 262.31. Ch. 2, ns. 7, 27
40 CFR Section 262.34(a)(1). Ch. 2, n. 19
40 CFR Section 262.34(a)(2), (3). Ch. 2, n. 18
40 CFR Section 262.34(a)(4). Ch. 2, n. 20
40 CFR Section 262.34(c)(1). Ch. 2, n. 22
40 CFR Section 262.34(d). Ch. 2, ns. 23, 24
40 CFR Section 262.34(d)(1)-(5). Ch. 2, n. 25
40 CFR Section 262.40(a)-(c). Ch. 2, n. 28
40 CFR Section 262.40 - 262.44. Ch. 6, n. 22
40 CFR Section 262.41. Ch. 2, n. 29
40 CFR Section 262.42. Ch. 2, n. 30
40 CFR Section 262.43. Ch. 2, n. 31
40 CFR Section 262.44. Ch. 2, n. 31
40 CFR Section 265 Subpart C. Ch. 2, n. 21
40 CFR Section 265 Subpart D. Ch. 2, n. 21
40 CFR Section 265.201. Ch. 5, n. 2
40 CFR Section 268. Ch. 5, n. 10
40 CFR Section 268.44. Ch. 2, n. 32
40 CFR Section 280.34(b)(1)-(5). Ch. 6, n. 24
40 CFR Section 280.34(c). Ch. 6, n. 23
40 CFR Section 280.50 - 280.53. Ch. 6, n. 1
40 CFR Section 280.50. Ch. 6, n. 5
40 CFR Section 280.50(c)(2). Ch. 6, n. 2
40 CFR Section 280.52. Ch. 6, n. 4
40 CFR Section 280.53. Ch. 6, n. 3
40 CFR Section 280.61 - 280.63. Ch. 6, n. 6
40 CFR Section 280.65 - 280.66. Ch. 6, n. 7
40 CFR Section 280.97. Ch. 9, n. 1
40 CFR Section 302. Ch. 6, n. 28; Ch. 7, n. 10
40 CFR Section 302.4. Ch. 6, ns. 9, 14, 25
40 CFR Section 302.5. Ch. 6., n. 9
40 CFR Section 304. Ch. 6, n. 28; Ch. 7, n. 10

49 CFR Section 172. Ch. 8, n. 12
49 CFR Section 172.101. Ch. 8, ns. 10, 11
49 CFR Section 173. Ch. 8, n. 12
49 CFR Section 177. Ch. 8, n. 17

REGULATIONS AND STATUTES

Federal Register

50 Fed. Reg. 614 (Jan. 4, 1985). Ch. 2, n. 4
51 Fed. Reg. 34547 (Sept. 29, 1986). Ch. 6., n. 9
54 Feg. Reg. 22538 (May 24, 1989). Ch. 6., n. 9
55 Fed. Reg. 11798 (Mar. 29, 1990). Ch. 2, n. 11

Clean Air Act (CAA) - 42 USC Section 7401 to 7642

CAA Section 108 - 109, 42 USC Section 7408 - 7409. Ch 1, n. 14
CAA Section 112, 42 USC Section 7412. Ch. 1, n. 15
CAA Section 113(c), 42 USC Section 7413(c). Ch. 10, n. 12

Comprehensive Environmental Response Compensation and Liability Act (CERCLA) - 42 USC Section 9601 to 9675

CERCLA, 42 USC Section 9601 - 9675. Ch. 1, n. 5; Ch. 6, n. 27; Ch. 7, n. 9
CERCLA Section 103(a) 42 USC Section 9603(a). Ch 6, ns. 9, 15; Ch. 7, n. 12
CERCLA Section 103(b)(3), 42 USC Section 9603(b)(3). Ch. 6, n. 11
CERCLA Section 107(c)(2), 42 USC Section 9607(c)(2). Ch. 10, n. 4
CERCLA Section 107(c)(3), 42 USC Section 9607(c)(3). Ch. 10, n. 5

Federal Water Pollution Control Act (FWPCA) - 33 USC Section 1251 to 1387

FWPCA Section 101, 33 USC Section 1251(a). Ch. 1, n. 10
FWPCA Section 309(c), 33 USC Section 1319(c). Ch. 10, n. 7
FWPCA Section 309(d), 33 USC Section 1319(d). Ch. 10, n. 6
FWPCA Section 311(b)(5), 33 USC Section 1321(b)(5). Ch. 6, n. 16

Hazardous Materials Transportation Act (HMTA) - 49 USC Section 1801 to 1812

HMTA, 49 USC Section 1801 - 1812. Ch. 8, n. 1
HMTA, 49 USC Section 1803. Ch. 8, n. 3
HMTA, 49 USC Section 1804. Ch. 8, n. 4
HMTA, 49 USC Section 1805. Ch. 8, n. 4
HMTA, 49 USC Section 1809. Ch. 8, n. 5
HMTA, 49 USC Section 1811(a). Ch. 8, n. 7
HMTA, 49 USC Section 1811(b). Ch. 8, n. 7

Occupational Safety and Health Act (OSHA) - 29 USC Section 651-678

OSHA, 29 USC Section 651 - 678. Ch. 1, n. 16; Ch. 3, n. 16; Ch. 7, n. 1
OSHA, 29 USC Section 666(b). Ch. 10, n. 10
OSHA, 29 USC Section 666(e). Ch. 10, n. 11

Pollution Prevention Act of 1990 (PPA) - 42 USC Section 13101 to 13109

PPA, 42 USC Section 13101 - 13109. Ch. 7, n. 16.

Resource Conservation and Recovery Act (RCRA) - 42 USC Section 6901 to 6991i

RCRA, 42 USC Section 6901 *et seq.* Ch. 1, n. 1; Ch. 8, ns. 1, 8
RCRA Section 1003(b), 42 USC Section 6902(b). Ch. 2, n. 1
RCRA Section 1004(5), 42 USC Section 6903(5). Ch. 3, n. 3
RCRA Section 1004(27), 42 USC Section 6903(27). Ch. 2, n. 3; Ch. 3, n. 5
RCRA Section 3002(a)(1), 42 USC Section 6922(a)(1). Ch. 6, n. 22
RCRA Section 3002(b), 42 USC Section 6922(b). Ch. 5, n. 13
RCRA Section 3008(a)(3), 42 USC Section 6928(a)(3). Ch. 10, n. 1
RCRA Section 3008(d), (e), 42 USC Section 6928(d), (e). Ch. 10, n. 2
RCRA, 42 USC Section 6991e(d). Ch. 10, n. 3

Superfund Amendments and Reauthorization Act of 1986 (SARA) - 42 USC Section 11001 to 11050

SARA Title III, 42 USC Section 11001 - 11050. Ch. 6, ns. 12, 26; Ch. 7, n. 8
SARA Title III, 42 USC Section 11004. Ch. 6, n. 13; Ch. 7, n. 11
SARA Title III, 42 USC Section 11023. Ch. 6, ns. 21, 30; Ch. 7, n. 13
SARA Title III, 42 USC Section 11023(b). Ch. 6, n. 31; Ch. 7, n. 14

Toxic Substances Control Act (TSCA) - 15 USC Section 2601 to 2654

TSCA, 15 USC Section 2601 *et seq.* Ch. 1, n. 12
TSCA Section 6(e), 15 USC Section 2605(e). Ch. 3, n. 10
TSCA Section 7, 15 USC Section 2606. Ch. 1, n. 13
TSCA Section 16(a), 17 USC Section 2615(a). Ch. 10, n. 8
TSCA Section 16(b), 17 USC Section 2615(b). Ch. 10, n. 9
TSCA, 15 USC Sections 2641-2654. Ch. 3, 15

Index

AGENCIES. See ENVIRONMENTAL AGENCIES AND REGULATIONS

AUDITS.
   Levels of Assessment, *Table* **121**
   Hazardous Waste Compliance Checklist **125**
   Hazardous Waste Audits **120**
   Performing an Audit **122**
      Pre-audit **123**
      Site Analysis **123**
      Post-audit **123**
   Preparing for an Audit **122**

COSTS.
   Hazardous Waste Management Costs **119**
   Hazardous Waste Management Costs, *Table* **120**

ENVIRONMENTAL AGENCIES AND REGULATIONS.
   Clean Air Act (CAA) **10**
   Comprehensive Environmental Response, Compensation and Liability Act (CERCLA) **6**
   Environmental Agencies, Generally **1**
   Environmental Statutes and Regulations **3**
   EPA Regional Offices **Appendix 1**
   Federal Environmental Agencies and Offices **Appendix 2**
   Federal Level **1**
   Federal Water Pollution Control Act (FWPCA), Generally **8**
   Occupational Safety and Health Act (OSHA), Generally **10**
   Resource Conservation and Recovery Act (RCRA), Generally **4**
   State Hazardous and Solid Waste Contacts **Appendix 3**
   State Level **2**
   Toxic Substances Control Act (TSCA), Generally **9**

DISPOSAL. See HAZARDOUS WASTE, HANDLING OF

GENERATORS.
   Introduction **37**
   Categories of Hazardous Waste Generators **38**
   CESQG Limits, *Table* **39**
   Conditionally Exempt Small Quantity Generator **38**

Determining the Waste Generator Category 40
EPA Notification of Hazardous Waste Activity **Appendix 5**
Large Quantity Generator (LQG) 40
Monthly Volume Generated 40
Small Quantity Generator (SQG) 39

HAZARD COMMUNICATION.
Introduction 71
Annual Reporting 76
Appendix E to Hazard Communication Standard (Advisory) - Guidelines for Employer Compliance **Appendix 8**
Emergency Planning and Community Right-to-Know Hotline 79
Emergency Planning and Community Right-to-Know Requirements 75
   Extremely Hazardous Substances 76
   Threshold Planning Quantity 76
Employee Training Program 74
Exempt Substances 75, 81
Form R **79-80**
Hazardous Chemical Inventory 76
Material Safety Data Sheets 72
   Sample Material Safety Data Sheet **Appendix 9**
OSHA Asbestos Standard 81
OSHA Asbestos Standard Requirements 82
   Air Sampling 82
   Employee Notification 82
   Record of Exposure Measurements 82
   Medical Surveillance 83
OSHA Hazard Communication Standard Requirements 72
Pollution Prevention Act of 1990 79
Reporting Requirements Added to Form R 80
Reporting Spills or Releases 76
Revised Form R 80
Tier One Form 77, **Appendix 6**
Tier Two Form 78, **Appendix 7**
Toxic Release Inventory (Form R) **Appendix 10**
Written Hazard Communication Plan 73
   Training Program Components, *Table* 74

HAZARDOUS WASTE ASSESSMENTS. See AUDITS

HAZARDOUS WASTE CLASSIFICATION. See HAZARDOUS WASTE IDENTIFICATION

HAZARDOUS WASTE DISPOSAL. See HAZARDOUS WASTE, HANDLING OF

HAZARDOUS WASTE GENERATORS. See GENERATORS

HAZARDOUS WASTE, HANDLING OF
  Contaminated Soils 50
  Emergency Response Plans 44
  Labeling 44
  Land Disposal Restrictions 46
    Best Demonstrated Available Technology (BDAT) 47, **Appendix 11**
    Schedule for Land Disposal Restrictions, *Table* 48
    Treatment Standards for Land Disposal of Wastes 49
    Variances 50
  Land Disposal Sites 46
  RCRA Hazardous and Solid Waste Amendments 47
  Recordkeeping Requirements 45
  Reporting Requirements 45
  Safe Storage Practices 45
    Safe Hazardous Waste Storage Checklist 45
  Storing Hazardous Wastes 43
  Waste Minimization 51
    Minimization Benefits, *Table* 52
    Minimization Program Implementation 52
  Waste Mixtures 50

HAZARDOUS WASTE IDENTIFICATION,
  Asbestos 33
  Chlorinated Hydrocarbons 31
  Definitions of Hazardous Waste 23
  EPA Hazard Classes 29
    Acute Hazardous Waste 30
    Corrosive Waste 29
    Ignitable Waste 29
    Reactive Waste 29
    Toxicity Characteristic Waste 30
    Toxic Wastes 30
  Federal Laws Governing Hazardous Substances 24
    Comprehensive Environmental Response, Compensation and Liability
      Act (CERCLA) 24
    Hazardous Materials Transportation Act 24
    Resource Conservation and Recovery Act 24
    Toxic Substances Control Act 24
  Hazardous Waste Classification 29
  Heavy Metals 35
  Listed and Characteristic Hazardous Wastes 28
  Petroleum Products 31
  Polychlorinated biphenyls (PCBs) 32
  Resource Conservation and Recovery Act (RCRA) 25
    RCRA Hazardous Wastes **Appendix 4**
    RCRA/Superfund Hotline 26
  Sulfuric Acid 34

Terminology Relating to Hazardous Waste 24
　Hazardous Material 24
　Hazardous Substance 25
　Hazardous Waste 24
　Toxic Waste 25

HAZARDOUS WASTE LIABILITY INSURANCE. See INSURANCE

HAZARDOUS WASTE PENALTIES. See PENALTIES

HAZARDOUS WASTE RECORDKEEPING. See RECORDKEEPING

HAZARDOUS WASTE REPORTING
　Clean Water Act 62
　　Designated Hazardous Substances 62
　　Penalties for Failure to Report 62
　Comprehensive Environmental Response, Compensation and Liability Act (CERCLA) 60
　　Reportable Quantities 60
　Developing a Contingency Plan 63
　　Contingency Plan Components, *Table* 63
　Emergency Coordinator 63
　Resource Conservation and Recovery Act (RCRA) 55
　　How to Respond to a Spill or Release, *Table* 54
　Underground Storage Tank Releases 57
　　Confirmation of Suspected Releases 57
　　Corrective Actions 58
　　UST Reporting, *Table* 57

IDENTIFICATION OF HAZARDOUS WASTES. See HAZARDOUS WASTE IDENTIFICATION

INSURANCE.
　Comprehensive General Liability (CGL) Policy 104
　　Duty to Defend 105
　　Damages Covered 106
　　Definition of Occurrence 108
　　Triggers of Coverage 108
　　Sudden and Accidental Pollution Exclusion 110
　　Absolute Pollution Exclusion 112
　　Owned Property Exclusion 114
　Other Types of Coverage 102
　　Environmental Impairment Liability (EIL) Insurance 103
　　Pollution Cleanup Coverage 103
　　Pollution Legal Liability Insurance 103
　　Underground Storage Tank Liability Insurance 103

PENALTIES.
  Introduction 117
  Hazardous Waste Penalties 117
  Penalties for Noncompliance, *Table* 118

RECORDKEEPING.
  CERCLA Recordkeeping 67
  Hazardous Waste Recordkeeping Checklist 68
  Record Retention Period and Filing Frequency, *Table* 65
  RCRA Recordkeeping 64
  Underground Storage Tank Recordkeeping 65
    UST Recordkeeping Checklist 66

REGULATIONS. See ENVIRONMENTAL AGENCIES AND REGULATIONS, RESOURCE CONSERVATION AND RECOVERY ACT

REPORTING. See HAZARDOUS WASTE REPORTING, HAZARD COMMUNICATION

RESOURCE CONSERVATION AND RECOVERY ACT.
  Characteristic Hazardous Wastes 14, 28
    Corrosivity 15, 29
    Ignitability 15, 29
    Reactivity 15, 29
    Toxicity 15, 30
  Common Hazardous Wastes 30
  Derived From Rule 15
  Exempt Substances 15
  Generators 16, 37
    Conditionally Exempt Small Quantity Generators 38
    Satellite Accumulation 18
    Small Quantity Generators 18, 39
    Large Quantity Generators 40
  Land Disposal Restrictions 19, 46
  Listed Hazardous Wastes 14, 28
  Recordkeeping and Reporting 19, 55
  Regulated Wastes 12, 25
    Hazardous Waste 14
    Solid Waste 12
    Mixture Rule 15
  RCRA Checklist 20
  Scope and Purpose 11
  Toxicity Characteristic Leachate Procedure 30
  Transporting Hazardous Wastes 18, 87
  Treatment, Storage and Disposal Facilities 16

SPILLS AND RELEASES. See HAZARDOUS WASTE REPORTING, HAZARD COMMUNICATION

TRANSPORTING HAZARDOUS WASTES.
  Choosing a Reputable Transporter 95
  DOT Shipment Number 91
  Hazard Class 91
  Hazardous Materials Transportation Act 88
  Hazardous Waste Shipment Preparation 90
  Labels and Placards 93
    Label/Placard Colors and Class Numbers, *Table* 94
  Manifest Requirements 96
    Completing the Hazardous Waste Manifest 96
    How to Decide Which Manifest Form to Use, *Table* 98
    Manifest Exception Reports 98
    Manifest Exemptions 98
  Markings 94
  Packaging 91
  Resource Conservation and Recovery Act 89
  Transporting Hazardous Wastes in Company Vehicles 99
  Used Batteries 88

DISCARDED

JUN 19 2025

ASHEVILLE-BUNCOMBE
TECHNICAL COMMUNITY COLLEGE

3 3312 00043 6998

TD 1030 .D45 1993

Dennison, Mark S.

RCRA regulatory compliance
 guide